T0292935

Symmetric Functions and Polynomials (Mathematics Essentials)

Symmetric Functions and Polynomials (Mathematics Essentials)

Editor: Alma Adams

New York

Published by NY Research Press
118-35 Queens Blvd., Suite 400,
Forest Hills, NY 11375, USA
www.nyresearchpress.com

Symmetric Functions and Polynomials (Mathematics Essentials)
Edited by Alma Adams

International Standard Book Number: 978-1-64725-462-9 (Hardback)

Cataloging-in-Publication Data

Symmetric functions and polynomials : mathematics essentials / edited by Alma Adams.
p. cm.
Includes bibliographical references and index.
ISBN 978-1-64725-462-9
1. Symmetric functions. 2. Polynomials. 3. Equations, Theory of. 4. Mathematics. I. Adams, Alma.
QA212 .S96 2023
515.22--dc23

Contents

Preface

This book aims to highlight the current researches and provides a platform to further the scope of innovations in this area. This book is a product of the combined efforts of many researchers and scientists, after going through thorough studies and analysis from different parts of the world. The objective of this book is to provide the readers with the latest information of the field.

A function containing several variables that remains unchanged for any permutation of the variables is called a symmetric function. Polynomials are a type of function. A symmetric polynomial refers to a type of polynomial P in n variables such that if any of the variables are swapped with each other, it remains the same polynomial. There are various types of symmetric polynomials including power-sum symmetric polynomials, elementary symmetric polynomials, complete homogeneous symmetric polynomials, monomial symmetric polynomials, and Schur polynomials. Symmetric polynomials have numerous applications in various areas of combinatorics, representation theory, mathematical physics, and mathematics. They are frequently found in Newton's identities and Vieta's formula. This book includes some of the vital pieces of works being conducted across the world, on various topics related to symmetric functions and polynomials, and their applications. It will serve as a valuable source of reference for graduate and postgraduate students.

I would like to express my sincere thanks to the authors for their dedicated efforts in the completion of this book. I acknowledge the efforts of the publisher for providing constant support. Lastly, I would like to thank my family for their support in all academic endeavors.

Editor

1

A Note on Degenerate Euler and Bernoulli Polynomials of Complex Variable

Dae San Kim [1], Taekyun Kim [2,3]* and Hyunseok Lee [3]

1 Department of Mathematics, Sogang University, Seoul 121-742, Korea; dskim@sogang.ac.kr
2 School of Science, Xi'an Technological University, Xi'an 710021, China
3 Department of Mathematics, Kwangwoon University, Seoul 139-701, Korea; luciasconstant@gmail.com
* Correspondence: tkkim@kw.ac.kr

Abstract: Recently, the so-called new type Euler polynomials have been studied without considering Euler polynomials of a complex variable. Here we study degenerate versions of these new type Euler polynomials. This has been done by considering the degenerate Euler polynomials of a complex variable. We also investigate corresponding ones for Bernoulli polynomials in the same manner. We derive some properties and identities for those new polynomials. Here we note that our result gives an affirmative answer to the question raised by the reviewer of the paper.

Keywords: degenerate cosine-Euler polynomials; degenerate sine-Euler polynomials; degenerate cosine-Bernoulli polynomials; degenerate sine-Bernoulli polynomials; degenerate cosine-polynomials; degenerate sine-polynomials

---

1. Introduction

The ordinary Bernoulli polynomials $B_n(x)$ and Euler polynomials $E_n(x)$ are respectively defined by

$$\frac{t}{e^t-1}e^{xt}=\sum_{n=0}^{\infty}B_n(x)\frac{t^n}{n!}, \tag{1}$$

and

$$\frac{2}{e^t+1}e^{xt}=\sum_{n=0}^{\infty}E_n(x)\frac{t^n}{n!}, \tag{2}$$

(see [1–20]).

For any nonzero $\lambda\in\mathbb{R}$, the degenerate exponential function is defined by

$$e_\lambda^x(t)=(1+\lambda t)^{\frac{x}{\lambda}},\quad e_\lambda(t)=e_\lambda^1(t), \tag{3}$$

(see [8]).

In [1,2], Carlitz considered the degenerate Bernoulli and Euler polynomials which are given by

$$\frac{t}{e_\lambda(t)-1}e_\lambda^x(t)=\frac{t}{(1+\lambda t)^{\frac{1}{\lambda}}-1}(1+\lambda t)^{\frac{x}{\lambda}}=\sum_{n=0}^{\infty}\beta_{n,\lambda}(x)\frac{t^n}{n!}, \tag{4}$$

and

$$\frac{2}{e_\lambda(t)+1}e_\lambda^x(t)=\frac{2}{(1+\lambda t)^{\frac{1}{\lambda}}+1}(1+\lambda t)^{\frac{x}{\lambda}}=\sum_{n=0}^{\infty}\mathcal{E}_{n,\lambda}(x)\frac{t^n}{n!}. \tag{5}$$

Note that

$$\lim_{\lambda\to0}\beta_{n,\lambda}(x)=B_n(x),\quad \lim_{\lambda\to0}\mathcal{E}_{n,\lambda}(x)=E_n(x).$$

The falling factorial sequence is defined as

$$(x)_0 = 1, \quad (x)_n = x(x-1)\cdots(x-n+1), \ (n \geq 1),$$

(see [17]).

The Stirling numbers of the first kind are defined by the coefficients in the expansion of $(x)_n$ in terms of powers of x as follows:

$$(x)_n = \sum_{l=0}^{n} S^{(1)}(n,l)x^l, \tag{6}$$

(see, [7,11,17]).

The Stirling numbers of the second kind are defined by

$$x^n = \sum_{l=0}^{n} S^{(2)}(n,l)(x)_l, \ (n \geq 0), \tag{7}$$

(see [9,10,17]).

In [9], the degenerate stirling numbers of the second kind are defined by

$$\frac{1}{k!}\left(e_\lambda(t)-1\right)^k = \sum_{n=k}^{\infty} S_\lambda^{(2)}(n,k)\frac{t^n}{n!}, \ (k \geq 0). \tag{8}$$

Note that $\lim_{\lambda\to 0} S_\lambda^{(2)}(n,k) = S^{(2)}(n,k), \ (n,k \geq 0)$.

Recently, Masjed–Jamei, Beyki and Koepf introduced the new type Euler polynomials which are given by

$$\frac{2e^{pt}}{e^t+1}\cos qt = \sum_{n=0}^{\infty} E_n^{(c)}(p,q)\frac{t^n}{n!}, \tag{9}$$

$$\frac{2e^{pt}}{e^t+1}\sin qt = \sum_{n=0}^{\infty} E_n^{(s)}(p,q)\frac{t^n}{n!}, \tag{10}$$

(see [15]).

They also considered the cosine-polynomials and sine-polynomials defined by

$$e^{pt}\cos qt = \sum_{n=0}^{\infty} C_n(p,q)\frac{t^n}{n!}, \tag{11}$$

and

$$e^{pt}\sin qt = \sum_{n=0}^{\infty} S_n(p,q)\frac{t^n}{n!}, \quad \text{(see [15])}. \tag{12}$$

In [15], the authors deduced many interesting identities and properties for those polynomials. It is well known that

$$e^{ix} = \cos x + i\sin x, \quad \text{where } x \in \mathbb{R}, \ i = \sqrt{-1}, \tag{13}$$

(see [20]).

From (1) and (2), we note that

$$\frac{t}{e^t-1}e^{(x+iy)t} = \sum_{n=0}^{\infty} B_n(x+iy)\frac{t^n}{n!}, \tag{14}$$

and

$$\frac{2}{e^t+1}e^{(x+iy)t} = \sum_{n=0}^{\infty} E_n(x+iy)\frac{t^n}{n!}. \tag{15}$$

By (14) and (15), we get

$$\begin{aligned}
\frac{t}{e^t-1}e^{xt}\cos yt &= \sum_{n=0}^{\infty} \frac{B_n(x+iy)+B_n(x-iy)}{2}\frac{t^n}{n!} = \sum_{n=0}^{\infty} B_n^{(c)}(x,y)\frac{t^n}{n!}, \\
\frac{t}{e^t-1}e^{xt}\sin yt &= \sum_{n=0}^{\infty} \frac{B_n(x+iy)-B_n(x-iy)}{2i}\frac{t^n}{n!} = \sum_{n=0}^{\infty} B_n^{(s)}(x,y)\frac{t^n}{n!}, \\
\frac{2}{e^t+1}e^{xt}\cos yt &= \sum_{n=0}^{\infty} \frac{E_n(x+iy)+E_n(x-iy)}{2}\frac{t^n}{n!} = \sum_{n=0}^{\infty} E_n^{(c)}(x,y)\frac{t^n}{n!},
\end{aligned} \tag{16}$$

and

$$\frac{2}{e^t+1}e^{xt}\sin yt = \sum_{n=0}^{\infty} \frac{E_n(x+iy)-E_n(x-iy)}{2i}\frac{t^n}{n!} = \sum_{n=0}^{\infty} E_n^{(s)}(x,y)\frac{t^n}{n!},$$

(see [12]).

In view of (4) and (5), we study the degenerate Bernoulli and Euler polynomials with complex variables and investigate some identities and properties for those polynomials. The outline of this paper is as follows. In Section 1, we will briefly recall the degenerate Bernoulli and Euler polynomials of Carlitz and the degenerate Stirling numbers of the second kind. Then we will introduce the so-called the new type Euler polynomials, and the cosine-polynomials and sine-polynomials recently introduced in [15]. Then we indicate that the new type Euler polynomials and the corresponding Bernoulli polynomials can be expressed by considering Euler and Bernoulli polynomials of a complex variable and treating the real and imaginary parts separately. In Section 2, the degenerate cosine-polynomials and degenerate sine-polynomials were introduced and their explicit expressions were derived. The degenerate cosine-Euler polynomials and degenerate sine-Euler polynomials were expressed in terms of degenerate cosine-polynomials and degenerate sine-polynomials and vice versa. Further, some reflection identities were found for the degenerate cosine-Euler polynomials and degenerate sine-Euler polynomials. In Section 3, the degenerate cosine-Bernoulli polynomials and degenerate sine-Bernoulli polynomials were introduced. They were expressed in terms of degenerate cosine-polynomials and degenerate sine-polynomials and vice versa. Reflection symmetries were deduced for the degenerate cosine-Bernoulli polynomials and degenerate sine-Bernoulli polynomials.

2. Degenerate Euler Polynomials of Complex Variable

Here we will consider the degenerate Euler polynomials of complex variable and, by treating the real and imaginary parts separately, introduce the degenerate cosine-Euler polynomials and degenerate sine-Euler polynomials. They are degenerate versions of the new type Euler polynomials studied in [15].

The degenerate sine and cosine functions are defined by

$$\cos_\lambda t = \frac{e_\lambda^i(t)+e_\lambda^{-i}(t)}{2}, \quad \sin_\lambda t = \frac{e_\lambda^i(t)-e_\lambda^{-i}(t)}{2i}. \tag{17}$$

From (13), we note that

$$\lim_{\lambda\to 0}\cos_\lambda t = \cos t, \quad \lim_{\lambda\to 0}\sin_\lambda t = \sin t.$$

By (5), we get

$$\frac{2}{e_\lambda(t)+1}e_\lambda^{x+iy}(t) = \sum_{n=0}^{\infty} \mathcal{E}_{n,\lambda}(x+iy)\frac{t^n}{n!}. \tag{18}$$

and

$$\frac{2}{e_\lambda(t)+1}e_\lambda^{x-iy}(t) = \sum_{n=0}^{\infty} \mathcal{E}_{n,\lambda}(x-iy)\frac{t^n}{n!}. \tag{19}$$

Now, we define the degenerate cosine and degenerate sine function as

$$\cos_\lambda^{(y)}(t) = \frac{e_\lambda^{iy}(t)+e_\lambda^{-iy}(t)}{2} = \cos\left(\frac{y}{\lambda}\log(1+\lambda t)\right), \tag{20}$$

$$\sin_\lambda^{(y)}(t) = \frac{e_\lambda^{iy}(t)-e_\lambda^{-iy}(t)}{2i} = \sin\left(\frac{y}{\lambda}\log(1+\lambda t)\right). \tag{21}$$

Note that $\lim_{\lambda\to 0}\cos_\lambda^{(y)}(t) = \cos yt$, $\lim_{\lambda\to 0}\sin_\lambda^{(y)}(t) = \sin yt$.
From (18) and (19), we note that

$$\frac{2}{e_\lambda(t)+1}e_\lambda^{x}(t)\cos_\lambda^{(y)}(t) = \sum_{n=0}^{\infty}\left(\frac{\mathcal{E}_{n,\lambda}(x+iy)+\mathcal{E}_{n,\lambda}(x-iy)}{2}\right)\frac{t^n}{n!}, \tag{22}$$

and

$$\frac{2}{e_\lambda(t)+1}e_\lambda^{x}(t)\sin_\lambda^{(y)}(t) = \sum_{n=0}^{\infty}\left(\frac{\mathcal{E}_{n,\lambda}(x+iy)-\mathcal{E}_{n,\lambda}(x-iy)}{2i}\right)\frac{t^n}{n!}. \tag{23}$$

In view of (9) and (10), we define the degenerate cosine-Euler polynomials and degenerate sine-Euler polynomials respectively by

$$\frac{2}{e_\lambda(t)+1}e_\lambda^{x}(t)\cos_\lambda^{(y)}(t) = \sum_{n=0}^{\infty}\mathcal{E}_{n,\lambda}^{(c)}(x,y)\frac{t^n}{n!}, \tag{24}$$

and

$$\frac{2}{e_\lambda(t)+1}e_\lambda^{x}(t)\sin_\lambda^{(y)}(t) = \sum_{n=0}^{\infty}\mathcal{E}_{n,\lambda}^{(s)}(x,y)\frac{t^n}{n!}. \tag{25}$$

Note that $\lim_{\lambda\to 0}\mathcal{E}_{n,\lambda}^{(c)}(x,y) = E_n^{(c)}(x,y)$, $\lim_{\lambda\to 0}\mathcal{E}_{n,\lambda}^{(s)}(x,y) = E_n^{(s)}(x,y)$, $(n\geq 0)$, where $E_n^{(c)}(x,y)$ and $E_n^{(s)}(x,y)$ are the new type of Euler polynomials of Masjed-Jamei, Beyki and Koepf (see [15]).
From (22)–(25), we note that

$$\mathcal{E}_{n,\lambda}^{(c)}(x,y) = \frac{\mathcal{E}_{n,\lambda}(x+iy)+\mathcal{E}_{n,\lambda}(x-iy)}{2}, \tag{26}$$

and

$$\mathcal{E}_{n,\lambda}^{(s)}(x,y) = \frac{\mathcal{E}_{n,\lambda}(x+iy)-\mathcal{E}_{n,\lambda}(x-iy)}{2i}, \quad (n\geq 0). \tag{27}$$

We recall here that the generalized falling factorial sequence is defined by

$$(x)_{0,\lambda} = 1, \quad (x)_{n,\lambda} = x(x-\lambda)(x-2\lambda)\cdots(x-(n-1)\lambda), \quad (n\geq 1).$$

Note that $\lim_{\lambda\to 1}(x)_{n,\lambda} = (x)_n$, $\lim_{\lambda\to 0}(x)_{n,\lambda} = x^n$.

We observe that

$$\begin{aligned} e_\lambda^{iy}(t) &= (1+\lambda t)^{\frac{iy}{\lambda}} = e^{\frac{iy}{\lambda}\log(1+\lambda t)} \\ &= \sum_{k=0}^{\infty}\left(\frac{iy}{\lambda}\right)^k \frac{1}{k!}\big(\log(1+\lambda t)\big)^k \\ &= \sum_{k=0}^{\infty}\lambda^{-k}(iy)^k \sum_{n=k}^{\infty} S^{(1)}(n,k)\frac{\lambda^n}{n!}t^n \\ &= \sum_{n=0}^{\infty}\left(\sum_{k=0}^{n}\lambda^{n-k} i^k y^k S^{(1)}(n,k)\right)\frac{t^n}{n!}. \end{aligned} \tag{28}$$

From (20), we can derive the following equation.

$$\begin{aligned} \cos_\lambda^{(y)}(t) &= \frac{e_\lambda^{iy}(t)+e_\lambda^{-iy}(t)}{2} \\ &= \frac{1}{2}\sum_{n=0}^{\infty}\left(\sum_{k=0}^{n}\lambda^{n-k}(i^k+(-i)^k)y^k S^{(1)}(n,k)\right)\frac{t^n}{n!} \\ &= \sum_{n=0}^{\infty}\left(\sum_{k=0}^{[\frac{n}{2}]}\lambda^{n-2k}(-1)^k y^{2k} S^{(1)}(n,2k)\right)\frac{t^n}{n!} \\ &= \sum_{k=0}^{\infty}\left(\sum_{n=2k}^{\infty}\lambda^{n-2k}(-1)^k y^{2k} S^{(1)}(n,2k)\right)\frac{t^n}{n!}. \end{aligned} \tag{29}$$

Note that

$$\lim_{\lambda\to 0}\cos_\lambda^{(y)}(t) = \sum_{k=0}^{\infty}(-1)^k y^{2k}\frac{t^{2k}}{(2k)!} = \cos yt.$$

By (21), we get

$$\begin{aligned} \sin_\lambda^{(y)}(t) &= \frac{e_\lambda^{iy}(t)-e_\lambda^{-iy}(t)}{2i} \\ &= \frac{1}{2i}\sum_{n=0}^{\infty}\left(\sum_{k=0}^{n}\lambda^{n-k}(i^k-(-i)^k)y^k S^{(1)}(n,k)\right)\frac{t^n}{n!} \\ &= \sum_{n=1}^{\infty}\left(\sum_{k=0}^{[\frac{n-1}{2}]}\lambda^{n-2k-1}(-1)^k y^{2k+1} S^{(1)}(n,2k+1)\right)\frac{t^n}{n!} \\ &= \sum_{k=0}^{\infty}\left(\sum_{n=2k+1}^{\infty}(-1)^k\lambda^{n-2k-1} S^{(1)}(n,2k+1)\frac{t^n}{n!}\right)y^{2k+1}, \end{aligned} \tag{30}$$

where $[x]$ denotes the greatest integer $\leq x$.

Note that

$$\lim_{\lambda\to 0}\sin_\lambda^{(y)}(t) = \sum_{k=0}^{\infty}(-1)^k y^{2k+1}\frac{t^{2k+1}}{(2k+1)!} = \sin(yt).$$

From (18), we note that

$$\sum_{n=0}^{\infty}\mathcal{E}_{n,\lambda}(x+iy)\frac{t^n}{n!}=\frac{2}{e_\lambda(t)+1}e_\lambda^x(t)\cdot e_\lambda^{iy}(t) \tag{31}$$
$$=\sum_{l=0}^{\infty}\mathcal{E}_{l,\lambda}(x)\frac{t^l}{l!}\sum_{j=0}^{\infty}(iy)_{j,\lambda}\frac{t^j}{j!}$$
$$=\sum_{n=0}^{\infty}\left(\sum_{l=0}^{n}\binom{n}{l}(iy)_{n-l,\lambda}\mathcal{E}_{l,\lambda}(x)\right)\frac{t^n}{n!}.$$

On the other hand

$$\frac{2}{e_\lambda(t)+1}e_\lambda^{x+iy}(t)=\sum_{n=0}^{\infty}\mathcal{E}_{l,\lambda}\frac{t^l}{l!}\sum_{j=0}^{\infty}(x+iy)_{j,\lambda}\frac{t^j}{j!} \tag{32}$$
$$=\sum_{n=0}^{\infty}\left(\sum_{l=0}^{n}\binom{n}{l}(x+iy)_{n-l,\lambda}\mathcal{E}_{l,\lambda}\right)\frac{t^n}{n!}.$$

Therefore, by (31) and (32), we obtain the following theorem,

Theorem 1. *For $n \geq 0$, we have*

$$\mathcal{E}_{n,\lambda}(x+iy)=\sum_{l=0}^{n}\binom{n}{l}(iy)_{n-l,\lambda}\mathcal{E}_{l,\lambda}(x)$$
$$=\sum_{l=0}^{n}\binom{n}{l}(x+iy)_{n-l,\lambda}\mathcal{E}_{l,\lambda}.$$

Also, we have

$$\mathcal{E}_{n,\lambda}(x-iy)=\sum_{l=0}^{n}\binom{n}{l}(-1)^{n-l}\langle iy\rangle_{n-l,\lambda}\mathcal{E}_{l,\lambda}(x)$$
$$=\sum_{l=0}^{n}\binom{n}{l}(-1)^{n-l}\langle iy-x\rangle_{n-l,\lambda}\mathcal{E}_{l,\lambda},$$

where $\langle x\rangle_{0,\lambda}=1,\ \langle x\rangle_{n,\lambda}=x(x+\lambda)\cdots(x+\lambda(n-1)),\ (n\geq 1).$

By (29), we get

$$e_\lambda^x(t)\cos_\lambda^{(y)}(t)=\sum_{l=0}^{\infty}(x)_{l,\lambda}\frac{t^l}{l!}\sum_{m=0}^{\infty}\sum_{k=0}^{[\frac{m}{2}]}\lambda^{m-2k}(-1)^k y^{2k}S^{(1)}(m,2k)\frac{t^m}{m!} \tag{33}$$
$$=\sum_{n=0}^{\infty}\left(\sum_{m=0}^{n}\sum_{k=0}^{[\frac{m}{2}]}\binom{n}{m}\lambda^{m-2k}(-1)^k y^{2k}S^{(1)}(m,2k)(x)_{n-m,\lambda}\right)\frac{t^n}{n!},$$

and

$$e_\lambda^x(t)\sin_\lambda^{(y)}(t)=\sum_{l=0}^{\infty}(x)_{\lambda,l}\frac{t^l}{l!}\sum_{m=1}^{\infty}\sum_{k=0}^{[\frac{m-1}{2}]}\lambda^{m-2k-1}(-1)^k y^{2k+1}S^{(1)}(m,2k+1)\frac{t^m}{m!} \tag{34}$$
$$=\sum_{n=1}^{\infty}\left(\sum_{m=1}^{n}\sum_{k=0}^{[\frac{m-1}{2}]}\binom{n}{m}\lambda^{m-2k-1}(-1)^k y^{2k+1}S^{(1)}(m,2k+1)(x)_{n-m,\lambda}\right)\frac{t^n}{n!}.$$

Now, we define the degenerate cosine-polynomials and degenerate sine-polynomials respectively by

$$e_\lambda^x(t)\cos_\lambda^{(y)}(t)=\sum_{k=0}^{\infty}C_{k,\lambda}(x,y)\frac{t^k}{k!},\tag{35}$$

and

$$e_\lambda^x(t)\sin_\lambda^{(y)}(t)=\sum_{k=0}^{\infty}S_{k,\lambda}(x,y)\frac{t^k}{k!}.\tag{36}$$

Note that

$$\lim_{\lambda\to0}C_{k,\lambda}(x,y)=C_k(x,y),\quad \lim_{\lambda\to0}S_{k,\lambda}(x,y)=S_k(x,y),$$

where $C_k(x,y)$ and $S_k(x,y)$ are the cosine-polynomials and sine-polynomials of Masijed–Jamei, Beyki and Koepf.

Therefore, by (33)–(36), we obtain the following theorem.

Theorem 2. *For $n\geq 0$, we have*

$$\begin{aligned}C_{n,\lambda}(x,y)&=\sum_{m=0}^{n}\sum_{k=0}^{[\frac{m}{2}]}\binom{n}{m}\lambda^{m-2k}(-1)^k y^{2k}S^{(1)}(m,2k)(x)_{n-m,\lambda}\\&=\sum_{k=0}^{[\frac{n}{2}]}\sum_{m=2k}^{n}\binom{n}{m}\lambda^{m-2k}(-1)^k y^{2k}S^{(1)}(m,2k)(x)_{n-m,\lambda}.\end{aligned}$$

Also, for $n\in\mathbb{N}$, we have

$$\begin{aligned}S_{n,\lambda}(x,y)&=\sum_{m=1}^{n}\sum_{k=0}^{[\frac{m-1}{2}]}\binom{n}{m}\lambda^{m-2k-1}(-1)^k y^{2k+1}S^{(1)}(m,2k+1)(x)_{n-m,\lambda}\\&=\sum_{k=0}^{[\frac{n-1}{2}]}\sum_{m=2k+1}^{n}\binom{n}{m}\lambda^{m-2k-1}(-1)^k y^{2k+1}S^{(1)}(m,2k+1)(x)_{n-m,\lambda}.\end{aligned}$$

and $S_{0,\lambda}(x,y)=0$.

From (24), we note that

$$\begin{aligned}\sum_{n=0}^{\infty}\mathcal{E}_{n,\lambda}^{(c)}(x,y)\frac{t^n}{n!}&=\frac{2}{e_\lambda(t)+1}e_\lambda^x(t)\cos_\lambda^{(y)}(t)\\&=\sum_{m=0}^{\infty}\mathcal{E}_{m,\lambda}\frac{t^m}{m!}\sum_{l=0}^{\infty}C_{l,\lambda}(x,y)\frac{t^l}{l!}\\&=\sum_{n=0}^{\infty}\left(\sum_{m=0}^{n}\binom{n}{m}\mathcal{E}_{m,\lambda}C_{n-m,\lambda}(x,y)\right)\frac{t^n}{n!}.\end{aligned}\tag{37}$$

On the other hand,

$$\begin{aligned}\frac{2}{e_\lambda(t)+1}e_\lambda^x(t)\cos_\lambda^{(y)}(t)&=\sum_{m=0}^{\infty}\mathcal{E}_{m,\lambda}(x)\frac{t^m}{m!}\sum_{l=0}^{\infty}\sum_{k=0}^{[\frac{l}{2}]}\lambda^{l-2k}(-1)^k y^{2k}S^{(1)}(l,2k)\frac{t^l}{l!}\\&=\sum_{n=0}^{\infty}\left(\sum_{l=0}^{n}\sum_{k=0}^{[\frac{l}{2}]}\binom{n}{l}\lambda^{l-2k}(-1)^k y^{2k}S^{(1)}(l,2k)\mathcal{E}_{n-l,\lambda}(x)\right)\frac{t^n}{n!}\\&=\sum_{n=0}^{\infty}\left(\sum_{k=0}^{[\frac{n}{2}]}\sum_{l=2k}^{n}\binom{n}{l}\lambda^{l-2k}(-1)^k y^{2k}S^{(1)}(l,2k)\mathcal{E}_{n-l,\lambda}(x)\right)\frac{t^n}{n!}.\end{aligned}\tag{38}$$

By (30), we get

$$\frac{2}{e_\lambda(t)+1}e_\lambda^x(t)\sin_\lambda^{(y)}(t) = \sum_{m=0}^{\infty}\mathcal{E}_{m,\lambda}(x)\frac{t^m}{m!}\sum_{l=1}^{n}\sum_{k=0}^{[\frac{l-1}{2}]}(-1)^k\lambda^{l-2k-1}y^{2k+1}S^{(1)}(l,2k+1)\frac{t^l}{l!} \tag{39}$$
$$= \sum_{n=1}^{\infty}\left(\sum_{l=1}^{n}\sum_{k=0}^{[\frac{l-1}{2}]}\binom{n}{l}\lambda^{l-2k-1}(-1)^k y^{2k+1}S^{(1)}(l,2k+1)\mathcal{E}_{n-l,\lambda}(x)\right)\frac{t^n}{n!}$$
$$= \sum_{n=1}^{\infty}\left(\sum_{k=0}^{[\frac{n-1}{2}]}\sum_{l=2k+1}^{n}\binom{n}{l}\lambda^{l-2k-1}(-1)^k y^{2k+1}S^{(1)}(l,2k+1)\mathcal{E}_{n-l,\lambda}(x)\right)\frac{t^n}{n!}.$$

Therefore, by (24), (25), and (37)–(39), we obtain the following theorem.

Theorem 3. *For $n \geq 0$, we have*

$$\mathcal{E}_{n,\lambda}^{(c)}(x,y) = \sum_{k=0}^{n}\binom{n}{k}\mathcal{E}_{k,\lambda}C_{n-k,\lambda}(x,y)$$
$$= \sum_{k=0}^{[\frac{n}{2}]}\sum_{l=2k}^{n}\binom{n}{l}\lambda^{l-2k}(-1)^k y^{2k}S^{(1)}(l,2k)\mathcal{E}_{n-l,\lambda}(x).$$

Also, for $n \in \mathbb{N}$, we obtain

$$\mathcal{E}_{n,\lambda}^{(s)}(x,y) = \sum_{k=0}^{n}\binom{n}{k}\mathcal{E}_{k,\lambda}S_{n-k,\lambda}(x,y)$$
$$= \sum_{k=0}^{[\frac{n-1}{2}]}\sum_{l=2k+1}^{n}\binom{n}{l}\lambda^{l-2k-1}(-1)^k y^{2k+1}S^{(1)}(l,2k+1)\mathcal{E}_{n-l,\lambda}(x).$$

By (24), we get

$$2e_\lambda^x(t)\cos_\lambda^{(y)}(t) = \sum_{l=0}^{\infty}\mathcal{E}_{l,\lambda}^{(c)}(x,y)\frac{t^l}{l!}(e_\lambda(t)+1) \tag{40}$$
$$= \sum_{l=0}^{\infty}\mathcal{E}_{l,\lambda}^{(c)}(x,y)\frac{t^l}{l!}\sum_{m=0}^{\infty}(1)_{m,\lambda}\frac{t^m}{m!} + \sum_{n=0}^{\infty}\mathcal{E}_{n,\lambda}^{(c)}(x,y)\frac{t^n}{n!}$$
$$= \sum_{n=0}^{\infty}\left(\sum_{l=0}^{n}\binom{n}{l}(1)_{n-l,\lambda}\mathcal{E}_{l,\lambda}^{(c)}(x,y) + \mathcal{E}_{n,\lambda}^{(c)}(x,y)\right)\frac{t^n}{n!}.$$

Therefore by comparing the coefficients on both sides of (35) and (40), we obtain the following theorem.

Theorem 4. *For $n \geq 0$, we have*

$$C_{n,\lambda}(x,y) = \frac{1}{2}\left(\sum_{l=0}^{n}\binom{n}{l}(1)_{n-l,\lambda}\mathcal{E}_{l,\lambda}^{(c)}(x,y) + \mathcal{E}_{n,\lambda}^{(c)}(x,y)\right),$$

and

$$S_{n,\lambda}(x,y) = \frac{1}{2}\left(\sum_{l=0}^{n}\binom{n}{l}(1)_{n-l,\lambda}\mathcal{E}_{l,\lambda}^{(s)}(x,y) + \mathcal{E}_{n,\lambda}^{(s)}(x,y)\right).$$

From (24), we have

$$\begin{aligned}\sum_{n=0}^{\infty}\mathcal{E}_{n,\lambda}^{(c)}(x+r,y)\frac{t^n}{n!} &= \frac{2}{e_\lambda(t)+1}e_\lambda^{x+r}(t)\cos_\lambda^{(y)}(t) \\ &= \frac{2}{e_\lambda(t)+1}e_\lambda^{x}(t)\cos_\lambda^{(y)}(t)e_\lambda^{r}(t) \\ &= \sum_{l=0}^{\infty}\mathcal{E}_{l,\lambda}^{(c)}(x,y)\frac{t^l}{l!}\sum_{m=0}^{\infty}(r)_{m,\lambda}\frac{t^m}{m!} \\ &= \sum_{n=0}^{\infty}\left(\sum_{l=0}^{n}\binom{n}{l}\mathcal{E}_{l,\lambda}^{(c)}(x,y)(r)_{n-l,\lambda}\right)\frac{t^n}{n!}.\end{aligned} \tag{41}$$

Therefore, by comparing the coefficients on both sides of (41), we obtain the following proposition.

Proposition 1. *For $n \geq 0$, we have*

$$\mathcal{E}_{n,\lambda}^{(c)}(x+r,y) = \sum_{l=0}^{n}\binom{n}{l}\mathcal{E}_{l,\lambda}^{(c)}(x,y)(r)_{n-l,\lambda},$$

and

$$\mathcal{E}_{n,\lambda}^{(s)}(x+r,y) = \sum_{l=0}^{n}\binom{n}{l}\mathcal{E}_{l,\lambda}^{(s)}(x,y)(r)_{n-l,\lambda},$$

where r is a fixed real (or complex) number.

Now, we consider the reflection symmetric identities for the degenerate cosine-Euler polynomials. By (24), we get

$$\begin{aligned}\sum_{n=0}^{\infty}\mathcal{E}_{n,\lambda}^{(c)}(1-x,y)\frac{t^n}{n!} &= \frac{2}{e_\lambda(t)+1}e_\lambda^{1-x}(t)\cos_\lambda^{(y)}(t) \\ &= \frac{2}{1+e_\lambda^{-1}(t)}e_\lambda^{-x}(t)\cos_\lambda^{(y)}(t) \\ &= \frac{2}{e_{-\lambda}(-t)+1}e_{-\lambda}^{x}(-t)\cos_{-\lambda}^{(y)}(-t) \\ &= \sum_{n=0}^{\infty}\mathcal{E}_{n,-\lambda}^{(c)}(x,y)\frac{(-1)^n t^n}{n!},\end{aligned} \tag{42}$$

and

$$\begin{aligned}\sum_{n=0}^{\infty}\mathcal{E}_{n,\lambda}^{(s)}(1-x,y)\frac{t^n}{n!} &= \frac{2}{e_\lambda(t)+1}e_\lambda^{1-x}(t)\sin_\lambda^{(y)}(t) \\ &= \frac{2}{1+e_\lambda^{-1}(t)}e_\lambda^{-x}(t)\sin_\lambda^{(y)}(t) \\ &= \frac{2}{e_{-\lambda}(-t)+1}e_{-\lambda}^{x}(-t)\sin_{-\lambda}^{(y)}(-t) \\ &= -\sum_{n=0}^{\infty}\mathcal{E}_{n,-\lambda}^{(s)}(x,y)\frac{(-1)^n t^n}{n!}.\end{aligned} \tag{43}$$

Therefore, by (42) and (43), we obtain the following theorem.

Theorem 5. *For $n \geq 0$, we have*

$$\mathcal{E}_{n,\lambda}^{(c)}(1-x,y) = (-1)^n\mathcal{E}_{n,-\lambda}^{(c)}(x,y),$$

and

$$\mathcal{E}_{n,\lambda}^{(s)}(1-x,y) = (-1)^{n+1}\mathcal{E}_{n,-\lambda}^{(s)}(x,y),$$

Now, we observe that

$$\begin{aligned}
\sum_{n=0}^{\infty} \mathcal{E}_{n,\lambda}^{(c)}(x,y)\frac{t^n}{n!} &= \frac{2}{e_\lambda(t)+1}(e_\lambda(t)-1+1)^x \cos_\lambda^{(y)}(t) \\
&= \frac{2}{e_\lambda(t)+1}\sum_{l=0}^{\infty}\binom{x}{l}(e_\lambda(t)-1)^l \cos_\lambda^{(y)}(t) \\
&= \frac{2}{e_\lambda(t)+1}\cos_\lambda^{(y)}(t)\sum_{l=0}^{\infty}(x)_l\sum_{k=l}^{\infty}S_\lambda^{(2)}(k,l)\frac{t^k}{k!} \\
&= \sum_{j=0}^{\infty}\mathcal{E}_{j,\lambda}^{(c)}(y)\frac{t^j}{j!}\sum_{k=0}^{\infty}\left(\sum_{l=0}^{k}(x)_l S_\lambda^{(2)}(k,l)\right)\frac{t^k}{k!} \\
&= \sum_{n=0}^{\infty}\left(\sum_{k=0}^{n}\sum_{l=0}^{k}\binom{n}{k}(x)_l S_\lambda^{(2)}(k,l)\mathcal{E}_{n-k}^{(c)}(y)\right)\frac{t^n}{n!}.
\end{aligned} \tag{44}$$

Therefore, by (44), we obtain the following theorem.

Theorem 6. *For $n \geq 0$, we have*

$$\mathcal{E}_{n,\lambda}^{(c)}(x,y) = \sum_{k=0}^{n}\sum_{l=0}^{k}\binom{n}{l}(x)_l S_\lambda^{(2)}(k,l)\mathcal{E}_{n-k,\lambda}^{(c)}(y).$$

Also, for $n \in \mathbb{N}$, we have

$$\mathcal{E}_{n,\lambda}^{(s)}(x,y) = \sum_{k=0}^{n}\sum_{l=0}^{k}\binom{n}{k}(x)_l S_\lambda^{(2)}(k,l)\mathcal{E}_{n-k,\lambda}^{(s)}(y).$$

3. Degenerate Bernoulli Polynomials of Complex Variable

In this section, we will consider the degenerate Bernoulli polynomials of complex variable and, by treating the real and imaginary parts separately, introduce the degenerate cosine-Bernoulli polynomials and degenerate sine-Bernoulli polynomials.

From (4), we have

$$\frac{t}{e_\lambda(t)-1}e_\lambda^{x+iy}(t) = \sum_{n=0}^{\infty}\beta_{n,\lambda}(x+iy)\frac{t^n}{n!}, \tag{45}$$

and

$$\frac{t}{e_\lambda(t)-1}e_\lambda^{x-iy} = \sum_{n=0}^{\infty}\beta_{n,\lambda}(x-iy)\frac{t^n}{n!}. \tag{46}$$

Thus, by (45) and (46), we get

$$\sum_{n=0}^{\infty}\left(\beta_{n,\lambda}(x+iy)+\beta_{n,\lambda}(x-iy)\right)\frac{t^n}{n!} = 2\frac{t}{e_\lambda(t)-1}e_\lambda^x(t)\cos_\lambda^{(y)}(t), \tag{47}$$

and

$$\sum_{n=0}^{\infty}\left(\beta_{n,\lambda}(x+iy)-\beta_{n,\lambda}(x-iy)\right)\frac{t^n}{n!} = 2i\frac{t}{e_\lambda(t)-1}e_\lambda^x(t)\sin_\lambda^{(y)}(t). \tag{48}$$

In view of (24) and (25), we define the degenerate cosine-Bernoulli polynomials and degenerate sine-Bernoulli polynomials respectively by

$$\frac{t}{e_\lambda(t)-1}e_\lambda^x(t)\cos_\lambda^{(y)}(t)=\sum_{n=0}^{\infty}\beta_{n,\lambda}^{(c)}(x,y)\frac{t^n}{n!}, \tag{49}$$

and

$$\frac{t}{e_\lambda(t)-1}e_\lambda^x(t)\sin_\lambda^{(y)}(t)=\sum_{n=0}^{\infty}\beta_{n,\lambda}^{(s)}(x,y)\frac{t^n}{n!}. \tag{50}$$

Note that $\beta_{0,\lambda}^{(s)}(x,y)=0$.
From (47)–(50), we have

$$\beta_{n,\lambda}^{(c)}(x,y)=\frac{\beta_{n,\lambda}(x+iy)+\beta_{n,\lambda}(x-iy)}{2}, \tag{51}$$

and

$$\beta_{n,\lambda}^{(s)}(x,y)=\frac{\beta_{n,\lambda}(x+iy)-\beta_{n,\lambda}(x-iy)}{2i}, \quad (n\geq 0). \tag{52}$$

Note that

$$\lim_{\lambda\to 0}\beta_{n,\lambda}^{(c)}(x,y)=B_n^{(c)}(x,y), \quad \lim_{\lambda\to 0}\beta_{n,\lambda}^{(s)}(x,y)=B_n^{(s)}(x,y),$$

where $B_n^{(c)}(x,y)$, $B_n^{(s)}(x,y)$ are cosine-Bernoulli polynomials, and sine-Bernoulli polynomials (see [12,16]).

By (49), we get

$$\begin{aligned}\sum_{n=0}^{\infty}\beta_{n,\lambda}^{(c)}(x,y)\frac{t^n}{n!}&=\frac{t}{e_\lambda(t)-1}e_\lambda^x(t)\cos_\lambda^{(y)}(t)\\&=\sum_{l=0}^{\infty}\beta_{l,\lambda}\frac{t^l}{l!}\sum_{m=0}^{\infty}C_{m,\lambda}(x,y)\frac{t^m}{m!}\\&=\sum_{n=0}^{\infty}\left(\sum_{l=0}^{n}\binom{n}{l}\beta_{l,\lambda}C_{n-l,\lambda}(x,y)\right)\frac{t^n}{n!}.\end{aligned} \tag{53}$$

On the other hand,

$$\begin{aligned}&\frac{t}{e_\lambda(t)-1}e_\lambda^x(t)\cos_\lambda^{(y)}(t)\\&=\sum_{m=0}^{\infty}\beta_{m,\lambda}(x)\frac{t^m}{m!}\sum_{l=0}^{n}\sum_{k=0}^{[\frac{l}{2}]}\lambda^{l-2k}(-1)^k y^{2k}S^{(1)}(l,2k)\frac{t^l}{l!}\\&=\sum_{n=0}^{\infty}\left(\sum_{l=0}^{n}\sum_{k=0}^{[\frac{l}{2}]}\binom{n}{l}\lambda^{l-2k}(-1)^k y^{2k}S^{(1)}(l,2k)\beta_{n-l,\lambda}(x)\right)\frac{t^n}{n!}\\&=\sum_{n=0}^{\infty}\left(\sum_{k=0}^{[\frac{n}{2}]}\sum_{l=2k}^{n}\binom{n}{l}\lambda^{l-2k}(-1)^k y^{2k}S^{(1)}(l,2k)\beta_{n-l,\lambda}(x)\right)\frac{t^n}{n!}.\end{aligned} \tag{54}$$

Therefore, by (53) and (54), we obtain the following theorem.

Theorem 7. *For $n \geq 0$, we have*

$$\beta^{(c)}_{n,\lambda}(x,y) = \sum_{k=0}^{n}\binom{n}{k}\beta_{k,\lambda}C_{n-k,\lambda}(x,y)$$
$$= \sum_{k=0}^{[\frac{n}{2}]}\sum_{l=2k}^{n}\binom{n}{l}\lambda^{l-2k}(-1)^k y^{2k} S^{(1)}(l,2k)\beta_{n-l,\lambda}(x).$$

Also, for $n \in \mathbb{N}$, we have

$$\beta^{(s)}_{n,\lambda}(x,y) = \sum_{k=0}^{n}\binom{n}{k}\beta_{k,\lambda}S_{n-k,\lambda}(x,y)$$
$$= \sum_{k=0}^{[\frac{n-1}{2}]}\sum_{l=2k+1}^{n}\binom{n}{l}\lambda^{l-2k-1}(-1)^k y^{2k+1} S^{(1)}(l,2k+1)\beta_{n-l,\lambda}(x).$$

and

$$\beta^{(s)}_{0,\lambda}(x,y) = 0.$$

From (49), we have

$$\sum_{n=0}^{\infty}\beta^{(c)}_{n,\lambda}(1-x,y)\frac{t^n}{n!} = \frac{t}{1-e_\lambda^{-1}(t)}e_\lambda^{-x}(t)\cos^{(y)}_\lambda(t) \tag{55}$$
$$= \frac{-t}{e_{-\lambda}(-t)-1}e^x_{-\lambda}(-t)\cos^{(y)}_{-\lambda}(-t)$$
$$= \sum_{n=0}^{\infty}\beta^{(c)}_{n,-\lambda}(x,y)\frac{(-1)^n}{n!}t^n.$$

Therefore, by (55), we obtain the following theorem.

Theorem 8. *For $n \geq 0$, we have*

$$\beta^{(c)}_{n,\lambda}(1-x,y) = (-1)^n\beta^{(c)}_{n,-\lambda}(x,y),$$

and

$$\beta^{(s)}_{n,\lambda}(1-x,y) = (-1)^{n+1}\beta^{(s)}_{n,-\lambda}(x,y).$$

By (49), we easily get

$$\sum_{n=0}^{\infty}\beta^{(c)}_{n,\lambda}(x+r,y)\frac{t^n}{n!} = \frac{t}{e_\lambda(t)-1}e_\lambda^{x+r}(t)\cos^{(y)}_\lambda(t) \tag{56}$$
$$= \frac{t}{e_\lambda(t)-1}e^x_\lambda(t)\cos^{(y)}_\lambda(t)e^r_\lambda(t)$$
$$= \sum_{l=0}^{\infty}\beta^{(c)}_{l,\lambda}(x,y)\frac{t^l}{l!}\sum_{m=0}^{\infty}(r)_{m,\lambda}\frac{t^m}{m!}$$
$$= \sum_{n=0}^{\infty}\left(\sum_{l=0}^{n}\binom{n}{l}\beta^{(c)}_{l,\lambda}(x,y)(r)_{n-l,\lambda}\right)\frac{t^n}{n!}.$$

By comparing the coefficients on both sides of (56), we get

$$\beta_{n\lambda}^{(c)}(x+r,y)=\sum_{l=0}^{n}\binom{n}{l}\beta_{l,\lambda}^{(c)}(x,y)(r)_{n-l,\lambda},\tag{57}$$

and

$$\beta_{n,\lambda}^{(s)}(x+r,y)=\sum_{l=0}^{n}\binom{n}{l}\beta_{l,\lambda}^{(s)}(x,y)(r)_{n-l,\lambda},\tag{58}$$

where r is a fixed real (or complex) number.

From (49), we note that

$$\begin{aligned}te_{\lambda}^{x}(t)\cos_{\lambda}^{(y)}(t)&=\sum_{l=0}^{\infty}\beta_{l,\lambda}^{(c)}(x,y)\frac{t^{l}}{l!}(e_{\lambda}(t)-1)\\&=\sum_{l=0}^{\infty}\beta_{l,\lambda}^{(c)}(x,y)\frac{t^{l}}{l!}\sum_{m=0}^{\infty}(1)_{m,\lambda}\frac{t^{m}}{m!}-\sum_{n=0}^{\infty}\beta_{n,\lambda}^{(c)}(x,y)\frac{t^{n}}{n!}\\&=\sum_{n=0}^{\infty}\left(\sum_{l=0}^{n}\binom{n}{l}\beta_{l,\lambda}^{(c)}(x,y)(1)_{n-l,\lambda}-\beta_{n,\lambda}^{(c)}(x,y)\right)\frac{t^{n}}{n!}\\&=\sum_{n=1}^{\infty}\left(\beta_{n,\lambda}^{(c)}(x+1,y)-\beta_{n,\lambda}^{(c)}(x,y)\right)\frac{t^{n}}{n!}\\&=\sum_{n=0}^{\infty}\left(\frac{\beta_{n+1,\lambda}^{(c)}(x+1,y)-\beta_{n+1,\lambda}^{(c)}(x,y)}{n+1}\right)\frac{t^{n+1}}{n!}.\end{aligned}\tag{59}$$

By (59), we get

$$\sum_{n=0}^{\infty}\left(\frac{\beta_{n+1,\lambda}^{(c)}(x+1,y)-\beta_{n+1,\lambda}^{(c)}(x,y)}{n+1}\right)\frac{t^{n}}{n!}=e_{\lambda}^{x}(t)\cos_{\lambda}^{(y)}(t)=\sum_{n=0}^{\infty}C_{n,\lambda}(x,y)\frac{t^{n}}{n!}.\tag{60}$$

Therefore, by comparing the coefficients on both sides of (60), we obtain the following theorem.

Theorem 9. *For $n\geq 0$, we have*

$$C_{n,\lambda}(x,y)=\frac{1}{n+1}\{\beta_{n+1,\lambda}^{(c)}(x+1,y)-\beta_{n+1,\lambda}^{(c)}(x,y)\},$$

and

$$S_{n,\lambda}(x,y)=\frac{1}{n+1}\{\beta_{n+1,\lambda}^{(s)}(x+1,y)-\beta_{n+1,\lambda}^{(s)}(x,y)\}.$$

Corollary 1. *For $n\geq 1$, we have*

$$C_{n,\lambda}(x,y)=\frac{1}{n+1}\sum_{l=0}^{n}\binom{n+1}{l}\beta_{l,\lambda}^{(c)}(x,y)(1)_{n+1-l,\lambda},$$

and

$$S_{n,\lambda}(x,y)=\frac{1}{n+1}\sum_{l=0}^{n}\binom{n+1}{l}\beta_{l,\lambda}^{(s)}(x,y)(1)_{n+1-l,\lambda}.$$

When $x=0$, let $\beta_{n,\lambda}^{(c)}(0,y)=\beta_{n,\lambda}^{(c)}(y)$, $\beta_{n,\lambda}^{(s)}(0,y)=\beta_{n,\lambda}^{(s)}(y)$, $\mathcal{E}_{n,\lambda}^{(c)}(0,y)=\mathcal{E}_{n,\lambda}^{(c)}(y)$, and $\mathcal{E}_{n,\lambda}^{(s)}(0,y)=\mathcal{E}_{n,\lambda}^{(s)}(y)$.

For $n \geq 0$, we have

$$\beta_{n,\lambda}^{(c)}(y) = \sum_{k=0}^{[\frac{n}{2}]} \sum_{l=2k}^{n} \binom{n}{l} \lambda^{l-2k}(-1)^k y^{2k} S^{(1)}(l,2k)\beta_{n-l,\lambda}. \tag{61}$$

Also, for $n \in \mathbb{N}$, we get

$$\beta_{n,\lambda}^{(s)}(y) = \sum_{k=0}^{[\frac{n-1}{2}]} \sum_{l=2k+1}^{n} \binom{n}{l} \lambda^{l-2k-1}(-1)^k y^{2k+1} S^{(1)}(l,2k+1)\beta_{n-l,\lambda}. \tag{62}$$

By (49), we get

$$\begin{aligned} \sum_{n=0}^{\infty} \beta_{n,\lambda}^{(c)}(x,y)\frac{t^n}{n!} &= \frac{t}{e_\lambda(t)-1}\cos_\lambda^{(y)}(t)(e_\lambda(t)-1+1)^x \\ &= \sum_{m=0}^{\infty} \beta_{m,\lambda}^{(c)}(y)\frac{t^m}{m!}\sum_{l=0}^{\infty}(x)_l \sum_{k=l}^{\infty} S_\lambda^{(2)}(k,l)\frac{t^k}{k!} \\ &= \sum_{m=0}^{\infty} \beta_{m,\lambda}^{(c)}(y)\frac{t^m}{m!}\sum_{k=0}^{\infty}\sum_{l=0}^{k}(x)_l S_\lambda^{(2)}(k,l)\frac{t^k}{k!} \\ &= \sum_{n=0}^{\infty}\left(\sum_{k=0}^{n}\sum_{l=0}^{k}\binom{n}{k}(x)_l S_\lambda^{(2)}(k,l)\beta_{n-k,\lambda}^{(c)}(y)\right)\frac{t^n}{n!}. \end{aligned} \tag{63}$$

Comparing the coefficients on both sides of (63), we have

$$\beta_{n,\lambda}^{(c)}(x,y) = \sum_{k=0}^{n}\sum_{l=0}^{k}\binom{n}{k}(x)_l S_\lambda^{(2)}(k,l)\beta_{n-k,\lambda}^{(c)}(y).$$

Also, for $n \in \mathbb{N}$, we get

$$\beta_{n,\lambda}^{(s)}(x,y) = \sum_{k=0}^{n}\sum_{l=0}^{k}\binom{n}{k}(x)_l S_\lambda^{(2)}(k,l)\beta_{n-k,\lambda}^{(s)},$$

and

$$\beta_{0,\lambda}^{(s)}(x,y) = 0.$$

4. Conclusions

In [15], the authors introduced the so-called the new type Euler polynomials by means of generating functions (see (9) and (10)) and deduced several properties and identities for these polynomials. Hacène Belbachir, the reviewer of the paper [15], asked the following question in Mathematical Reviews (MR3808565) of the American Mathematical Society: Is it possible to obtain their results by considering the classical Euler polynomials of complex variable z, and treating the real part and the imaginary part separately?

Our result gives an affirmative answer to the question (see (16)). In this paper, we considered the degenerate Euler and Bernoulli polynomials of a complex variable and, by treating the real and imaginary parts separately, were able to introduce degenerate cosine-Euler polynomials, degenerate sine-Euler polynomials, degenerate cosine-Bernoulli polynomials, and degenerate sine-Bernoulli polynomials. They are degenerate versions of the new type Euler polynomials studied by Masjed–Jamei, Beyki and Koepf [15] and of the 'new type Bernoulli polynomials.'

In Section 2, the degenerate cosine-polynomials and degenerate sine-polynomials were introduced and their explicit expressions were derived. The degenerate cosine-Euler polynomials and degenerate sine-Euler polynomials were expressed in terms of degenerate cosine-polynomials and degenerate sine-polynomials and vice versa. Further, some reflection identities were found for the degenerate cosine-Euler polynomials and degenerate sine-Euler polynomials. In Section 3, the degenerate cosine-Bernoulli polynomials and degenerate sine-Bernoulli polynomials were introduced. They were expressed in terms of degenerate cosine-polynomials and degenerate sine-polynomials and vice versa. Reflection symmetries were deduced for the degenerate cosine-Bernoulli polynomials and degenerate sine-Bernoulli polynomials. Further, some expressions involving the degenerate Stirling numbers of the second kind were derived for them.

It was Carlitz [1,2] who initiated the study of degenerate versions of some special polynomials, namely the degenerate Bernoulli and Euler polynomials. Studying degenerate versions of some special polynomials and numbers have turned out to be very fruitful and promising (see [3,5–11,13,14,19] and references therein). In fact, this idea of considering degenerate versions of some special polynomials are not limited just to polynomials but can be extended even to transcendental functions like gamma functions [8].

Author Contributions: All authors contributed equally to the manuscript and typed, read, and approved the final manuscript.

Acknowledgments: The authors would like to thank the referees for their valuable comments and suggestions.

References

1. Carlitz, L. Degenerate Stirling, Bernoulli and Eulerian numbers. *Utilitas Math.* **1979**, *15*, 51–88.
2. Carlitz, L., A degenerate Staud-Clausen theorem. *Arch. Math.* **1956**, *7*, 28–33. [CrossRef]
3. Dolgy, D.V.; Kim, T. Some explicit formulas of degenerate Stirling numbers associated with the degenerate special numbers and polynomials. *Proc. Jangjeon Math. Soc.* **2018**, *21*, 309–317.
4. Haroon, H.; Khan, W.A. Degenerate Bernoulli numbers and polynomials associated with degenerate Hermite polynomials. *Commum Korean Math. Soc.* **2018**, *33*, 651–669.
5. Jang, G.-W.; Kim, T.; Kwon, H.-I. On the extension of degenerate Stirling polynomials of the second kind and degenerate Bell polynomials. *Adv. Stud. Contemp. Math.* **2018**, *28*, 305–316.
6. Kim, D.S.; Kim, T. A note on polyexponential and unipoly functions. *Russ. J. Math. Phys.* **2019**, *26*, 40–49. [CrossRef]
7. Kim, T.; Jang, G.-W. A note on degenerate gamma function and degenerate Stirling number of the second kind. *Adv. Stud. Contemp. Math.* **2018**, *28*, 207–214.
8. Kim, T.; Kim, D.S. Degenerate Laplace transform and degenerate gamma function. *Russ. J. Math. Phys.* **2017**, *24*, 241–248. [CrossRef]
9. Kim, T. A note on degenerate Stirling polynomials of the second kind. *Proc. Jangjeon Math. Soc.* **2017**, *20*, 319–331.
10. Kim, T.; Yao, Y.; Kim, D.S.; Jang, G.-W. Degenerate r-Stirling numbers and r-Bell polynomials. *Russ. J. Math. Phys.* **2018**, *25*, 44–58. [CrossRef]
11. Kim, T.; Kim, D.S.; Kwon, H.-I. A note on degenerate Stirling numbers and their applications. *Proc. Jangjeon Math. Soc.* **2018**, *21*, 195–203.
12. Kim, T.; Ryoo, C.S. Some identities for Euler and Bernoulli polynomials and their zeros. *Axioms* **2018**, *7*, 56. [CrossRef]
13. Lee, J.G.; Jang, L.-C. On modified degenerate Carlitz q-Bernoulli numbers and polynomials. *Adv. Differ. Equ.* **2017**, *2017*, 22. [CrossRef]
14. Lee, J.G.; Kwon, J. The modified degenerate q-Bernoulli polynomials arising from p-adic invariant integral on $\mathbb{Z}_p$. *Adv. Difference Equ.* **2017**, *2017*, 29. [CrossRef]

15. Masjed-Jamei, M.; Beyki, M.R.; Koepf, W. A new type of Euler polynomials and numbers. *Mediterr. J. Math.* **2018**, *15*, 138. [CrossRef]
16. Masjed-Jamei, M.; Beyki, M.R.; Koepf, W. An extension of the Euler-Maclauin quadrature formula using a parametric type of Bernoulli polynomials. 2019, in press.
17. Roman, S. The umbral calculus. In *Pure and Applied Mathematics*; Academic Press, Inc. [Harcourt Brace Jovanovich, Publishers]: New York, NY, USA.
18. Simsek, Y. Identities on the Changhee numbers and Apostol-type Daehee polynomials. *Adv. Stud. Contemp. Math.* **2017**, *27*, 199–212.
19. Zhang, Z.; Yang, J. On sums of products of the degenerate Bernoulli numbers. *Integral Transforms Spec. Funct.* **2009**, *20*, 751–755. [CrossRef]
20. Zill, D.G.; Cullen, M.R. *Advanced Engineering Mathematics*; Jones and Barrtlett: Mississauga, ON, Canada, 2006.

2

Supersymmetric Polynomials on the Space of Absolutely Convergent Series

Farah Jawad and Andriy Zagorodnyuk *

Department of Mathematical and Functional Analysis, Vasyl Stefanyk Precarpathian National University, 57 Shevchenka Str., 76018 Ivano-Frankivsk, Ukraine

* Correspondence: azagorodn@gmail.com

Abstract: We consider an algebra H_b^{sup} of analytic functions on the Banach space of two-sided absolutely summing sequences which is generated by so-called supersymmetric polynomials. Our purpose is to investigate H_b^{sup} and its spectrum with using methods of infinite dimensional complex analysis and the theory of Fréchet algebras. Some algebraic bases of H_b^{sup} are described. Also, we show that the spectrum of the algebra of supersymmetric analytic functions of bounded type contains a metric ring $\mathcal{M}$. We prove that $\mathcal{M}$ is a complete metric (nonlinear) space and investigate homomorphisms and additive operators on this ring. Some possible applications are discussed.

Keywords: symmetric and supersymmetric polynomials on Banach spaces; algebras of analytic functions on Banach spaces; spectra algebras of analytic functions

---

1. Introduction and Preliminaries

Let X be a complex Banach space. A (continuous) map $P\colon X \to \mathbb{C}$ is said to be a (continuous) n-homogeneous polynomial if there exists a (continuous) n-linear map $B_P\colon X^n \to \mathbb{C}$ such that $P(x) = B_P(x, \ldots, x)$. 0-homogeneous polynomial is just a constant function. A finite sum of homogeneous polynomials is a polynomial. We denote by $\mathcal{P}(^nX)$ the space of all continuous n-homogeneous polynomials on X and by $\mathcal{P}(X)$ the space of all polynomials on X. Note that $\mathcal{P}(^nX)$ is a Banach space with respect to any of the norms

$$\|P\|_r = \sup_{\|x\| \le r} |P(x)|, \qquad r > 0. \tag{1}$$

Let τ_b be the topology on $\mathcal{P}(X)$ of uniform convergence on bounded subsets of X. This topology is generated by the countable family of norms (1) for positive rational numbers r and so is metrisable. We denote by $H_b(X)$ the completion of $(\mathcal{P}(X), \tau_b)$. So $H_b(X)$ is a Fréchet algebra which consists of entire analytic functions on X which are bounded on all bounded subsets (so-called *entire functions of bounded type*). For details on polynomials and analytic functions on Banach spaces we refer the reader to [1]. The spectra (sets of continuous complex homomorphisms = sets of characters) of $H_b(X)$ and its subalgebras were investigated by many authors (see e.g., [2–5]).

Let G be a group of isometric operators on X. We denote by $H_{bG}(X)$ the subalgebra of $H_b(X)$ which consists of G-invariant analytic functions. Such algebras were considered in the general case in [6,7]. For some special cases of G there is a sequence of G-symmetric homogeneous polynomials

$\{P_1, P_2, \ldots, P_n, \ldots\}$, $\deg P_n = n$ which forms an algebraic basis in the algebra of G-symmetric polynomials $\mathcal{P}_G(X)$. For example, if $G = s$ is the group of all permutations of the basis vectors in ℓ_1, then the functions

$$F_k(x) = \sum_{i=1}^{\infty} x_i^k, \qquad k \in \mathbb{N}$$

form an algebraic basis in $\mathcal{P}_s(\ell_1)$ [8]. The following bases in $\mathcal{P}_s(\ell_1)$ also are important

$$G_n(x) = \sum_{i_1 < \cdots < i_n} x_{i_1} \cdots x_{i_n}$$

and

$$H_n(x) = \sum_{i_1 \leq \cdots \leq i_n} x_{i_1} \cdots x_{i_n}.$$

Let $\mathcal{F}(x)(t)$, $\mathcal{G}(x)(t)$ and $\mathcal{H}(x)(t)$ be formal series

$$\mathcal{F}(x)(t) = \sum_{n=1}^{\infty} t^{n-1} F_n(x),$$

$$\mathcal{G}(x)(t) = \sum_{n=0}^{\infty} t^n G_n(x), \qquad G_0 = 1$$

and

$$\mathcal{H}(x)(t) = \sum_{n=0}^{\infty} t^n H_n(x), \qquad H_0 = 1$$

which also are called generating functions. From combinatorial considerations it is known ([9] p. 3) that

$$\mathcal{G}(x)(t) = \frac{1}{H(-x)(t)} \tag{2}$$

and

$$\mathcal{G}(x)(t) = \exp\left(-\sum_{n=1}^{\infty} t^n \frac{F_n(-x)}{n}\right) = \exp\left(-\int_0^t \mathcal{F}(-x)(\xi)d\xi\right), \tag{3}$$

where the equality holds for every $x \in \ell_1$ and every t in the common domain of convergence. In [10] it is shown that every complex homomorphism φ of $H_{bs}(\ell_1)$ is completely defined by its value on $\mathcal{G}(x)(t)$ and

$$g(t) = \varphi(\mathcal{G}(t)) = \sum_{n=0}^{\infty} t^n \varphi(G_n) \tag{4}$$

is a function of exponential type with $g(0) = 1$. Moreover, if $\varphi = \delta_x$ is the point evaluation functional at $x \in \ell_1$ (that is $\delta_x(f) = f(x)$, $f \in H_b(\ell_1)$), then

$$\delta_x(\mathcal{G}(t)) = \mathcal{G}(x)(t) = \prod_{k=1}^{\infty}(1 + x_k t). \tag{5}$$

Note that (5) is an absolutely convergent Hadamard Product—the entire function defined by its zeros $a_n = 1/(-x_n)$ for $x_n \neq 0$. Also [10,11], there is a family ψ_λ, $\lambda \in \mathbb{C}$ in the spectrum of $H_b(\ell_1)$ such that

$$\psi_\lambda(\mathcal{G}(t)) = e^{\lambda t}.$$

In [12] it is shown that there is a function of exponential type γ with $\gamma(0) = 1$ but which cannot be represented as in (4). Spectra of algebras $H_{bs}(\ell_p)$ were investigated also in [13,14]. Polynomials which are symmetric with respect to some other representations of the group of permutations of natural numbers were considered in [15–17].

In this paper we consider a subalgebra of entire functions of bounded type which is generated by so-called supersymmetric polynomials. Algebras of supersymmetric polynomials on finite-dimensional spaces were considered in [18–20]. In Section 2.1 we consider some important bases in the algebra of supersymmetric polynomials. Section 2.2 is devoted to the spectrum of the algebra of supersymmetric analytic functions of bounded type. In particular, we show that the set of point evaluation functionals on the algebra can be described as a metric ring which is not a linear space. Some operators on this ring are investigated.

2. Results

2.1. Bases of Supersymmetric Polynomials

We will use $\mathbb{N}$ for natural numbers and $\mathbb{Z}$ for integers. Also, we set $\mathbb{Z}_0 = \mathbb{Z} \setminus 0$ and denote by $\ell_1(\mathbb{Z}_0)$ the Banach space of all absolutely summing complex sequences indexed by numbers in $\mathbb{Z}_0$. The symbol $\ell_1 = \ell_1(\mathbb{N})$ means the classical Banach space of absolutely summing complex sequences. Any element z in $\ell_1(\mathbb{Z}_0)$ has the representation

$$z = (\ldots, z_{-n}, \ldots, z_{-2}, z_{-1}, z_1, z_2, \ldots, z_n, \ldots)$$
$$= (y|x) = (\ldots, y_n, \ldots, y_2, y_1|x_1, x_2, \ldots, x_n, \ldots)$$

with

$$\|z\| = \sum_{i=-\infty}^{\infty} |z_i|,$$

where $x = (x_1, x_2, \ldots, x_n, \ldots)$ and $y = (y_1, y_2, \ldots, y_n, \ldots)$ are in ℓ_1, $z_n = x_n$, $z_{-n} = y_n$ for $n \in \mathbb{N}$ and

$$x \longmapsto (0|x_1, x_2, \ldots, x_n, \ldots) \text{ and } y \longmapsto (\ldots, y_{-n}, \ldots, y_{-2}, y_{-1}|0)$$

are natural isometric embeddings of the copies of ℓ_1 into $\ell_1(\mathbb{Z}_0)$.

Let us define the following polynomials on $\ell_1(\mathbb{Z}_0)$:

$$T_k(z) = F_k(x) - F_k(y) = \sum_{i=1}^{\infty} x_i^k - \sum_{i=1}^{\infty} y_i^k, \quad k \in \mathbb{N}.$$

Definition 1. *A polynomial P on $\ell_1(\mathbb{Z}_0)$ is said to be supersymmetric if it can be represented as an algebraic combination of polynomials $\{T_k\}_{k=1}^{\infty}$. In other words, P is a finite sum of finite products of polynomials in $\{T_k\}_{k=1}^{\infty}$ and constants. We denote by $\mathcal{P}_{sup}$ the algebra of all supersymmetric polynomials on $\ell_1(\mathbb{Z}_0)$.*

Note first that polynomials T_k are algebraically independent because F_k are so. Hence $\{T_k\}_{k=1}^{\infty}$ forms an algebraic basis in $\mathcal{P}_{sup}$.

We say that $z \sim w$, for some $z, w \in \ell_1(\mathbb{Z}_0)$ if $T_k(z) = T_k(w)$ for every $k \in \mathbb{N}$. Let us denote by $\mathcal{M}$ the quotient set $\ell_1(\mathbb{Z}_0)/\sim$ which is a natural domain for supersymmetric polynomials. For a given $z \in \ell_1(\mathbb{Z}_0)$, let $[z] \in \mathcal{M}$ be the class of equivalence which contains z.

Similarly like in [10] we introduce an operation "$\bullet$" on $\ell_1(\mathbb{Z}_0)$:

$$z \bullet w = (y \bullet v|x \bullet u) = (\ldots, v_n, y_n, \ldots, v_1, y_1|x_1, u_1, \ldots, x_n, u_n, \ldots),$$

where $z = (y|x)$ and $w = (v|u)$. Also, we denote $z^{-} = (y|x)^{-} = (x|y)$. Clearly, $(z^{-})^{-} = z$ and $z \bullet z^{-} \sim (0|0)$. These operations can be naturally defined on $\mathcal{M}$ by

$$[z] \bullet [w] = [z \bullet w] \text{ and } [z]^{-} = [z^{-}]. \tag{6}$$

Theorem 1. *The following statements hold:*

1. $T_k(z \bullet w) = T_k(z) + T_k(w)$ *for every* $k \in \mathbb{N}$.
2. *The operations in (6) are well defined, that is, they do not depend on the choice of representatives.*
3. $(\mathcal{M}, \bullet, [z] \mapsto [z]^{-})$ *is a commutative group with zero* $0 = (0|0)$.
4. $z \sim 0$ *if and only if there are* $d, s \in \ell_1$ *such that* $z = (d|s)$ *and* $F_k(d) = F_k(s)$ *for all* $k \in \mathbb{N}$. *Equivalently, all nonzero coordinates of* d *coincides with nonzero coordinates of* s *up to a permutation.*

Proof. Assertions (1)–(3) are straightforward consequences of definitions. In [13] is proved that for given $d, s \in \ell_1$, $F_k(d) = F_k(s)$ for all $k \in \mathbb{N}$ if and only if all nonzero coordinates of d coincides with nonzero coordinates of s up to a permutation. □

Let $\mathcal{P}_1$ and $\mathcal{P}_2$ be some algebras of polynomials on linear spaces X and Y respectively such that $\mathcal{P}_1$ is generated by an algebraic basis $\{P_1, P_2, \ldots, P_n, \ldots\}$ and $\mathcal{P}_2$ is generated by an algebraic basis $\{Q_1, Q_2, \ldots, Q_n, \ldots\}$ with $\deg P_n = \deg Q_n = n$, $n \in \mathbb{N}$. Then the map, defined on the basic vectors by $P_n \longmapsto Q_n$ and extended to $\mathcal{P}_1$ by linearity and multiplicativity, obviously is an algebraic isomorphism from $\mathcal{P}_1$ onto $\mathcal{P}_2$ which preserves degrees of polynomials.

Let us denote by Λ the isomorphism from $\mathcal{P}_s = \mathcal{P}_s(\ell_1)$ to $\mathcal{P}_{sup}$ such that

$$\Lambda \colon F_n \longmapsto T_n, \qquad n \in \mathbb{N}.$$

Proposition 1. *If* $\{P_n\}_{n=1}^{\infty}$ *is an algebraic basis in* $\mathcal{P}_s$, *then* $\{\Lambda(P_n)\}_{n=1}^{\infty}$ *is an algebraic basis in* $\mathcal{P}_{sup}$.

Proof. The proof follows from the general fact that the range of any algebraic basis under an isomorphism is an algebraic basis. Indeed, $\Lambda(P_n)$, $n \in \mathbb{N}$ are algebraically independent because P_n, $n \in \mathbb{N}$ are so and Λ is injective. Also, every $Q \in \mathcal{P}_{sup}$ belongs to the algebraic combination of $\{\Lambda(P_n)\}_{n=1}^{\infty}$ because $\Lambda^{-1}(Q)$ belongs to the algebraic combination of $\{P_n\}_{n=1}^{\infty}$ and Λ is surjective (cf. [13]). □

Let H_b^{sup} be the completion of $\mathcal{P}_{sup}$ with respect to the topology of uniform convergence on bounded subsets. In other words, H_b^{sup} is the minimal closed subspace of $H_b(\ell_1(\mathbb{Z}_0))$ which contains $\mathcal{P}_{sup}$. Elements of H_b^{sup} will be called *supersymmetric analytic* or *entire* functions on $\ell_1(\mathbb{Z}_0)$.

Proposition 2. *The map* Λ^{-1} *is continuous and can be extended to a continuous homomorphism from* H_b^{sup} *to* $H_{bs} = H_{bs}(\ell_1)$ *with a dense range. The map* Λ *is discontinuous and densely defined on* H_{bs}.

Proof. Let us observe first that $\Lambda^{-1}(P)$ is the restriction of $P \in \mathcal{P}_{sup}$ onto the closed subspace $\{(0|x) \colon x \in \ell_1\}$. The operator of the restriction is obviously continuous on H_b^{sup} and is the extension of Λ^{-1}. The range of Λ^{-1} is dense because it contains all symmetric polynomials on ℓ_1. On the other hand, in [10] it is proved that the homomorphism $\Lambda_{-} \colon \mathcal{P}_s \to \mathcal{P}_s$ such that $\Lambda_{-} F_k = -F_k$, $k \in \mathbb{N}$ is discontinuous on $(\mathcal{P}_s, \tau_b)$. Moreover, in [21] a function $g(x) \in H_{bs}$ such that $\Lambda_{-}(g) \notin H_{bs}$ was constructed. If Λ is continuous, it can be extended to the whole space H_{bs} and so $\Lambda g(x|0) = (\Lambda_{-} g)(x)$. It leads to a contradiction because on the left side we have a bounded function on all bounded subsets but on the right side, it is not so. □

For a given $y \in \ell_1$ we denote by $\Lambda_y P(x) = (\Lambda P)(y|x)$, $P \in \mathcal{P}_s$. It is easy to see that

$$\Lambda_y P(x \bullet y) = (\Lambda P)(y|x \bullet y) = (\Lambda P)(0|x) = P(x).$$

Theorem 2. *Let $\Lambda G_n = W_n$. Then*

$$W_n(y|x) = \sum_{k=0}^{n} G_k(x) H_{n-k}(-y), \qquad n \in \mathbb{N} \tag{7}$$

and

$$\mathcal{W}(y|x)(t) = \sum_{n=0}^{\infty} t^n W_n(y|x) = \frac{\mathcal{G}(x)(t)}{\mathcal{G}(y)(t)}, \tag{8}$$

where the equality is true on the common domains of convergence.

Proof. In [10] it is proved that

$$\mathcal{G}(x \bullet y)(t) = \mathcal{G}(x)(t)\mathcal{G}(y)(t), \qquad x, y \in \ell_1. \tag{9}$$

Hence, for a fixed $y \in \ell_1$

$$\Lambda_y(\mathcal{G}(x \bullet y)(t)) = \sum_{n=0}^{\infty} t^n (\Lambda G_n)(y|x) \sum_{n=0}^{\infty} t^n G_n(y) = \sum_{n=0}^{\infty} t^n W_n(y|x) \sum_{n=0}^{\infty} t^n G_n(y).$$

On the other hand,

$$\Lambda_y(\mathcal{G}(x \bullet y)(t)) = \sum_{n=0}^{\infty} t^n G_n(x).$$

So

$$\sum_{n=0}^{\infty} t^n W_n(y|x) \sum_{n=0}^{\infty} t^n G_n(y) = \sum_{n=0}^{\infty} t^n G_n(x). \tag{10}$$

From (10) we have

$$\mathcal{W}(y|x)(t)\mathcal{G}(y)(t) = \mathcal{G}(x)(t)$$

and so (8) holds. Taking into account Formula (2) we have

$$\sum_{n=0}^{\infty} t^n W_n(y|x) = \sum_{n=0}^{\infty} t^n G_n(x) \sum_{n=0}^{\infty} t^n H_n(-y) = \sum_{n=0}^{\infty} t^n \sum_{k=0}^{n} G_k(x) H_{n-k}(-y).$$

From here we have (7). □

Corollary 1.

$$\mathcal{W}((y|x) \bullet (d|b))(t) = \mathcal{W}(y|x)(t)\mathcal{W}(d|b)(t), \quad x, y, b, d \in \ell_1.$$

Proof. The required statement immediately follows if we combine Formulas (9) and (8). □

Corollary 2. *For every $n \in \mathbb{N}$ and $x, y, b, d \in \ell_1$ we have*

$$W_n((y|x) \bullet (d|b)) = \sum_{k=0}^{n} W_k(y|x) W_{n-k}(d|b),$$

$$G_n(x \bullet b) = \sum_{k=0}^{n} G_k(x) G_{n-k}(b)$$

and

$$H_n(y \bullet d) = \sum_{k=0}^{n} H_k(y) H_{n-k}(d).$$

Proof. From Corollary 1 we have

$$\sum_{n=0}^{\infty} t^n W_n((y|x) \bullet (d|b)) = \sum_{k=0}^{\infty} t^k W_k(y|x) \sum_{j=0}^{\infty} t^j W_j(d|b).$$

Taking coefficients of t^n we have the first equality. The second and thirds equalities we can obtain by the same reasoning. □

It is clear that $(y \bullet a | x \bullet a) \sim (y|x)$ for all $x, y, a \in \ell_1$. We say that $(y|x)$ is an *irreducible* representative of $u \in \mathcal{M}$ if $[(y|x)] = u$ and for every $x_n \neq 0$ and every k, $x_n \neq y_k$.

Proposition 3. *$(y|x)$ is irreducible if and only if $\mathcal{G}(x)(t)$ and $\mathcal{G}(y)(t)$ have no common zeros.*

Proof. According to (5), for nonzero elements x_k and y_k the numbers $(-x_k)^{-1}$ and $(-y_k)^{-1}$ are zeros of $\mathcal{G}(x)(t)$ and $\mathcal{G}(y)(t)$ respectively. □

Corollary 3. *Let $u \in \mathcal{M}$. Then u is completely defined by $\mathcal{W}(u)(t) = \mathcal{W}(y|x)(t)$ and $\mathcal{W}(y|x)(t)$ is a meromorphic functions of the form $f(t)/g(t)$ such that f, g are entire functions of exponential type with $f(0) = 1$ and $g(0) = 1$, where $(y|x) \in u$. Moreover, let (α_k) and β_k be zeros of f and g respectively. Then both $(1/\alpha_k)$ and $(1/\beta_k)$ belong to ℓ_1,*

$$f(t) = \prod_{k=1}^{\infty} \left(1 - \frac{t}{\alpha_k}\right) \quad \text{and} \quad g(t) = \prod_{k=1}^{\infty} \left(1 - \frac{t}{\beta_k}\right),$$

and

$$\left(\ldots, -\frac{1}{\beta_n}, \ldots, -\frac{1}{\beta_2}, -\frac{1}{\beta_1} \,\middle|\, -\frac{1}{\alpha_1}, -\frac{1}{\alpha_2}, \ldots, -\frac{1}{\alpha_n}, \ldots\right)$$

is an irreducible representation of u.

Let $x \in \ell_1$. We denote by $\operatorname{supp} x$ the support of x, that is,

$$\operatorname{supp} x = \{n \in \mathbb{N} \colon x_n \neq 0\}.$$

Corollary 4. *Let $(y|x)$ and $(y'|x')$ be two irreducible representatives of u. Then there are bijections $i \colon \operatorname{supp} x \to \operatorname{supp} x'$ and $j \colon \operatorname{supp} y \to \operatorname{supp} y'$ such that $x_n = x'_{i(n)}$ and $y_m = y'_{j(m)}$ for all $n \in \operatorname{supp} x$ and $m \in \operatorname{supp} y$.*

Proposition 4. *For every $u = [(y|x)] \in \mathcal{M}$ the following equality holds on the common domain of convergence*

$$\mathcal{W}(u)(t) = \exp\left(-\sum_{n=1}^{\infty} t^n \frac{T_n(-u)}{n}\right) = \exp\left(-\int_0^t \mathcal{T}(-u)(\xi)d\xi\right), \tag{11}$$

where $-u = [(-y| - x)]$.

Proof. From (8) and (5) it follows that $\mathcal{W}(u)(t)$ converges for every $t \in \mathbb{C}$ if $y = 0$ and in the ball $|t| < r$, where

$$r = \min_{|y_n| \neq 0} |y_n|^{-1}$$

if $y \neq 0$. Since Λ^{-1} is a continuous homomorphism, from (3) we have that for each $t \in \mathbb{C}$ such that $\mathcal{W}(u)(t)$ converges

$$\Lambda^{-1}\mathcal{W}(u)(t) = \mathcal{G}(x)(t) = \exp\left(-\sum_{n=1}^{\infty} t^n \frac{F_n(-x)}{n}\right).$$

Since $\|F_n\| = 1$, the series

$$\sum_{n=1}^{\infty} t^n \frac{F_n(-x)}{n}$$

converges if $|t| < \|x\|$. Also, $\|T_n\| = 1$ and the series

$$\sum_{n=1}^{\infty} t^n \frac{T_n(-u)}{n}$$

converges if $|t| < \|u\|$. So in the common domain of convergence

$$\exp\left(-\sum_{n=1}^{\infty} t^n \frac{F_n(-x)}{n}\right)$$

is in the domain of Λ and

$$\mathcal{W}(u)(t) = \Lambda\mathcal{G}(x)(t) = \Lambda \exp\left(-\sum_{n=1}^{\infty} t^n \frac{F_n(-x)}{n}\right).$$

Also, in the domain

$$\Lambda^{-1} \exp\left(-\sum_{n=1}^{\infty} t^n \frac{T_n(-u)}{n}\right) = \exp\left(-\sum_{n=1}^{\infty} t^n \frac{F_n(-x)}{n}\right).$$

□

Theorem 3. *Let $u \in \mathcal{M}$ and $u \neq 0$. For a given $\lambda \in \mathbb{C}$ there is $v \in \mathcal{M}$ such that $T_k(v) = \lambda T_k(u)$ if and only if λ is an integer number.*

Proof. Let $\lambda = m \in \mathbb{Z}$. If $n = 0$, then $v = 0$. If $n > 0$, then $v = \underbrace{u \bullet \cdots \bullet u}_{n}$. If $n < 0$, then $v = \underbrace{u^- \bullet \cdots \bullet u^-}_{n}$.

Let now $\lambda \notin \mathbb{Z}$. According to (11)

$$\mathcal{W}(v)(t) = (\mathcal{W}(u)(t))^{\lambda}.$$

But it contradicts representation (8) for v. □

2.2. *The Spectrum of H_B^{Sup} and the Nonlinear Normed Ring $\mathcal{M}$*

2.2.1. The Spectrum

Let us denote by M_b^{sup} the spectrum of H_b^{sup}, that is, the set of all continuous nonzero complex homomorphisms (characters) of H_b^{sup}. Clearly for every point $u \in \mathcal{M}$ there is a character $\delta_u \in M_b^{sup}$ (so-called point evaluation functional) such that $\delta_u(f) = f(u)$, $f \in H_b^{sup}$. Moreover, if $u \neq v$, then $\delta_u \neq \delta_v$. In this sense, we can say that $\mathcal{M} \subset M_b^{sup}$.

Since polynomials $\{W_n\}$ form an algebraic basis in H_b^{sup}, any character $\varphi \in M_b^{sup}$ is completely defined by its values on W_n, $n \in \mathbb{N}$. In other words, every character φ can be represented by the function

$$\varphi(\mathcal{W}(\iota)) = \sum_{n=0}^{\infty} \iota^n \varphi(W_n). \tag{12}$$

Note that if $\varphi = \delta_u$ for some $u \in \mathcal{M}$, then $\varphi(\mathcal{W}(t))$ can be described by Corollary 3. Using ideas in [11,13] it is possible to construct a character which is not a point-evaluation functional. Let λ and μ be complex numbers. Consider

$$u_n = \left(0,\dots,0,\frac{\mu}{n},\dots,\frac{\mu}{n}\Big|\frac{\lambda}{n},\dots,\frac{\lambda}{n},0,\dots,0\right).$$

From the compactness reasons, we have that $\{\delta_{u_n}\}$ has a cluster point $\psi_{\lambda,\mu}$ in M_b^{sup}. So

$$\psi_{\lambda,\mu}(\mathcal{W}(t)) = \lim_{n\to\infty} \frac{\sum_{k=0}^{\infty} t^k G_k(\lambda/n,\dots,\lambda/n,0,\dots,0)}{\sum_{k=0}^{\infty} t^k G_k(\mu/n,\dots,\mu/n,0,\dots,0)}.$$

Taking into account [10] that

$$\lim_{n\to\infty} \sum_{k=0}^{\infty} t^k G_k(\lambda/n,\dots,\lambda/n,0,\dots,0) = e^{\lambda t}$$

we have

$$\psi_{\lambda,\mu}(\mathcal{W}(t)) = e^{(\lambda-\mu)t}.$$

Comparing the representation with Corollary 3, we can see that $\psi_{\lambda,\mu}$ cannot be equal to a point evaluation functional.

2.2.2. The Normed Ring Structure of $\mathcal{M}$

We consider the set $\mathcal{M}$ more detailed. Let $\mathcal{M}_+ = \{u \in \mathcal{M} \colon u = [(0|x)], x \in \ell_1\}$. According to [12] we introduce an operation '$\diamond$' on $\mathcal{M}_+$ and extend it to $\mathcal{M}$.

Let $x, y \in \ell_1$. Then $x \diamond y$, we mean the resulting sequence of ordering the set $\{x_i y_j \colon i,j \in \mathbb{N}\}$ with one single index in some fixed order. If $u = [(0|x)]$ and $v = [(0|y)]$, then $u \diamond v = [(0|x \diamond y)]$. From [12,22] we know that the operation on $\mathcal{M}_+$ is commutative, associative and $[y \diamond (x \bullet d)] = [(y \diamond x) \bullet (y \diamond d)]$. Finally, let $u = [(y|x)]$ and $v = [(d|b)]$ are in $\mathcal{M}$. We define

$$u \diamond v = [((y \diamond b) \bullet (x \diamond d)|(y \diamond d) \bullet (x \diamond b))].$$

Proposition 5. *For every $k \in \mathbb{N}$, $T_k(u \diamond v) = T_k(u)T_k(v)$, $u, v \in \mathcal{M}$.*

Proof. From [12] we know that for all $x, z \in \ell_1$, $F_k(x \diamond z) = F_k(x)F_k(z)$. Let $u = [(y|x)]$ and $v = [(d|b)]$. Then

$$T_k(u \diamond v) = F_k((y \diamond d) \bullet (x \diamond b)) - F_k((y \diamond b) \bullet (x \diamond d))$$
$$= F_k(y)F_k(d) + F_k(x)F_k(b) - F_k(y)F_k(b) - F_k(x)F_k(d) = T_k(u)T_k(v).$$

□

Theorem 4. *$(\mathcal{M}, \bullet, \diamond)$ is a commutative ring with zero $0 = [(0|0)]$ and unity $\mathbb{I} = [(0|1, 0, \ldots)]$.*

Proof. Note first that $(\mathcal{M}, \bullet)$ is a commutative group and if $u = [(y|x)] \in \mathcal{M}$, then $u^- = [(y|x)^-] = [(x|y)]$ is the inverse of u. The associativity and commutativity of the multiplication and the distributive low were proved in [12] for the case $\mathcal{M}_+$ and can be checked for the general case by simple computations. □

Note that there is an operation of multiplication by a constant on $\mathbb{C} \times \mathcal{M}$:

$$\lambda[(y|x)] = [(\lambda y|\lambda x)], \quad \lambda \in \mathbb{C}, \quad [(y|x)] \in \mathcal{M}.$$

Clearly,

$$\lambda(u \bullet v) = \lambda u \bullet \lambda v \quad \text{and} \quad \lambda(u \diamond v) = (\lambda u) \diamond v = u \diamond (\lambda v) \quad \lambda \in \mathbb{C}, \quad u, v \in \mathcal{M}.$$

But, in the general case,

$$(\lambda_1 + \lambda_2)u \neq \lambda_1 u \bullet \lambda_2 u.$$

So $(\mathcal{M}, \bullet, (\lambda, u) \mapsto \lambda u)$ is not a linear space over $\mathbb{C}$. Hence $(\mathcal{M}, \bullet, \diamond)$ is not an algebra. In order to topologise $\mathcal{M}$, we can use the standard norm on $\ell_1(\mathbb{Z}_0)$.

Definition 2. *Let $u \in \mathcal{M}$. We define a norm of u by the following way:*

$$\|u\| = \|x\| + \|y\| = \sum_{n=1}^{\infty} |x_n| + \sum_{n=1}^{\infty} |y_n|,$$

where $(y|x)$ is an irreducible representative of u.

From Corollary 4 it follows that the definition of norm $\|\cdot\|$ does not depend on the irreducible representative. The next proposition shows that, like in a linear space, the norm has natural properties.

Proposition 6. *Let $u, v \in \mathcal{M}$, $\lambda \in \mathbb{C}$ The following properties hold:*

1. $\|u\| \geq 0$ *and* $\|u\| = 0$ *if and only if* $u = 0$.
2. $\|\lambda u\| = |\lambda| \|u\|$.
3. $\|u \bullet v\| \leq \|u\| + \|v\|$.
4. $\|u \diamond v\| \leq \|u\| \|v\|$.
5. $\|u^-\| = \|u\|$.
6. $\|u\| = \min_{(y|x) \in u} (\|x\| + \|y\|)$.

Proof. We need to prove just item (6). Let $(y|x)$ be a representation of u. We can write up to a permutation that $(y|x) = (y' \bullet a|x' \bullet a)$ for some $a \in \ell_1$ and irreducible $(y'|x')$. So

$$\|u\| = \|x'\| + \|y'\| \leq \|x\| + \|y\|$$

for ever $(y|x) \in u$. □

We define a metric ρ on $\mathcal{M}$, associated with the norm by the natural way. Let $u, v \in \mathcal{M}$. Set

$$\rho(u,v) = \|u \bullet v^-\|.$$

It is easy to check that ρ is a metric using the same arguments as in the classical case of linear normed spaces.

Proposition 7. *The multiplication by $\lambda \in \mathbb{C}$, $\lambda \mapsto \lambda u$ for a fixed $u \in \mathcal{M}$ is discontinuous in general at each nonzero point in $\mathbb{C}$ and continuous at zero. Here we consider the standard topology on $\mathbb{C}$ and the topology on $\mathcal{M}$, generated by ρ.*

Proof. Let ε_n be a sequence in $\mathbb{C}$ such that $\varepsilon_n \neq 0$, $\varepsilon_n \to 0$ as $n \to \infty$, $\lambda \neq 0$ and $u = [(\ldots,0,y_1|x_1,0,\ldots)]$, where $x_1 \neq 0$ or $y_1 \neq 0$ and $x_1 \neq y_1$. Then

$$\rho(\lambda(1+\varepsilon_n)u, \lambda u) = \|[(\ldots,0,\lambda y_1,\lambda(1+\varepsilon_n)x_1|\lambda x_1,\lambda(1+\varepsilon_n)y_1,0,\ldots)]\|$$

$$= \|\lambda x_1\| + \|\lambda y_1\| + \|\lambda(1+\varepsilon_n)x_1\| + \|\lambda(1+\varepsilon_n)y_1\| > |\lambda|\|u\| > 0$$

while $\lambda(1+\varepsilon_n) \to \lambda$ as $n \to \infty$.

Let now $\lambda = 0$, $u \in \mathcal{M}$ and $(y|x)$ be an irreducible representation of u. Then

$$\rho(\varepsilon_n u, 0) = \|\varepsilon_n u\| = |\varepsilon_n|\|x\| + |\varepsilon_n|\|y\| \to 0.$$

□

Theorem 5. *The operations '•' and '⋄' are jointly continuous on $(\mathcal{M},\rho)$.*

Proof. It is easy to check that if $\rho(u,u') < \varepsilon_1$ and $\rho(v,v') < \varepsilon_2$, then

$$\rho(u \bullet v, u' \bullet v') < \|(u \bullet v) \bullet (u' \bullet v')^-\| < \varepsilon_1 + \varepsilon_2$$

and

$$\rho(u \diamond v, u' \diamond v') < \varepsilon_1\|u\| + \varepsilon_2\|v\| + \varepsilon_1\varepsilon_2.$$

□

Proposition 8. *The metric space $(\mathcal{M},\rho)$ is nonseparable.*

Proof. Let us consider the following set

$$S_1 = \{u_\lambda = \lambda\mathbb{I} = (0|\lambda,0,0\ldots) : \lambda \in \mathbb{C}, |\lambda| = 1\}.$$

If $\lambda_1 \neq \lambda_2$, then

$$\rho(u_{\lambda_1}, u_{\lambda_2}) = \|(\ldots,0,0,\lambda_2|\lambda_1,0,0\ldots)\| = 2.$$

So the unit sphere of $(\mathcal{M},\rho)$ contains an uncountable set S_1 such that the distance between each pair of distinct points of S_1 is equal to 2. □

Theorem 6. *The metric space $(\mathcal{M},\rho)$ is complete.*

Proof. Let u and v be in $\mathcal{M}$ and $\rho(u,v) = \|u \bullet v^-\| < \varepsilon$ and $(y|x)$ be an irreducible representations of u. Then there is an irreducible representation $(d|b)$ of v such that $\|(y|x) - (d|b)\| < \varepsilon$. Indeed the inequality $\|u \bullet v^-\| < \varepsilon$ implies that there is $w \in \mathcal{M}$ such that $u = u' \bullet w$, $v = v' \bullet w$ and $\|u'\| + \|v'\| < \varepsilon$. Let us consider a representation $(d|b)$ of v such that the element w in $(d|b)$ is represented by the same vector that in $(y|x)$. Let $(y'|x')$ be the irreducible representation of u' in $(y|x)$ and $(d'|b')$ be the irreducible representation of v' in $(d|b)$. Then

$$\|(y|x) - (d|b)\| = \|(y'|x') - (d'|b')\| \leq \|u'\| + \|v'\| < \varepsilon.$$

Let $u^{(m)}$, $m \in \mathbb{N}$ be a Cauchy sequence in $(\mathcal{M}, \rho)$. Taking a subsequence, if necessary, we can assume that if $n \geq N$ and $m \geq N$, then $\rho(u^{(m)}, u^{(n)}) < 1/2^{N+1}$. Let us chose irreducible representations $(y^{(m)}|x^{(m)})$ of $u^{(m)}$ such that

$$\|(y^{(m+1)}|x^{(m+1)}) - (y^{(m)}|x^{(m)})\| = \rho(u^{(m+1)}, u^{(m)}) < 1/2^{m+1}.$$

So if $n \geq N$ and $m \geq N$, then

$$\|(y^{(m)}|x^{(m)}) - (y^{(n)}|x^{(n)})\| < 1/2^N.$$

Hence, $(y^{(m)}|x^{(m)})$, $m \in \mathbb{N}$ is a Cauchy sequence in $\ell_1(\mathbb{Z})$ and so it has a limit point $z^{(0)} = (y^{(0)}|x^{(0)})$. Let $z_i^{(m)}$ be the ith coordinate of $z^{(m)} = (y^{(m)}|x^{(m)})$, $i \in \mathbb{Z}_0$, that is, $z_i^{(m)} = x_i^{(m)}$ if $i > 0$ and $z_i^{(m)} = y_{-i}^{(m)}$ if $i < 0$. Clearly that $z_i^{(m)} \to z_i^{(0)}$ as $m \to \infty$. We claim that if $z_i^{(0)} = c \neq 0$ then there is a number N such that for every $m > N$, $z_i^{(m)} = c$. Indeed, it it is not so, then for every $n, m > N$, $\rho(u^{(m)}, u^{(n)}) > c$ and we have a contradiction.

For a given $\varepsilon > 0$ we denote by z^ε a vector in $\ell_1(\mathbb{Z}_0)$ such that z^ε has a finite support, $z_i^\varepsilon = z_i^{(0)}$ or $z_i^\varepsilon = 0$ and

$$\rho(z^\varepsilon, z^{(0)}) < \frac{\varepsilon}{3}.$$

Note that for this case $\rho(z^\varepsilon, z^{(0)}) = \|z^\varepsilon - z^{(0)}\|$. Let N be a number such that for every $n > N$, $z_i^\varepsilon = z_i^{(n)}$ for all $i \in \operatorname{supp} z^\varepsilon$ and $\|z^{(n)} - z^{(0)}\| < \frac{\varepsilon}{3}$. So

$$\rho(z^{(n)}, z^\varepsilon) = \|z^\varepsilon - z^{(n)}\| \leq \|z^\varepsilon - z^{(0)}\| + \|z^{(n)} - z^{(0)}\| < \frac{2}{3}\varepsilon.$$

Thus

$$\rho(z^{(n)}, z^{(0)}) \leq \rho(z^{(n)}, z^\varepsilon) + \rho(z^\varepsilon, z^{(0)}) < \varepsilon.$$

□

2.2.3. Invertibility and Homomorphisms

If $u \in \mathcal{M}$ has an inverse with respect to the multiplication '$\diamond$' we denote it by $u^{-1} = u^{\diamond(-1)}$, that is,

$$u \diamond u^{-1} = u^{-1} \diamond u = \mathbb{I}.$$

Proposition 9. *Let $u \in \mathcal{M}$ and $\|u\| < 1$. Then $\mathbb{I} \bullet u^-$ is invertible in $\mathcal{M}$.*

Proof. It is easy to check that the proof for classical Banach algebras can be literally repeated for this case. In particular,

$$(\mathbb{I} \bullet u^-)^{-1} = \mathop{\bullet}_{n=0}^{\infty} u^{\diamond n},$$

where $u^{\diamond 0} = \mathbb{I}$, $u^{\diamond n} = \underbrace{u \diamond \cdots \diamond u}_{n}$ and the series on the right converges in $\mathcal{M}$. □

Next we consider ring homomorphisms and subrings of $\mathcal{M}$. In sequel we do not assume that ring homomorphisms preserve the multiplication by constants. Note that an element x of a commutative Banach algebra A is invertible if and only if $\varphi(x) \neq 0$ for every character φ of A. The situation in $\mathcal{M}$ is different. Let $\mathbb{I}^{\bullet n} = \underbrace{\mathbb{I} \bullet \cdots \bullet \mathbb{I}}_{n} = (0|\underbrace{1, \ldots, 1}_{n}, 0, \ldots)$.

Proposition 10. *Let φ be a nonzero ring homomorphism from $\mathcal{M}$ to $\mathbb{C}$. Then $\varphi(\mathbb{I}^{\bullet n}) = n$ but $\mathbb{I}^{\bullet n}$ is non invertible for $n > 1$.*

Proof. Clearly, $\varphi(\mathbb{I}^{\bullet n}) = n\varphi(\mathbb{I}) = n$. On the other hand, $\mathbb{I}^{\bullet n} \diamond u = u^{\bullet n} \neq \mathbb{I}$ for every $u \in \mathcal{M}$. □

Example 1. *The following maps are ring homomorphisms from $\mathcal{M}$ to $\mathbb{C}$.*

1. *Polynomials T_n, $n \in \mathbb{N}$ are (continuous) complex valued ring homomorphism of $\mathcal{M}$ but only T_1 preserves the multiplication by constants.*
2. *Let $u = [(y|x)] \in \mathcal{M}$. We define*

$$\Theta(u) = \sum_{n=1}^{\infty} |x_n| - \sum_{n=1}^{\infty} |y_n|.$$

Clearly, Θ is well defined. The additivity and multiplicativity will be proved for more general case.

As usual $\mathcal{R}$ is a *subring* of $\mathcal{M}$ if it is a subset of $\mathcal{M}$ and a ring with respect to '$\bullet$' and '$\diamond$'. For example, let $\mathcal{M}_{00}$ consists of all elements $u = [(y|x)]$ such that if $(y|x)$ is irreducible, then $\operatorname{supp} x$ and $\operatorname{supp} y$ are finite sets. Then $\mathcal{M}_{00}$ is a dense subring of $\mathcal{M}$. We consider some nontrivial examples of closed subrings of $\mathcal{M}$.

Example 2. *Let $\mathcal{M}_\Delta$, $\mathcal{M}_S$ and $\mathcal{M}_1$ be defined by*

$$\mathcal{M}_\Delta = \{u \in \mathcal{M}\colon |x_j| \le 1,\ |y_j| \le 1\, \forall \text{ irreducible representations } (y|x) \in u\},$$

$$\mathcal{M}_S = \{u \in \mathcal{M}\colon |x_j| = 1 \text{ or } 0,\ |y_j| = 1 \text{ or } 0\, \forall \text{ irreducible representations } (y|x) \in u\} \cup \{0\},$$

$$\mathcal{M}_1 = \{u \in \mathcal{M}\colon x_j = 1 \text{ or } 0,\ y_j = 1 \text{ or } 0\, \forall \text{ irreducible representations } (y|x) \in u\} \cup \{0\}.$$

Clearly, $\mathcal{M}_\Delta$, $\mathcal{M}_S$ and $\mathcal{M}_1$ are subrings of $\mathcal{M}$ and

$$\mathcal{M}_\Delta \supset \mathcal{M}_S \supset \mathcal{M}_1.$$

Also, $\mathcal{M}_1$ is isomorphic to the ring $\mathbb{Z}$ of integer numbers and the restriction of the topology of $(\mathcal{M}, \rho)$ to $\mathcal{M}_S$ and $\mathcal{M}_1$ coincides with the discrete topology. In the general case, let U be a subset of $\mathbb{C}$. We denote by

$$\mathcal{M}_U = \{u \in \mathcal{M}\colon x_j \in U,\ y_j \in U\, \forall \text{ irreducible representations } (y|x) \in u\} \cup \{0\}.$$

Then $\mathcal{M}_U$ is a subring of $\mathcal{M}$ if U is closed with respect to the multiplication in $\mathbb{C}$ and $1 \in U$.

Proposition 11. *Let $\gamma(t)$ be a function of one variable which is well defined and multiplicative on a subset $U \in \mathbb{C}$. We define*

$$\Theta_\gamma(u) = \sum_{n=1}^{\infty} \gamma(x_n) - \sum_{n=1}^{\infty} \gamma(y_n), \qquad u \in \mathcal{M},$$

where $(y|x) \in u$. If U is closed with respect to the multiplication and $1 \in U$, then Θ_γ is a complex valued ring homomorphism of $\mathcal{M}_U$.

Proof. Note first that $\Theta_\gamma(u)$ does not depend of the choice of a representative. Thus

$$\Theta_\gamma((y|x) \bullet (d|b)) = \Theta_\gamma(y \bullet d|x \bullet b)$$

$$= \sum_{n=1}^{\infty} \gamma(x_n) + \sum_{n=1}^{\infty} \gamma(b_n) - \sum_{n=1}^{\infty} \gamma(y_n) - \sum_{n=1}^{\infty} \gamma(d_n) = \Theta_\gamma(y|x) + \Theta_\gamma(d|b).$$

By the multiplicativity of γ we have

$$\Theta_\gamma((0|x) \diamond (0|b)) = \sum_{n=i,j}^{\infty} \gamma(x_i)\gamma(b_j) = \sum_{n=i}^{\infty} \gamma(x_i) \sum_{n=j}^{\infty} \gamma(b_j)$$

and $\Theta_\gamma(x|0) = -\Theta_\gamma(0|x)$. So

$$\Theta_\gamma((y|x) \diamond (d|b)) = \Theta_\gamma((y \diamond b) \bullet (x \diamond d)|(x \diamond b) \bullet (y \diamond d))$$

$$= \sum_{n=i}^{\infty} \gamma(x_i) \sum_{n=j}^{\infty} \gamma(b_j) + \sum_{n=i}^{\infty} \gamma(y_i) \sum_{n=j}^{\infty} \gamma(d_j) - \sum_{n=i}^{\infty} \gamma(y_i) \sum_{n=j}^{\infty} \gamma(b_j) - \sum_{n=i}^{\infty} \gamma(x_i) \sum_{n=j}^{\infty} \gamma(d_j)$$

$$= \Theta_\gamma(y|x)\Theta_\gamma(d|b).$$

□

Example 3. *Let us consider some examples of complex valued homomorphisms of subrings of $\mathcal{M}$.*

1. *Let g be a multiplicative function from $\mathbb{N} \to \mathbb{C}$. In Number Theory such functions are called completely multiplicative arithmetic functions. Then for $\gamma = |g|$, Θ_γ is a complex valued ring homomorphisms of $\mathcal{M}_S$ and $\mathcal{M}_1$.*
2. *Let $\varepsilon < 1$ and $\varepsilon\Delta$ be the closed disk in $\mathbb{C}$ of radius ε, centered at zero. Then $\mathcal{M}_{\varepsilon\Delta}$ is an ideal in $\mathcal{M}_1$. Let*

$$\chi_{\mathbb{C}\setminus\varepsilon\Delta}(t) = \begin{cases} 0 & \text{if } |t| \le \varepsilon \\ 1 & \text{if } |t| > \varepsilon, \end{cases}$$

then $\Theta_{\chi_{\mathbb{C}\setminus\varepsilon\Delta}}$ is a complex valued ring homomorphisms of $\mathcal{M}_1$. Note that if $u \in \mathcal{M}_1 \setminus \mathcal{M}_{\varepsilon\Delta}$ and $v \in \mathcal{M}_{\varepsilon\Delta}$, then $\rho(u,v) \ge \varepsilon$. From here we have that $\Theta_{\chi_{\mathbb{C}\setminus\varepsilon\Delta}}$ is continuous.

We do not know whether or not every complex valued homomorphism of $\mathcal{M}$ or its closed subring is continuous.

2.2.4. Additive Operator Calculus

Let $\Phi\colon \mathcal{M} \to \mathcal{M}$ be an additive map. Since it is a homomorphism of the additive group $(\mathcal{M}, \bullet)$ to itself, Φ is continuous at every point if and only if it is continuous at a point in $\mathcal{M}$. Let $\gamma\colon \mathbb{C} \to \mathbb{C}$ be an arbitrary function. Then it is well defined the following additive map from $\mathcal{M}_{00}$ to itself:

$$\Phi_\gamma(u) = (\ldots, \gamma(y_n), \ldots, \gamma(y_1)|\gamma(x_1), \ldots, \gamma(x_n), \ldots). \tag{13}$$

Proposition 12. *If there are constants $C > 0$ and $m \in \mathbb{N}$ such that $|\gamma(t)| \le C|t|^m$, then Φ_γ is continuous, additive and well defined on $\mathcal{M}$.*

Proof. For every $u \in \mathcal{M}$

$$\|\Phi_\gamma(u)\| = \sum_{n=1}^{\infty} \|\gamma(x_n)\| + \sum_{n=1}^{\infty} \|\gamma(y_n)\| \le C \sum_{n=1}^{\infty} (|x_n|^m + |y_n|^m) < \infty.$$

If $\|u\| < \varepsilon < 1$, then $\|\Phi_\gamma(u)\| < C\varepsilon$ and so Φ_γ is continuous at zero. Thus it is continuous. □

Example 4. *(Power operators.) Let $m \in \mathbb{N}$. Then $\gamma(t) = t^m$ satisfies Proposition 12 and so the map $\Phi_m : u \mapsto u^m$, where $u = [(y|x)] \in \mathcal{M}$ and*

$$u^m = (\ldots, y_n^m, \ldots, y_1^m | x_1^m, \ldots, x_n^m, \ldots)$$

is a continuous additive operator on $\mathcal{M}$.

Let $k \in \mathbb{N}$ and

$$\sqrt[k]{a} = (a)^{1/k} = (a^{(1/k,1)}, a^{(1/k,2)}, \ldots, a^{(1/k,k)}), \qquad a \in \mathbb{C}$$

be the multi-valued kth power root function. Let us consider

$$(a^{(1/k,1)}, a^{(1/k,2)}, \ldots, a^{(1/k,k)}) = (a^{(1/k,1)}, a^{(1/k,2)}, \ldots, a^{(1/k,k)}, 0, 0 \ldots)$$

as an element in ℓ_1. Then, for every $u \in \mathcal{M}$ such that for an irreducible representation $(y|x)$ of u

$$\sum_{n=1}^{\infty} (|x_n|^{1/k} + |y_n|^{1/k}) < \infty$$

we can define

$$\Phi_{1/k}(u) = u^{1/k} = (\cdots \bullet y_n^{1/k} \bullet \cdots \bullet y_1^{1/k} | x_1^{1/k} \bullet \cdots \bullet x_n^{1/k} \bullet \cdots).$$

The map $\Phi_{1/k}$ for $k > 1$ is a discontinuous additive operator, defined on a dense subset of $\mathcal{M}$. But if $m > k$, then we can define an additive operator

$$\Phi_{1/k} \circ \Phi_m$$

which is continuous on $\mathcal{M}$. Note that $\Phi_{1/k} \circ \Phi_k \ne \Phi_1$ if $k > 1$ because Φ_1 is the identical operator while

$$\Phi_{1/k} \circ \Phi_k(u) = \underbrace{u \bullet \cdots \bullet u}_{k} = u \diamond \mathbb{I}^{\bullet k}.$$

We say that a map $A : \mathcal{M} \to \mathcal{M}$ is a *linear operator* if it is additive, preserves multiplications by constants, that is, $A(\lambda u) = \lambda A(u)$, $\lambda \in \mathbb{C}$ and if $A(u^-) = (A(u))^-$ for all $u \in \mathcal{M}$. From Proposition 12 it follows that there are a lot of additive operators. Linear operators, in contrast, can be described in a simple way.

Theorem 7. *Let A be a continuous linear operator from $\mathcal{M}$ to itself. Then there exists an element $v \in \mathcal{M}$ such that*

$$A(u) = v \diamond u, \qquad u \in \mathcal{M}.$$

Proof. Let $u = \mathbb{I} = [(0|1,0,\ldots)]$ and $A(u) = [(b|a)] \in \mathcal{M}$. Set $v = [(b|a)]$. Let now u be an element in $\mathcal{M}$ which can be represented by a vector $(y|x)$ with finite support

$$(y|x) = (\ldots,0,y_m,\ldots,y_1|x_1,\ldots,x_n,0,\ldots).$$

Then we can write

$$[(y|x)] = y_1\mathbb{I}^- \bullet \cdots \bullet y_m\mathbb{I}^- \bullet x_1\mathbb{I} \bullet \cdots x_n\mathbb{I}$$

and so

$$A(u) = y_1u^- \bullet \cdots \bullet y_mu^- \bullet x_1u \bullet \cdots x_nu = v \diamond u.$$

Since the set $\mathcal{M}_{00}$ of elements with finite supports is dense in $\mathcal{M}$ and A is continuous, $A = v \diamond u$ for every $u \in \mathcal{M}$. □

We denote by $A_v(u)$ the operator $u \mapsto v \diamond u$, $u \in \mathcal{M}$. Let us prove some natural properties of operators A_v.

Proposition 13. *1. The operator A_v is bijective if and only if v is invertible in $\mathcal{M}$.*
2. *If the operator A_v is surjective, then it is bijective.*
3. *The operator A_v is injective if and only if* $\ker A_v = 0$.
4. *If $u \in \ker A_v$ for some $u \neq 0$, then $T_n(v) = 0$ for some $n \in \mathbb{N}$.*
5. *If $T_n(v) = 0$ for some $n \in \mathbb{N}$, then A_v is not surjective.*

Proof. (1) If v is invertible, then $A_{v^{-1}} = A_v^{-1}$ so A_v is a bijection. Let now $B = A_v^{-1}$. Then from the Open Map theorem for complete metric groups (see [23]) it follows that B is continuous. From Theorem 7 we have that $B = A_w$ for some $w \in \mathcal{M}$. Since

$$A_v \circ A_w = A_{v \diamond w} = A_{\mathbb{I}},$$

$w = v^{-1}$.

(2) Let A_v be surjective. Then there exists $u \in \mathcal{M}$ such that $A_v(u) = v \diamond u = \mathbb{I}$. So v is invertible and $A_u = A_v^{-1}$.

(3) If A_v is injective, then $\ker A_v = 0$. Conversely, If there are $u, w \in \mathcal{M}$, $u \neq w$ such that $A_v(u) = A_v(w)$, then $A_v(u \bullet w^-) = 0$ and so $\ker A_v$ is nontrivial.

(4) If $u \in \ker A_v$, then $v \diamond u = 0$ and so

$$T_k(v \diamond u) = T_k(v)T_k(u) = 0, \qquad k \in \mathbb{N}.$$

Since $u \neq 0$, there exists a number $n \in \mathbb{N}$ such that $T_n(u) \neq 0$. So $T_n(v) = 0$.

(5) If A_v is surjective, then it is bijective and so v is invertible. But

$$1 = T_n(\mathbb{I}) = T_n(v \diamond v^{-1}) = T_n(v)T_n(v^{-1}) = 0,$$

a contradiction. □

Note that for $v = \mathbb{I}^m$ the operator A_v is not surjective but it is injective because

$$A_v(u) = \underbrace{u \bullet \cdots \bullet u}_{m}, \qquad u \in \mathcal{M}$$

and $T_k(v) = m > 0$ for every k. On the other hand for $v = (\ldots,0,1,2|3,0\ldots)$, $T_1(v) = 0$ and so A_v is not surjective but it is injective. Indeed, it is easy to check that $T_k(v) \neq 0$ for $k > 1$. So, if $v \diamond u = 0$ for some $u = (y|x) \in \mathcal{M}$, then $F_k(x) = F_k(y)$ for $k > 1$. But from [13] it follows that also $F_1(x) = F_1(y)$ and so $T_k(u) = 0$ for all $k \in \mathbb{N}$, that is $u = 0$. Finally, for $v = (\ldots,0,-1|1,0\ldots)$, A_v has a nontrivial kernel which contains $u = (\ldots,0|1,-1,0\ldots)$.

3. Discussion

According to Gelfand's theory, every commutative semi-simple algebra Fréchet $\mathcal{A}$ can be represented as an algebra of continuous functions on its spectrum $M(\mathcal{A})$ (see e.g., [24] p. 217, p. 231). If $\mathcal{A}$ consists of analytic functions on a Banach space X, then for every $x \in X$ the point evaluation functional δ_x belongs to $M(\mathcal{A})$. The map $x \mapsto \delta_x$ is one-to-one if and only if $\mathcal{A}$ separates points of X, for example, if $\mathcal{A} = H_b(X)$ is the algebra of all analytic functions of bounded type on X. Investigations of the spectrum of $H_b(X)$ were started by Aron, Cole and Gamelin in their fundamental work [2]. Note that, in the general case, $M_b = M(H_b(X))$ has complicated topological and algebraic structures (see [5,25]) which can be described only implicitly involving such tools as the Aron-Berner extension, topological tensor products, StoneČech compactification, ect. On the other hand, it is convenient for applications to have algebras of analytic functions of infinite many variables whose spectra admit explicit descriptions. If a subalgebra $\mathcal{A}$ of $H_b(X)$ has an algebraic basis of polynomials $P_1, P_2, \ldots, P_n, \ldots$, then every $\varphi \in M(\mathcal{A})$ is completely defined by its values on this basis, $\xi_1 = \varphi(P_1), \xi_2 = \varphi(P_2), \ldots, \xi_n = \varphi(P_n), \ldots$. So we can describe $M(\mathcal{A})$ as a subset of a sequence space $\{(\xi_1, \ldots, \xi_n, \ldots) \colon \xi_j \in \mathbb{C}\}$. Moreover, if $\|P_n\| = 1$ and $\deg P_n = n$, then it is not difficult to check that sequences (ξ_n) should satisfy the following condition

$$\sup_n |\xi_n|^{1/n} < \infty. \tag{14}$$

Note that for the algebra of symmetric analytic functions of bounded type on $L_\infty[0,1]$ condition (14) is sufficient [14] but for the algebra $H_{bs}(\ell_1)$ is not [12]. In the present paper we use this approach for H_b^{sup} which is a subalgebra of $H_b(\ell_1(\mathbb{Z}_0))$ generated by polynomials $T_1, T_2, \ldots$. We can see that $H_b(\ell_1(\mathbb{Z}_0))$ is quite different than $H_{bs}(\ell_1)$. For example, the homomorphism defined by $T_k \mapsto -T_k$ is continuous in $H_b(\ell_1(\mathbb{Z}_0))$, while $F_k \mapsto -F_k$ is discontinuous in $H_{bs}(\ell_1)$. On the other hand, the homomorphism defined by $T_k \mapsto \lambda T_k$ is discontinuous for $\lambda \notin \mathbb{Z}$ and so the set of sequences $\xi_1 = \varphi(T_1), \xi_2 = \varphi(T_2), \ldots, \xi_n = \varphi(T_n), \ldots$, $\varphi \in M_b^{sup}$ does not support multiplications by constants. From here we have that condition (14) is not sufficient for description of M_b^{sup}.

The results of Section 2.2 show that the spectrum of H_b^{sup} admits an interesting algebraic structure of commutative ring with respect to operations '$\bullet$' and '$\diamond$' which play roles of addition and multiplication. Using these operations and the ℓ_1-norm we introduced a natural metric ρ on $\mathcal{M}$, and proved that $(\mathcal{M}, \rho)$ is a complete metric space. We studied homomorphisms of $\mathcal{M}$ and described all linear operators of $\mathcal{M}$ to itself. So obtained results may be interesting in the theory of commutative topological algebras and for algebras of analytic functions on Banach spaces as well.

Supersymmetric polynomials and analytic functions are applicable in other branches of Mathematics and in Physics. Note first that supersymmetric polynomials of several variables were studied by many authors and in [18–20] we can find analogs of Formulas (7) and (8) for these cases (with using some different notations). Here we proved such results for infinite many variables and due to ℓ_1-topology we can claim that $\mathcal{W}(y|x)(t)$ is a rational function, where the numerator and the denominator are functions of exponential type for every fixed $(y|x) \in \ell_1(\mathbb{Z}_0)$. But an important difference between finite- and infinite-dimensional case is that in the finite-dimensional case we can not to use the operations '$\bullet$' and '$\diamond$' because they do not preserve the dimension of the underlying space. Some applications of supersymmetric polynomials for Brauer groups are described in [26]. It seems to be that H_b^{sup} can be applied for infinite generated Brauer groups in a similar way. Another application can be obtained for Statistical Mechanics.

In [27] we can find an approach to how classical symmetric polynomials can be used to modeling the behavior of ideal gas. According to this approach and using our notations, independent variables $x_1, x_2, \ldots$ correspond to abstract energy levels which particles of ideal gas may occupy; symmetric monomials

$$\sum_{i_1<\cdots<i_n} x_{i_1}^{k^1} \cdots x_{i_n}^{k^n}$$

correspond to occupation these energy levels by particles; generating functions $\mathcal{G}(x)(t)$ and $\mathcal{H}(x)(t)$ correspond to grand canonical partition functions for bosons and fermions respectively, and Equation (2) is modeling the Bose-Fermi symmetry law. From this point of view and taking into account (7), supersymmetric polynomials may be useful for the description of ideal gas consisting of both type particles: bosons, and fermions. Moreover, the Bose-Fermi symmetry in our notations means just $[(x|x)] = 0$.

Note that Statistical Mechanics work with the situation when the number of particles, N tends to infinity. The fact that we consider the closure of polynomials in a metrizable topology allows us to proceed with limit values as $N \to \infty$. The ℓ_1-topology of the underlying space $\ell_1(\mathbb{Z}_0)$ is guarantying that all abstract supersymmetric polynomials are well defined on this space. For example, if we will use $\ell_2(\mathbb{Z}_0)$ instead of $\ell_1(\mathbb{Z}_0)$, then T_1 will be not defined. Finally, we can expect that the algebraic operations '$\bullet$' and '$\diamond$' may have a physical meaning in the proposed approach. But such kind of problems is outside of the topic of our article.

4. Conclusions

In this article, we considered the algebra H_b^{sup} of analytic functions of bounded type generated by supersymmetric polynomials on $\ell_1(\mathbb{Z}_0)$. We have described some algebraic bases of the subalgebra of supersymmetric polynomials and corresponding generating functions. Such a description is important in order to study the spectrum (the set of complex homomorphisms) of H_b^{sup}. In particular, it is shown that every point evaluation complex homomorphism can be represented as a ratio of two entire functions of exponential type. Also, we constructed an example of complex homomorphism which is not a point evaluation functional. However, we have not a complete description of the spectrum of H_b^{sup}. In particular, it is unclear under which conditions a meromorphic function is of the form (12) for some $\varphi \in M_b^{sup}$? Note that such kind of problem is also open for the algebra $H_{bs}(\ell_1)$ [10,12].

Our goal is establishing the structure of a complete metric commutative ring on the set $\mathcal{M}$ of point evaluation functionals of H_b^{sup}. The algebraic structure of $\mathcal{M}$ is very close to the Banach algebra structure but $\mathcal{M}$ is not a Banach algebra because it is not a linear space. So we have a natural question: which Banach algebras properties can be extended to the ring $\mathcal{M}$? For example, we can see that if an element is closed to the unity, then it is invertible. But we do not know: do $\mathcal{M}$ admits a discontinuous complex homomorphisms? Also, we investigated homomorphisms of $\mathcal{M}$, its subrings and additive operators of $\mathcal{M}$. The role of obtained results in the theory of algebras of analytic functions on Banach spaces and possible applications in Physics are discussed.

Author Contributions: These authors contributed equally to this work.

References

1. Dineen, S. *Complex Analysis in Infinite Dimensional Spaces*; Springer: London, UK, 1999.
2. Aron, R.M.; Cole, B.J.; Gamelin, T.W. Spectra of algebras of analytic functions on a Banach space. *J. Reine Angew. Math.* **1991**, *415*, 51–93.
3. Aron, R.M.; Cole, B.J.; Gamelin, T.W. Weak-star continuous analytic functions. *Can. J. Math.* **1995**, *47*, 673–683. [CrossRef]

4. Mujica, J. Ideals of holomorphic functions on Tsirelson's space. *Arch. Math.* **2001**, *76*, 292–298. [CrossRef]
5. Zagorodnyuk, A. Spectra of Algebras of Entire Functions on Banach Spaces. *Proc. Amer. Math. Soc.* **2006**, *134*, 2559–2569. [CrossRef]
6. Aron, R.; Galindo, P.; Pinasco, D.; Zalduendo, I. Group-symmetric holomorphic functions on a Banach space. *Bull. Lond. Math. Soc.* **2016**, *48*, 779–796. [CrossRef]
7. García, D.; Maestre, V.; Zalduendo, I. The spectra of algebras of group-symmetric functions. *Proc. Edinb. Math. Soc.* **2019**, *62*, 609–623. [CrossRef]
8. Gonzaléz, M.; Gonzalo, R.; Jaramillo, J. Symmetric polynomials on rearrangement invariant function spaces. *J. London Math. Soc.* **1999**, *59*, 681–697. [CrossRef]
9. Macdonald, I.G. *Symmetric Functions and Orthogonal Polynomials*; University Lecture Serie 12; AMS: Providence, RI, USA, 1997.
10. Chernega, I.; Galindo, P.; Zagorodnyuk, A. The convolution operation on the spectra of algebras of symmetric analytic functions. *J. Math. Anal. Appl.* **2012**, *395*, 569–577. [CrossRef]
11. Chernega, I.; Galindo, P.; Zagorodnyuk, A. Some algebras of symmetric analytic functions and their spectra. *Proc. Edinb. Math. Soc.* **2012**, *55*, 125–142. [CrossRef]
12. Chernega, I.; Galindo, P.; Zagorodnyuk, A. A multiplicative convolution on the spectra of algebras of symmetric analytic functions. *Rev. Mat. Complut.* **2014**, *27*, 575–585. [CrossRef]
13. Alencar, R.; Aron, R.; Galindo, P.; Zagorodnyuk, A. Algebra of symmetric holomorphic functions on ℓ_p. *Bull. Lond. Math. Soc.* **2003**, *35*, 55–64. [CrossRef]
14. Galindo, P.; Vasylyshyn, T.; Zagorodnyuk, A. The algebra of symmetric analytic functions on L_∞. *Proc. Roy. Soc. Edinburgh Sect. A.* **2017**, *147*, 743–761. [CrossRef]
15. Jawad, F. Note on separately symmetric polynomials on the Cartesian product of ℓ_1. *Mat. Stud.* **2018**, *50*, 204–210.
16. Kravtsiv, V. Algebraic basis of the algebra of block-symmetric polynomials on $\ell_1 \oplus \ell_\infty$. *Carpathian Math. Publ.* **2019**, *11*, 89–95. [CrossRef]
17. Kravtsiv, V.; Vasylyshyn, T.; Zagorodnyuk, A. On Algebraic Basis of the Algebra of Symmetric Polynomials on $\ell_p(\mathbf{C}^n)$. *J. Funct. Spaces* **2017**, *2017*, 4947925.
18. Olshanski, G.; Regev, A.; Vershik, A.; Ivanov, V. Frobenius-Schur Functions. In *Studies in Memory of Issai Schur. Progress in Mathematics*; Joseph, A., Melnikov, A., Rentschler, R., Eds.; Birkhäuser: Boston, MA, USA, 2003; Volume 210, pp. 251–299.
19. Sergeev, A.N. On rings of supersymmetric polynomials. *J. Algebra* **2019**, *517*, 336–364. [CrossRef]
20. Stembridge, J.R. A Characterization of Supersymmetric Polynomials. *J. Algebra* **1985**, *95*, 439–444. [CrossRef]
21. Chernega, I.; Zagorodnyuk, A. Unbounded symmetric analytic functions on ℓ_1. *Math. Scand.* **2018**, *122*, 84–90. [CrossRef]
22. Chernega, I.V. A semiring in the spectrum of the algebra of symmetric analytic functions in the space ℓ_1. *J. Math. Sci.* **2016**, *212*, 38–45. [CrossRef]
23. Brown, L.G. Topologically complete groups. *Proc. Amer. Math. Soc.* **1972**, *35*, 593–600. [CrossRef]
24. Mujica, J. *Complex Analysis in Banach Spaces*; Elsevier: Amsterdam, The Netherlands, 1986.
25. Aron, R.M.; Galindo, P.; Garcia, D.; Maestre, M. Regularity and algebras of analytic functions in infinite dimensions. *Trans. Am. Math. Soc.* **1996**, *348*, 543–559. [CrossRef]
26. Jung J.H.; Kim, M. Supersymmetric polynomials and the center of the walled Brauer algebra. *arXiv* **2017**, arXiv:1508.06469.
27. Schmidt, H.J.; Schnack, J. *Symmetric Polynomials in Physics*; Gazeau, J.-P., Kerner, R., Antoine, J.-P., Métens, S., Thibon., J.-Y., Eds.; IOP: Bristol, UK; Philadelphia, PA, USA, 2003; Volume 173, pp. 147–152.

3

Structure of Approximate Roots based on Symmetric Properties of (p,q)-Cosine and (p,q)-Sine Bernoulli Polynomials

Cheon Seoung Ryoo [1] and Jung Yoog Kang [2,*]

1 Department of Mathematics, Hanman University, Daejeon 10216, Korea; ryoocs@hnu.kr
2 Department of Mathematics Education, Silla University, Busan 469470, Korea
* Correspondence: jykang@silla.ac.kr

Abstract: This paper constructs and introduces (p,q)-cosine and (p,q)-sine Bernoulli polynomials using (p,q)-analogues of $(x+a)^n$. Based on these polynomials, we discover basic properties and identities. Moreover, we determine special properties using (p,q)-trigonometric functions and verify various symmetric properties. Finally, we check the symmetric structure of the approximate roots based on symmetric polynomials.

Keywords: (p,q)-cosine Bernoulli polynomials; (p,q)-sine Bernoulli polynomials; (p,q)-numbers; (p,q)-trigonometric functions

MSC: 11B68; 11B75; 33A10

---

1. Introduction

In 1991, (p,q)-calculus including (p,q)-number with two independent variables p and q, was first independently considered [1,2]. Throughout this paper, the sets of natural numbers, integers, real numbers and complex numbers are denoted by $\mathbb{N}, \mathbb{Z}, \mathbb{R}$ and $\mathbb{C}$, respectively.

For any $n \in \mathbb{N}$, the (p,q)-number is defined by the following:

$$[n]_{p,q} = \frac{p^n - q^n}{p-q}, \qquad \text{where} \quad |p/q| < 1, \tag{1}$$

which is a natural generalization of the q-number. From Equation (1), we note that $[n]_{p,q} = [n]_{q,p}$.

Many physical and mathematical problems have led to the necessity of studying (p,q)-calculus. Since 1991, many mathematicians and physicists have developed (p,q)-calculus in several different research areas. For example, in 1994, [3] introduced (p,q)-hypergeometric functions. Three years later, [3,4] derived related preliminary results by considering a more general (p,q)-hypergeometric series and Burban's (p,q)-hypergeometric series, respectively. In 2005, based on the (p,q)-numbers, [5] studied about (p,q)-hypergeometric series and discovered results corresponding to the (p,q)-extensions of known q-identities. Moreover, [6] established properties similar to the ordinary and q-binomial coefficients after developing the (p,q)-hypergeometric series in 2008. About seven years later, [7] introduced (p,q)-gamma and (p,q)-beta functions, which are generalizations of the gamma and beta functions.

The different variations of Bernoulli polynomials, q-Bernoulli polynomials and (p,q)-Bernoulli polynomials are illustrated in the diagram below. Kim, Ryoo and many mathematicians have studied the first and second rows of the polynomials in the diagram(see [8–12]). These studies began producing valuable results in areas related to number theory and combinatorics.

The main idea is to use property of (p,q)-numbers and combine (p,q)-trigonometric functions. From this idea, we construct (p,q)-cosine and (p,q)-sine Bernoulli polynomials. Investigating the various explicit identities for (p,q)-cosine and (p,q)-sine Bernoulli polynomials in the diagram's third row is the main goal of this paper.

$\frac{t}{e^t-1}e^{tx} = \sum_{n=0}^{\infty} B_n(x)\frac{t^n}{n!}$
(Bernoulli polynomials)

$\frac{t}{e^t-1}e^{tx}\cos(ty) = \sum_{n=0}^{\infty} B_n^{(C)}(x,y)\frac{t^n}{n!}$
(cosine Bernoulli polynomials)
$\frac{t}{e^t-1}e^{tx}\sin(ty) = \sum_{n=0}^{\infty} B_n^{(S)}(x,y)\frac{t^n}{n!}$
(sine Bernoulli polynomials)

$\frac{t}{e_q(t)-1}e_q(tx) = \sum_{n=0}^{\infty} B_{n,q}(x)\frac{t^n}{n!}$
(q-Bernoulli polynomials)

$\frac{t}{e_q(t)-1}e_q(tx)COS_q(ty) = \sum_{n=0}^{\infty} {}_CB_{n,q}(x,y)\frac{t^n}{n!}$
(q-cosine Bernoulli polynomials)
$\frac{t}{e_q(t)-1}e_q(tx)SIN_q(ty) = \sum_{n=0}^{\infty} {}_SB_{n,q}(x,y)\frac{t^n}{n!}$
(q-sine Bernoulli polynomials)

$\frac{t}{e_{p,q}(t)-1}e_{p,q}(tx) = \sum_{n=0}^{\infty} B_{n,p,q}(x)\frac{t^n}{n!}$
((p,q)-Bernoulli polynomials)

$\frac{t}{e_{p,q}(t)-1}e_{p,q}(tx)COS_{p,q}(ty) = \sum_{n=0}^{\infty} {}_CB_{n,p,q}(x,y)\frac{t^n}{n!}$
((p,q)-cosine Bernoulli polynomials)
$\frac{t}{e_{p,q}(t)-1}e_{p,q}(tx)SIN_{p,q}(ty) = \sum_{n=0}^{\infty} {}_SB_{n,p,q}(x,y)\frac{t^n}{n!}$
((p,q)-sine Bernoulli polynomials)

Due to their importance, the classical Bernoulli, Euler, and Genocchi polynomials have been studied extensively and are well-known. Mathematicians have studied these polynomials through various mathematical applications including finite difference calculus, p-adic analytic number theory, combinatorial analysis and number theory. Many mathematicians are familiar with the theorems and definitions of classical Bernoulli, Euler, and Genocchi polynomials. Based on the theorems and definitions, it is significant to study these properties in various ways by the combining with Bernoulli, Euler, and Genocchi polynomials. Mathematicians are studying the extended versions of these polynomials and are researching new polynomials by combining mathematics with other fields, such as physics or engineering (see [9–14]). The definition of Bernoulli polynomials combined with (p,q)-numbers follows:

Definition 1. *The (p,q)-Bernoulli numbers, $B_{n,p,q}$, and polynomials, $B_{n,p,q}(z)$, can be expressed as follows (see [8])*

$$\sum_{n=0}^{\infty} B_{n,p,q}\frac{t^n}{[n]_{p,q}!} = \frac{t}{e_{p,q}(t)-1}, \quad \sum_{n=0}^{\infty} B_{n,p,q}(z)\frac{t^n}{[n]_{p,q}!} = \frac{t}{e_{p,q}(t)-1}e_{p,q}(tz). \tag{2}$$

In [11], we confirmed the properties of q-cosine and q-sine Bernoulli polynomials. Their definitions and representative properties are as follows.

Definition 2. *The q-cosine Bernoulli polynomials ${}_CB_{n,q}(x,y)$ and q-sine Bernoulli polynomials ${}_SB_{n,q}(x,y)$ are defined by the following:*

$$\sum_{n=0}^{\infty} {}_CB_{n,q}(x,y)\frac{t^n}{n!} = \frac{t}{e_q(t)-1}e_q(tx)COS_q(ty), \quad \sum_{n=0}^{\infty} {}_SB_n(x,y)\frac{t^n}{n!} = \frac{t}{e_q(t)-1}e_q(tx)SIN_q(ty). \tag{3}$$

Theorem 1. *For $x, y \in \mathbb{R}$, we have the following:*

$$(i) \begin{cases} {}_CB_{n,q}((x \oplus r)_q, y) + {}_SB_{n,q}((x \ominus r)_q, y) \\ = \sum_{k=0}^{n} \begin{bmatrix} n \\ k \end{bmatrix}_q q^{\binom{n-k}{2}} r^{n-k} \left({}_CB_{k,q}(x,y) + (-1)^{n-k} {}_SB_{k,q}(x,y) \right) \\ {}_SB_{n,q}((x \oplus r)_q, y) + {}_CB_{n,q}((x \ominus r)_q, y) \\ = \sum_{k=0}^{n} \begin{bmatrix} n \\ k \end{bmatrix}_q q^{\binom{n-k}{2}} r^{n-k} \left({}_SB_{k,q}(x,y) + (-1)^{n-k} {}_CB_{k,q}(x,y) \right) \end{cases} \tag{4}$$

$$(ii) \begin{cases} \frac{\partial}{\partial x} {}_CB_{n,q}(x,y) = [n]_q {}_CB_{n-1,q}(x,y), & \frac{\partial}{\partial y} {}_CB_{n,q}(x,y) = -[n]_q {}_SB_{n-1,q}(x,qy) \\ \frac{\partial}{\partial x} {}_SB_{n,q}(x,y) = [n]_q {}_SB_{n-1,q}(x,y), & \frac{\partial}{\partial y} {}_SB_{n,q}(x,y) = [n]_q {}_CB_{n-1,q}(x,qy) \end{cases}$$

The main goal of this paper is to identify the properties of (p,q)-cosine and (p,q)-sine Bernoulli polynomials. In Section 2, we review some definitions and theorem of (p,q)-calculus. In Section 3, we introduce (p,q)-cosine and (p,q)-sine Bernoulli polynomials. Using the properties of exponential functions and trigonometric functions associated with (p,q)-numbers, we determine the various properties and identities of the polynomials. Section 4 presents the investigation of the symmetric properties of (p,q)-cosine and (p,q)-sine Bernoulli polynomials in different forms and based on the symmetric polynomials, we check the symmetric structure of the approximate roots.

2. Preliminaries

In this section, we introduce definitions and preliminary facts that are used throughout this paper (see [6,12–20]).

Definition 3. *For $n \geq k$, the Gaussian binomial coefficients are defined by the following:*

$$\begin{bmatrix} m \\ r \end{bmatrix}_{p,q} = \frac{[n]_{p,q}!}{[n-k]_{p,q}![k]_{p,q}!}, \tag{5}$$

where m and r are non-negative integers.

We note that $[n]_{p,q}! = [n]_{p,q}[n-1]_{p,q} \cdots [2]_{p,q}[1]_{p,q}$, where $n \in \mathbb{N}$. For $r = 0$, the value of the equation is 1, because both the numerator and denominator are empty products. Moreover, (p,q)-analogues of the binomial formula exist, and this definition has numerous properties.

Definition 4. *The (p,q)-analogues of $(x-a)^n$ and $(x+a)^n$ are defined by the following:*

$$(i)\ (x \ominus a)^n_{p,q} = \begin{cases} 1, & \text{if } n = 0 \\ (x-a)(px-qa)\cdots(p^{n-1}x - q^{n-1}a), & \text{if } n \geq 1 \end{cases}$$

$$\begin{aligned}(ii)\ (x \oplus a)^n_{p,q} &= \begin{cases} 1, & \text{if } n = 0 \\ (x+a)(px+qa)\cdots(p^{n-2}x + q^{n-2}a)(p^{n-1}x + q^{n-1}a), & \text{if } n \geq 1 \end{cases} \\ &= \sum_{k=0}^{n} \begin{bmatrix} n \\ k \end{bmatrix}_{p,q} p^{\binom{k}{2}} q^{\binom{n-k}{2}} x^k a^{n-k}. \end{aligned} \tag{6}$$

Definition 5. *We express the two forms of (p,q)-exponential functions as follows:*

$$e_{p,q}(x) = \sum_{n=0}^{\infty} p^{\binom{n}{2}} \frac{x^n}{[n]_{p,q}!}, \qquad E_{p,q}(x) = \sum_{n=0}^{\infty} q^{\binom{n}{2}} \frac{t^n}{[n]_{p,q}!}. \tag{7}$$

From Equation (7), we determine an important property, $e_{p,q}(x)E_{p,q}(-x) = 1$. Moreover, Duran, Acikgos, and Araci defined $\widetilde{e}_{p,q}(x)$ in [17] as follows:

$$\widetilde{e}_{p,q}(x) = \sum_{n=0}^{\infty} \frac{x^n}{[n]_{p,q}!}. \tag{8}$$

From Equations (8) and (6), we remark $e_{p,q}(x)E_{p,q}(y) = \widetilde{e}_{p,q}(x \oplus y)_{p,q}$.

Definition 6. *For $x \neq 0$, the (p,q)-derivative of a function f with respect to x is defined by the following:*

$$D_{p,q}f(x) = \frac{f(px) - f(qx)}{(p-q)x}, \tag{9}$$

where $(D_{p,q}f)(0) = f'(0)$, which prove that f is differentiable at 0. Moreover, it is evident that $D_{p,q}x^n = [n]_{p,q}x^{n-1}$.

Definition 7. *Let $i = \sqrt{-1} \in \mathbb{C}$. Then the (p,q)-trigonometric functions are defined by the following:*

$$\begin{aligned} sin_{p,q}(x) &= \frac{e_{p,q}(ix) - e_{p,q}(-ix)}{2i}, \qquad SIN_{p,q}(x) = \frac{E_{p,q}(ix) - E_{p,q}(-ix)}{2i} \\ cos_{p,q}(x) &= \frac{e_{p,q}(ix) + e_{p,q}(-ix)}{2}, \qquad COS_{p,q}(x) = \frac{E_{p,q}(ix) + E_{p,q}(-ix)}{2}, \end{aligned} \tag{10}$$

where, $SIN_{p,q}(x) = sin_{p^{-1},q^{-1}}(x)$ and $COS_{p,q}(x) = cos_{p^{-1},q^{-1}}(x)$.

In the same way as well-known Euler expressions using exponential functions, we define the (p,q)-analogues of hyperbolic functions and find several formulae (see [3,5,17]).

Theorem 2. *The following relationships hold:*

$$\begin{aligned} &(i) \quad sin_{p,q}(x)COS_{p,q}(x) = cos_{p,q}(x)SIN_{p,q}(x) \\ &(ii) \quad e_{p,q}(x) = cosh_{p,q}(x)sinh_{p,q}(x) \\ &(iii) \quad E_{p,q}(x) = COSH_{p,q}(x)SINH_{p,q}(x). \end{aligned} \tag{11}$$

From Definition 7 and Theorem 2, we note that $\cosh_{p,q}(x)\text{COSH}_{p,q}(x) - \sinh_{p,q}(x)$ $\text{SINH}_{p,q}(x) = 1$.

3. Several Basic Properties of (p,q)-Cosine and (p,q)-Sine Bernoulli Polynomials

We look for Lemma 1 and Theorem 3 in order to introduce (p,q)-cosine and (p,q)-sine Bernoulli polynomials. From the definitions of the (p,q)-cosine and (p,q)-sine Bernoulli polynomials, we search for a variety of properties. We also find relationships with other polynomials using properties of (p,q)-trigonometric functions or other methods.

Lemma 1. *For $y \in \mathbb{R}$ and $i = \sqrt{-1}$, we have the following:*

$$\begin{aligned} &(i) \quad E_{p,q}(ity) = COS_{p,q}(ty) + iSIN_{p,q}(ty), \\ &(ii) \quad E_{p,q}(-ity) = COS_{p,q}(ty) - iSIN_{p,q}(ty). \end{aligned} \tag{12}$$

Proof.

(i) $E_{p,q}(ity)$ can be expressed using the (p,q)-cosine and (p,q)-sine functions as

$$\begin{aligned} E_{p,q}(ity) &= \frac{E_{p,q}(ity)+E_{p,q}(-ity)}{2}+\frac{E_{p,q}(ity)-E_{p,q}(-ity)}{2} \\ &= COS_{p,q}(ty)+iSIN_{p,q}(ty). \end{aligned} \tag{13}$$

(ii) By substituting $-ity$ instead of (i), we obtain the following:

$$\begin{aligned} E_{p,q}(-ity) &= \frac{E_{p,q}(-ity)+E_{p,q}(-ity)}{2}-\frac{E_{p,q}(-ity)-E_{p,q}(-ity)}{2} \\ &= COS_{p,q}(ty)-iSIN_{p,q}(ty). \end{aligned} \tag{14}$$

Therefore, we complete the proof of Lemma 1. □

We note the following relations between $e_{p,q}$, $E_{p,q}$ and $\widetilde{e}_{p,q}$.

$$\begin{aligned} &(i) \quad e_{p,q}(x)E_{p,q}(y)=\sum_{n=0}^{\infty}\frac{(x\oplus y)_{p,q}^{n}}{[n]_{p,q}!}=\widetilde{e}_{p,q}(x\oplus y)_{p,q}, \\ &(ii) \quad e_{p,q}(x)E_{p,q}(-y)=\sum_{n=0}^{\infty}\frac{(x\ominus y)_{p,q}^{n}}{[n]_{p,q}!}=\widetilde{e}_{p,q}(x\ominus y)_{p,q}. \end{aligned} \tag{15}$$

Theorem 3. *Let $x,y\in\mathbb{R}$, $i=\sqrt{-1}$, and $|q/p|<1$. Then, we have*

$$\begin{aligned} &(i) \quad \sum_{n=0}^{\infty}\sum_{k=0}^{n}\begin{bmatrix} n \\ k \end{bmatrix}_{p,q}\left(\frac{(x\oplus iy)_{p,q}^{k}+(x\ominus iy)_{p,q}^{k}}{2}\right)B_{n-k,p,q}\frac{t^n}{[n]_{p,q}!} \\ &= \frac{t}{e_{p,q}(t)-1}e_{p,q}(tx)COS_{p,q}(ty), \\ &(ii) \quad \sum_{n=0}^{\infty}\sum_{k=0}^{n}\begin{bmatrix} n \\ k \end{bmatrix}_{p,q}\left(\frac{(x\oplus iy)_{p,q}^{k}-(x\ominus iy)_{p,q}^{k}}{2i}\right)B_{n-k,p,q}\frac{t^n}{[n]_{p,q}!} \\ &= \frac{t}{e_{p,q}(t)-1}e_{p,q}(tx)SIN_{p,q}(ty). \end{aligned} \tag{16}$$

Proof.

(i) We note that

$$\sum_{n=0}^{\infty}B_{n,p,q}\frac{t^n}{[n]_{p,q}!}=\frac{t}{e_{p,q}(t)-1}. \tag{17}$$

We find the following by multiplying $\widetilde{e}_{p,q}\left(t(x\oplus y)_{p,q}\right)$ in both sides of Equation (17).

$$\sum_{n=0}^{\infty}B_{n,p,q}\frac{t^n}{[n]_{p,q}!}\widetilde{e}_{p,q}\left(t(x\oplus y)_{p,q}\right)=\frac{t}{e_{p,q}(t)-1}\widetilde{e}_{p,q}\left(t(x\oplus y)_{p,q}\right). \tag{18}$$

The left-hand side of Equation (18) can be changed into

$$\begin{aligned} \sum_{n=0}^{\infty}B_{n,p,q}\frac{t^n}{[n]_{p,q}!}\widetilde{e}_{p,q}\left(t(x\oplus y)_{p,q}\right) &= \sum_{n=0}^{\infty}B_{n,p,q}\frac{t^n}{[n]_{p,q}!}\sum_{n=0}^{\infty}(x\oplus y)_{p,q}^{n}\frac{t^n}{[n]_{p,q}!} \\ &= \sum_{n=0}^{\infty}\left(\sum_{k=0}^{n}\begin{bmatrix} n \\ k \end{bmatrix}_{p,q}(x\oplus y)_{p,q}^{k}B_{n-k,p,q}\right)\frac{t^n}{[n]_{p,q}!}, \end{aligned} \tag{19}$$

and by using Lemma 1 (i) on the right-hand side of Equation (18), we yield

$$\begin{aligned}\frac{t}{e_{p,q}(t)-1}\widetilde{e}_{p,q}\left(t(x\oplus y)_{p,q}\right) &= \frac{t}{e_{p,q}(t)-1}e_{p,q}(x)E_{p,q}(y) \\ &= \frac{te_{p,q}(x)}{e_{p,q}(t)-1}\left(COS_{p,q}(ty)+iSIN_{p,q}(ty)\right).\end{aligned} \tag{20}$$

From Equations (19) and (20), we derive the following:

$$\sum_{n=0}^{\infty}\left(\sum_{k=0}^{n}\begin{bmatrix} n \\ k \end{bmatrix}_{p,q}(x\oplus y)_{p,q}^{k}B_{n-k,p,q}\right)\frac{t^n}{[n]_{p,q}!} = \frac{te_{p,q}(x)}{e_{p,q}(t)-1}\left(COS_{p,q}(ty)+iSIN_{p,q}(ty)\right). \tag{21}$$

We obtain the equation below for (p,q)-Bernoulli numbers using a similar method.

$$\sum_{n=0}^{\infty}\left(\sum_{k=0}^{n}\begin{bmatrix} n \\ k \end{bmatrix}_{p,q}(x\ominus iy)_{p,q}^{k}B_{n-k,p,q}\right)\frac{t^n}{[n]_{p,q}!} = \frac{te_{p,q}(tx)}{e_{p,q}(t)-1}\left(COS_{p,q}(ty)-iSIN_{p,q}(ty)\right). \tag{22}$$

By using Equations (21) and (22), we have

$$\sum_{n=0}^{\infty}\sum_{k=0}^{n}\begin{bmatrix} n \\ k \end{bmatrix}_{p,q}\left(\frac{(x\oplus iy)_{p,q}^{k}+(x\ominus iy)_{p,q}^{k}}{2}\right)B_{n-k,p,q}\frac{t^n}{[n]_{p,q}!} = \frac{t}{e_{p,q}(t)-1}e_{p,q}(tx)COS_{p,q}(ty) \tag{23}$$

and

$$\sum_{n=0}^{\infty}\sum_{k=0}^{n}\begin{bmatrix} n \\ k \end{bmatrix}_{p,q}\left(\frac{(x\oplus iy)_{p,q}^{k}-(x\ominus iy)_{p,q}^{k}}{2i}\right)B_{n-k,p,q}\frac{t^n}{[n]_{p,q}!} = \frac{t}{e_{p,q}(t)-1}e_{p,q}(tx)SIN_{p,q}(ty). \tag{24}$$

Therefore, we can conclude the required results. □

Thus, we are ready to introduce (p,q)-cosine and (p,q)-sine Bernoulli polynomials using Lemma 1 and Theorem 3.

Definition 8. *Let $|p/q|<1$ and $x,y\in\mathbb{R}$. Then (p,q)-cosine and (p,q)-sine Bernoulli polynomials are respectively defined by the following:*

$$\sum_{n=0}^{\infty}{}_{C}B_{n,p,q}(x,y)\frac{t^n}{[n]_{p,q}!} = \frac{t}{e_{p,q}(t)-1}e_{p,q}(tx)COS_{p,q}(ty),$$

and

$$\sum_{n=0}^{\infty}{}_{S}B_{n,p,q}(x,y)\frac{t^n}{[n]_{p,q}!} = \frac{t}{e_{p,q}(t)-1}e_{p,q}(tx)SIN_{p,q}(ty). \tag{25}$$

From Definition 8, we determine q-cosine and q-sine Bernoulli polynomials when $|q|<1$ and $p=1$. In addition, we observe cosine Bernoulli polynomials and sine Bernoulli polynomials for $q\to 1$ and $p=1$.

Corollary 1. *From Theorem 3 and Definition 8, the following holds*

$$\begin{aligned}
&(i)\quad {}_CB_{n,p,q}(x,y)=\sum_{k=0}^{n}\begin{bmatrix} n\\ k\end{bmatrix}_{p,q}\left(\frac{(x\oplus iy)_{p,q}^{k}+(x\ominus iy)_{p,q}^{k}}{2}\right)B_{n-k,p,q},\\
&(ii)\,{}_SB_{n,p,q}(x,y)=\sum_{k=0}^{n}\begin{bmatrix} n\\ k\end{bmatrix}_{p,q}\left(\frac{(x\oplus iy)_{p,q}^{k}-(x\ominus iy)_{p,q}^{k}}{2i}\right)B_{n-k,p,q},
\end{aligned}\tag{26}$$

where $B_{n,p,q}$ denotes the (p,q)-Bernoulli numbers.

Example 1. *From Definition 8, a few examples of ${}_CB_{n,p,q}(x,y)$ and ${}_SB_{n,p,q}(x,y)$ are the follows:*

$$\begin{aligned}
&{}_CB_{0,p,q}(x,y)=0\\
&{}_CB_{1,p,q}(x,y)=px\\
&{}_CB_{2,p,q}(x,y)=p^2x^2-qy^2\\
&{}_CB_{3,p,q}(x,y)=p^3x^3-pq(p^2+pq+q^2)xy^2\\
&{}_CB_{4,p,q}(x,y)=p^4x^4-p^2q(p^2+q^2)(p^2+pq+q^2)x^2y^2+q^6y^4,\\
&\cdots.
\end{aligned}\tag{27}$$

and

$$\begin{aligned}
&{}_SB_{0,p,q}(x,y)=0\\
&{}_SB_{1,p,q}(x,y)=\frac{y}{p+q}\\
&{}_SB_{2,p,q}(x,y)=\frac{pxy}{p^2+pq+q^2}\\
&{}_SB_{3,p,q}(x,y)=\frac{y\left(\frac{p^2x^2}{p^2+q^2}-q^3y^2\right)}{p+q}\\
&{}_SB_{4,p,q}(x,y)=p(p-q)xy\left(\frac{p^2x^2}{p^5-q^5}+\frac{q^3(p^2+q^2)y^2}{-p^3+q^3}\right),\\
&\cdots.
\end{aligned}\tag{28}$$

Definition 9. *Let $|p/q|<1$. Then, we define*

$$\sum_{n=0}^{\infty}C_{n,p,q}(x,y)\frac{t^n}{[n]_{p,q}!}=e_{p,q}(tx)COS_{p,q}(ty),\quad \sum_{n=0}^{\infty}S_{n,p,q}(x,y)\frac{t^n}{[n]_{p,q}!}=e_{p,q}(tx)SIN_{p,q}(ty).\tag{29}$$

Theorem 4. *Let k be a nonnegative integer and $|p/q|<1$. Then, we have*

$$\begin{aligned}
&(i)\quad {}_CB_{n,p,q}(x,y)=\sum_{k=0}^{n}\begin{bmatrix} n\\ k\end{bmatrix}_{p,q}B_{n-k,p,q}C_{k,p,q}(x,y),\\
&(ii)\quad {}_SB_{n,p,q}(x,y)=\sum_{k=0}^{n}\begin{bmatrix} n\\ k\end{bmatrix}_{p,q}B_{n-k,p,q}S_{k,p,q}(x,y),
\end{aligned}\tag{30}$$

where $B_{n,p,q}$ is the (p,q)-Bernoulli numbers.

Proof.

(i) Using the generating function of the (p,q)-cosine Bernoulli polynomials and Definition 9, we find

$$\begin{aligned}\sum_{n=0}^{\infty} {}_{C}B_{n,p,q}(x,y)\frac{t^n}{[n]_{p,q}!} &= \sum_{n=0}^{\infty} B_{n,p,q}\frac{t^n}{[n]_{p,q}!}\sum_{n=0}^{\infty} C_{n,p,q}(x,y)\frac{t^n}{[n]_{p,q}!} \\ &= \sum_{n=0}^{\infty}\left(\sum_{k=0}^{n}\begin{bmatrix} n \\ k \end{bmatrix}_{p,q} B_{n-k,p,q}C_{n,p,q}(x,y)\right)\frac{t^n}{[n]_{p,q}!}.\end{aligned} \tag{31}$$

Through comparison of the coefficients of both sides for Equation (31), we obtain the desired results immediately.

(ii) By applying a method similar to (i) in the generating function of the (p,q)-sine Bernoulli polynomials, we complete the proof of Theorem 4 (ii).

□

Theorem 5. *For a nonnegative integer n, we derive*

$$\begin{aligned}&(i)\quad [n]_{p,q}C_{n-1,p,q}(x,y) = \sum_{k=0}^{n}\begin{bmatrix} n \\ k \end{bmatrix}_{p,q} p^{\binom{n-k}{2}}{}_{C}B_{k,p,q}(x,y) - {}_{C}B_{n,p,q}(x,y), \\ &(ii)\quad [n]_{p,q}S_{n-1,p,q}(x,y) = \sum_{k=0}^{n}\begin{bmatrix} n \\ k \end{bmatrix}_{p,q} p^{\binom{n-k}{2}}{}_{S}B_{k,p,q}(x,y) - {}_{S}B_{n,p,q}(x,y).\end{aligned} \tag{32}$$

Proof.

(i) Suppose that $e_{p,q}(t) \neq 1$ in the generating function of the (p,q)-cosine Bernoulli polynomials. Then, we have

$$\sum_{n=0}^{\infty} {}_{C}B_{n,p,q}(x,y)\frac{t^n}{[n]_{p,q}!}(e_{p,q}(t)-1) = te_{p,q}(tx)COS_{p,q}(ty). \tag{33}$$

We write the left-hand side of Equation (33) as follows:

$$\begin{aligned}&\sum_{n=0}^{\infty} {}_{C}B_{n,p,q}(x,y)\frac{t^n}{[n]_{p,q}!}(e_{p,q}(t)-1) \\ &= \sum_{n=0}^{\infty} {}_{C}B_{n,p,q}(x,y)\frac{t^n}{[n]_{p,q}!}\left(\sum_{n=0}^{\infty} p^{\binom{n}{2}}\frac{t^n}{[n]_{p,q}!}-1\right) \\ &= \sum_{n=0}^{\infty}\left(\sum_{k=0}^{n}\begin{bmatrix} n \\ k \end{bmatrix}_{p,q} p^{\binom{n-k}{2}}{}_{C}B_{k,p,q}(x,y) - {}_{C}B_{n,p,q}(x,y)\right)\frac{t^n}{[n]_{p,q}!},\end{aligned} \tag{34}$$

and we transform the right-hand side into the following:

$$\begin{aligned}te_{p,q}(tx)COS_{p,q}(ty) &= \sum_{n=0}^{\infty} C_{n,p,q}(x,y)\frac{t^{n+1}}{[n]_{p,q}!} \\ &= \sum_{n=0}^{\infty} [n]_{p,q}C_{n-1,p,q}(x,y)\frac{t^n}{[n]_{p,q}!}.\end{aligned} \tag{35}$$

Therefore, we obtain the following:

$$\sum_{k=0}^{n}\begin{bmatrix} n \\ k \end{bmatrix}_{p,q} p^{\binom{n-k}{2}}{}_{C}B_{k,p,q}(x,y) - {}_{C}B_{n,p,q}(x,y) = [n]_{p,q}C_{n-1,p,q}(x,y). \tag{36}$$

By calculating the left-hand side of Equation (36), we investigate the required result.

(ii) We do not include the proof of Theorem 5 (ii) because the proving process is similar to that of Theorem 5 (i).

□

Corollary 2. *Setting $p = 1$ in Theorem 5, the following equations hold*

$$\begin{aligned}
&(i)\quad [n]_q C_{n-1,q}(x,y) = \sum_{k=0}^{n} \begin{bmatrix} n \\ k \end{bmatrix}_q {}_C B_{k,q}(x,y) - {}_C B_{n,q}(x,y) \\
&(ii)\quad [n]_q S_{n-1,q}(x,y) = \sum_{k=0}^{n} \begin{bmatrix} n \\ k \end{bmatrix}_q {}_S B_{k,q}(x,y) - {}_S B_{n,q}(x,y),
\end{aligned} \tag{37}$$

where ${}_C B_{n,q}(x,y)$ represents the q-cosine Bernoulli polynomials and ${}_S B_{n,q}(x,y)$ denotes the q-sine Bernoulli polynomials.

Corollary 3. *Assigning $p = 1$ and $q \to 1$ in Theorem 5, the following holds:*

$$\begin{aligned}
&(i)\quad nC_{n-1}(x,y) = \sum_{k=0}^{n-1} \binom{n}{k} {}_C B_n(x,y) \\
&(ii)\quad nS_{n-1}(x,y) = \sum_{k=0}^{n-1} \binom{n}{k} {}_S B_n(x,y),
\end{aligned} \tag{38}$$

where ${}_C B_n(x,y)$ represents the cosine Bernoulli polynomials and ${}_S B_n(x,y)$ represents the sine Bernoulli polynomials.

Theorem 6. *Let $|p/q| < 1$. Then, we have*

$$\begin{aligned}
&(i)\quad {}_C B_{n,p,q}(1,y) = \sum_{k=0}^{n} \begin{bmatrix} n \\ k \end{bmatrix}_{p,q} (-1)^{n-k} q^{\binom{n-k}{2}} \left([k]_{p,q} C_{k-1,p,q}(x,y) + {}_C B_{k,p,q}(x,y) \right) x^{n-k}, \\
&(ii)\quad {}_S B_{n,p,q}(1,y) = \sum_{k=0}^{n} \begin{bmatrix} n \\ k \end{bmatrix}_{p,q} (-1)^{n-k} q^{\binom{n-k}{2}} \left([k]_{p,q} S_{k-1,p,q}(x,y) + {}_S B_{k,p,q}(x,y) \right) x^{n-k}.
\end{aligned} \tag{39}$$

Proof.

(i) If we put 1 instead of x in the generating function of the (p,q)-cosine Bernoulli polynomials, we find the following:

$$\begin{aligned}
\sum_{n=0}^{\infty} {}_C B_{n,p,q}(1,y) \frac{t^n}{[n]_{p,q}!} &= \frac{t}{e_{p,q}(t) - 1} \left(e_{p,q}(t) - 1 + 1 \right) COS_{p,q}(ty) \\
&= tCOS_{p,q}(ty) + \frac{t}{e_{p,q}(t) - 1} COS_{p,q}(ty).
\end{aligned} \tag{40}$$

Using a property of the (p,q)-exponential function, $e_{p,q}(x)E_{p,q}(-x) = 1$, in Equation (40), we obtain the following:

$$\begin{aligned}
&\sum_{n=0}^{\infty} {}_{C}B_{n,p,q}(1,y)\frac{t^n}{[n]_{p,q}!}\\
&= \left(te_{p,q}(tx)COS_{p,q}(ty) + \frac{t}{e_{p,q}(t)-1}e_{p,q}(tx)COS_{p,q}(ty)\right)E_{p,q}(-tx)\\
&= \sum_{n=0}^{\infty}\left([n]_{p,q}C_{n-1,p,q}(x,y) + {}_{C}B_{n,p,q}(x,y)\right)\frac{t^n}{[n]_{p,q}!}\sum_{n=0}^{\infty} q^{\binom{n}{2}}(-x)^n\frac{t^n}{[n]_{p,q}!}\\
&= \sum_{n=0}^{\infty}\left(\sum_{k=0}^{n}\begin{bmatrix} n\\ k\end{bmatrix}_{p,q}(-1)^{n-k}q^{\binom{n-k}{2}}([k]_{p,q}C_{k-1,p,q}(x,y) + {}_{C}B_{k,p,q}(x,y))x^{n-k}\right)\frac{t^n}{[n]_{p,q}!},
\end{aligned} \tag{41}$$

and we immediately derive the results.

(ii) By applying a similar process for proving (i) to the (p,q)-sine Bernoulli polynomials, we find Theorem 6 (ii).

□

Corollary 4. *Setting $p = 1$ in Theorem 6, the following holds:*

$$\begin{aligned}
&(i)\quad {}_{C}B_{n,q}(1,y) = \sum_{k=0}^{n}\begin{bmatrix} n\\ k\end{bmatrix}_{q}(-1)^{n-k}q^{\binom{n-k}{2}}\left([k]_qC_{k-1,q}(x,y) + {}_{C}B_{k,q}(x,y)\right)x^{n-k}\\
&(ii)\quad {}_{S}B_{n,q}(1,y) = \sum_{k=0}^{n}\begin{bmatrix} n\\ k\end{bmatrix}_{q}(-1)^{n-k}q^{\binom{n-k}{2}}\left([k]_qS_{k-1,q}(x,y) + {}_{S}B_{k,q}(x,y)\right)x^{n-k},
\end{aligned} \tag{42}$$

where ${}_{C}B_{n,q}(x,y)$ denotes the q-cosine Bernoulli polynomials and ${}_{S}B_{n,q}(x,y)$ denotes the q-sine Bernoulli polynomials.

Corollary 5. *Setting $p = 1$ and $q \to 1$ in Theorem 6, the following holds:*

$$\begin{aligned}
&(i)\quad {}_{C}B_{n}(1,y) = \sum_{k=0}^{n}\binom{n}{k}(-1)^{n-k}\left(kC_{k-1}(x,y) + {}_{C}B_{k}(x,y)\right)x^{n-k}\\
&(ii)\quad {}_{S}B_{n}(1,y) = \sum_{k=0}^{n}\binom{n}{k}(-1)^{n-k}\left(kS_{k-1}(x,y) + {}_{S}B_{k}(x,y)\right)x^{n-k},
\end{aligned} \tag{43}$$

where ${}_{C}B_{n}(x,y)$ is the cosine Bernoulli polynomials and ${}_{S}B_{n}(x,y)$ is the sine Bernoulli polynomials.

Theorem 7. *For a nonnegative integer k and $|p/q| < 1$, we investigate*

$$\begin{aligned}
B_{n,p,q}(x) &= \sum_{k=0}^{[\frac{n}{2}]}\begin{bmatrix} n\\ 2k\end{bmatrix}_{p,q}(-1)^k p^{(2k-1)k}y^{2k}{}_{C}B_{n-k,p,q}(x,y)\\
&+ \sum_{k=0}^{[\frac{n-1}{2}]}\begin{bmatrix} n\\ 2k+1\end{bmatrix}_{p,q}(-1)^k p^{(2k+1)k}y^{2k+1}{}_{S}B_{n-(2k+1),p,q}(x,y),
\end{aligned} \tag{44}$$

where $B_{n,p,q}(x)$ is the (p,q)-Bernoulli polynomials and $[x]$ is the greatest integer not exceeding x.

Proof. In [9], we observe the power series of (p,q)-cosine and (p,q)-sine functions as follows:

$$cos_{p,q}(x)=\sum_{n=0}^{\infty}(-1)^n p^{(2n-1)n}\frac{x^{2n}}{[2n]_{p,q}!},\quad sin_{p,q}(x)=\sum_{n=0}^{\infty}(-1)^n p^{(2n+1)n}\frac{x^{2n+1}}{[2n+1]_{p,q}!}. \tag{45}$$

Let us consider (p,q)-cosine Bernoulli polynomials. If we multiply (p,q)-cosine Bernoulli polynomials and the (p,q)-cosine function to determine the relationship between (p,q)-Bernoulli polynomials and, combined (p,q)-cosine Bernoulli polynomials, and (p,q)-sine Bernoulli polynomials, we have

$$\sum_{n=0}^{\infty} {}_CB_{n,p,q}(x,y)\frac{t^n}{[n]_{p,q}!}cos_{p,q}(ty)=\frac{t}{e_{p,q}(t)-1}e_{p,q}(tx)COS_{p,q}(ty)cos_{p,q}(ty). \tag{46}$$

The left-hand side of Equation (46) is transformed as

$$\begin{aligned}
&\sum_{n=0}^{\infty} {}_CB_{n,p,q}(x,y)\frac{t^n}{[n]_{p,q}!}cos_{p,q}(ty)\\
&=\sum_{n=0}^{\infty} {}_CB_{n,p,q}(x,y)\frac{t^n}{[n]_{p,q}!}\sum_{n=0}^{\infty}(-1)^n p^{(2n-1)n}y^{2n}\frac{t^{2n}}{[n]_{p,q}!}\\
&=\sum_{n=0}^{\infty}\left(\sum_{k=0}^{n}\begin{bmatrix} n+k\\ 2k\end{bmatrix}_{p,q}(-1)^k p^{(2k-1)k}y^{2k}{}_CB_{n-k,p,q}(x,y)\right)\frac{t^{n+k}}{[n+k]_{p,q}!}\\
&=\sum_{n=0}^{\infty}\left(\sum_{k=0}^{[\frac{n}{2}]}\begin{bmatrix} n\\ 2k\end{bmatrix}_{p,q}(-1)^k p^{(2k-1)k}y^{2k}{}_CB_{n-k,p,q}(x,y)\right)\frac{t^{n}}{[n]_{p,q}!}.
\end{aligned} \tag{47}$$

From Equations (46) and (47), we derive the following:

$$\begin{aligned}
&\sum_{n=0}^{\infty}\left(\sum_{k=0}^{[\frac{n}{2}]}\begin{bmatrix} n\\ 2k\end{bmatrix}_{p,q}(-1)^k p^{(2k-1)k}y^{2k}{}_CB_{n-k,p,q}(x,y)\right)\frac{t^{n}}{[n]_{p,q}!}\\
&=\frac{t}{e_{p,q}(t)-1}e_{p,q}(tx)COS_{p,q}(ty)cos_{p,q}(ty),
\end{aligned} \tag{48}$$

where $[x]$ is the greatest integer that does not exceed x.

From now on, let us consider the (p,q)-sine Bernoulli polynomials in a same manner of (p,q)-cosine Bernoulli polynomials. If we multiply ${}_SB_{n,p,q}(x,y)$ and $sin_{p,q}(ty)$, we obtain

$$\sum_{n=0}^{\infty} {}_SB_{n,p,q}(x,y)\frac{t^n}{[n]_{p,q}!}sin_{p,q}(ty)=\frac{t}{e_{p,q}(t)-1}e_{p,q}(tx)SIN_{p,q}(ty)sin_{p,q}(ty). \tag{49}$$

The left-hand side of Equation (49) can be changed as the following.

$$\begin{aligned}
&\sum_{n=0}^{\infty} {}_SB_{n,p,q}(x,y)\frac{t^n}{[n]_{p,q}!}sin_{p,q}(ty)\\
&=\sum_{n=0}^{\infty} {}_SB_{n,p,q}(x,y)\frac{t^n}{[n]_{p,q}!}\sum_{n=0}^{\infty}(-1)^n p^{(2n+1)n}y^{2n+1}\frac{t^{2n+1}}{[n]_{p,q}!}\\
&=\sum_{n=0}^{\infty}\left(\sum_{k=0}^{[\frac{n-1}{2}]}\begin{bmatrix} n\\ 2k+1\end{bmatrix}_{p,q}(-1)^k p^{(2k+1)k}y^{2k+1}{}_SB_{n-(2k+1),p,q}(x,y)\right)\frac{t^{n}}{[n]_{p,q}!}.
\end{aligned} \tag{50}$$

From Equations (49) and (50), we have the following:

$$\sum_{n=0}^{\infty}\left(\sum_{k=0}^{[\frac{n-1}{2}]}\begin{bmatrix} n \\ 2k+1 \end{bmatrix}_{p,q}(-1)^k p^{(2k+1)k} y^{2k+1}{}_S B_{n-(2k+1),p,q}(x,y)\right)\frac{t^n}{[n]_{p,q}!}$$
$$= \frac{t}{e_{p,q}(t)-1} e_{p,q}(tx) SIN_{p,q}(ty) sin_{p,q}(ty), \tag{51}$$

where $[x]$ is the greatest integer that does not exceed x.

Here, we recall that

$$(COS_{p,q}(x)cos_{p,q}(x) + SIN_{p,q}(x)sin_{p,q}(x)) = 1. \tag{52}$$

Using Equations (48) and (51) and applying the property of (p,q)-trigonometric functions, we find (p,q)-Bernoulli polynomials as follows:

$$\sum_{n=0}^{\infty}\left(\sum_{k=0}^{[\frac{n}{2}]}\begin{bmatrix} n \\ 2k \end{bmatrix}_{p,q}(-1)^k p^{(2k-1)k} y^{2k}{}_C B_{n-k,p,q}(x,y)\right)\frac{t^n}{[n]_{p,q}!}$$
$$+\sum_{n=0}^{\infty}\left(\sum_{k=0}^{[\frac{n-1}{2}]}\begin{bmatrix} n \\ 2k+1 \end{bmatrix}_{p,q}(-1)^k p^{(2k+1)k} y^{2k+1}{}_S B_{n-(2k+1),p,q}(x,y)\right)\frac{t^n}{[n]_{p,q}!}$$
$$= \frac{t}{e_{p,q}(t)-1} e_{p,q}(tx)\left(COS_{p,q}(ty)cos_{p,q}(ty) + SIN_{p,q}(ty)sin_{p,q}(ty)\right)$$
$$= \sum_{n=0}^{\infty} B_{n,p,q}(x)\frac{t^n}{[n]_{p,q}!}, \tag{53}$$

where $B_{n,p,q}(x)$ is the (p,q)-Bernoulli polynomials.

By comparing the coefficients of both sides of t^n, we produce the desired result. □

Corollary 6. *Setting $p = 1$ in Theorem 7, the following holds:*

$$B_{n,q}(x) = \sum_{k=0}^{[\frac{n}{2}]}\begin{bmatrix} n \\ 2k \end{bmatrix}_{q}(-1)^k y^{2k}{}_C B_{n-k,q}(x,y)$$
$$+\sum_{k=0}^{[\frac{n-1}{2}]}\begin{bmatrix} n \\ 2k+1 \end{bmatrix}_{q}(-1)^k y^{2k+1}{}_S B_{n-(2k+1),q}(x,y), \tag{54}$$

where $B_{n,q}(x)$ is the q-Bernoulli polynomials, ${}_C B_{n,q}(x,y)$ denote the q-cosine Bernoulli polynomials, and ${}_S B_{n,q}(x)$ denote the q-sine Bernoulli polynomials.

Corollary 7. *Setting $y = 1$ in Theorem 7, one holds:*

$$B_{n,p,q}(x) = \sum_{k=0}^{[\frac{n}{2}]}\begin{bmatrix} n \\ 2k \end{bmatrix}_{p,q}(-1)^k p^{(2k-1)k}{}_C B_{n-k,p,q}(x,1)$$
$$+\sum_{k=0}^{[\frac{n-1}{2}]}\begin{bmatrix} n \\ 2k+1 \end{bmatrix}_{p,q}(-1)^k p^{(2k+1)k}{}_S B_{n-(2k+1),p,q}(x,1), \tag{55}$$

where $B_{n,p,q}(x)$ is the (p,q)-Bernoulli polynomials and $[x]$ is the greatest integers that does not exceed x.

Theorem 8. *For a nonnegative integer k and $|p/q| < 1$, we derive*

$$\begin{aligned}&\sum_{k=0}^{[\frac{n-1}{2}]}\begin{bmatrix} n \\ 2k+1 \end{bmatrix}_{p,q}(-1)^k p^{(2k+1)k} y^{2k+1} {}_CB_{n-(2k+1),p,q}(x,y)\\&=\sum_{k=0}^{[\frac{n}{2}]}\begin{bmatrix} n \\ 2k \end{bmatrix}_{p,q}(-1)^k p^{(2k-1)k} y^{2k} {}_SB_{n-k,p,q}(x,y),\end{aligned} \tag{56}$$

where $[x]$ is the greatest integer not exceeding x.

Proof. If we multiply ${}_CB_{n,p,q}(x,y)$ and $sin_{p,q}(ty)$, then we find

$$\sum_{n=0}^{\infty} {}_CB_{n,p,q}(x,y)\frac{t^n}{[n]_{p,q}!}sin_{p,q}(ty) = \frac{t}{e_{p,q}(t)-1}e_{p,q}(tx)COS_{p,q}(ty)sin_{p,q}(ty), \tag{57}$$

and the left-hand side of Equation (57) can be transformed as

$$\begin{aligned}&\sum_{n=0}^{\infty} {}_CB_{n,p,q}(x,y)\frac{t^n}{[n]_{p,q}!}sin_{p,q}(ty)\\&=\sum_{n=0}^{\infty} {}_CB_{n,p,q}(x,y)\frac{t^n}{[n]_{p,q}!}\sum_{n=0}^{\infty}(-1)^n p^{(2n+1)n} y^{2n+1}\frac{t^{2n+1}}{[2n+1]_{p,q}!}\\&=\sum_{n=0}^{\infty}\left(\sum_{k=0}^{[\frac{n-1}{2}]}\begin{bmatrix} n \\ 2k+1 \end{bmatrix}_{p,q}(-1)^k p^{(2k+1)k} y^{2k+1} {}_CB_{n-(2k+1),p,q}(x,y)\right)\frac{t^n}{[n]_{p,q}!}.\end{aligned} \tag{58}$$

Similarly, we multiply the (p,q)-sine Bernoulli polynomials and (p,q)-cosine function as the follows:

$$\sum_{n=0}^{\infty} {}_SB_{n,p,q}(x,y)\frac{t^n}{[n]_{p,q}!}cos_{p,q}(ty) = \frac{t}{e_{p,q}(t)-1}e_{p,q}(tx)SIN_{p,q}(ty)cos_{p,q}(ty). \tag{59}$$

The left-hand side of Equation (59) can be changed as

$$\begin{aligned}&\sum_{n=0}^{\infty} {}_SB_{n,p,q}(x,y)\frac{t^n}{[n]_{p,q}!}cos_{p,q}(ty)\\&=\sum_{n=0}^{\infty} {}_SB_{n,p,q}(x,y)\frac{t^n}{[n]_{p,q}!}\sum_{n=0}^{\infty}(-1)^n p^{(2n-1)n} y^{2n}\frac{t^{2n}}{[2n]_{p,q}!}\\&=\sum_{n=0}^{\infty}\left(\sum_{k=0}^{[\frac{n-1}{2}]}\begin{bmatrix} n \\ 2k \end{bmatrix}_{p,q}(-1)^k p^{(2k-1)k} y^{2k} {}_SB_{n-k,p,q}(x,y)\right)\frac{t^n}{[n]_{p,q}!}.\end{aligned} \tag{60}$$

In here, we recall that $sin_{p,q}(x)COS_{p,q}(x) = cos_{p,q}(x)SIN_{p,q}(x)$. From Equations (58) and (60), and the above property of (p,q)-trigonometric functions, we investigate

$$\begin{aligned}&\sum_{n=0}^{\infty}\left(\sum_{k=0}^{[\frac{n-1}{2}]}\begin{bmatrix} n \\ 2k+1 \end{bmatrix}_{p,q}(-1)^k p^{(2k+1)k} y^{2k+1} {}_CB_{n-(2k+1),p,q}(x,y)\right)\frac{t^n}{[n]_{p,q}!}\\&-\sum_{n=0}^{\infty}\left(\sum_{k=0}^{[\frac{n}{2}]}\begin{bmatrix} n \\ 2k \end{bmatrix}_{p,q}(-1)^k p^{(2k-1)k} y^{2k} {}_SB_{n-k,p,q}(x,y)\right)\frac{t^n}{[n]_{p,q}!}\\&=\frac{t}{e_{p,q}(t)-1}e_{p,q}(tx)\left(COS_{p,q}(ty)sin_{p,q}(ty)-SIN_{p,q}(ty)cos_{p,q}(ty)\right)\end{aligned} \tag{61}$$

From Equation (61), we complete the proof of Theorem 8. □

Corollary 8. *Putting $y = 1$ in Theorem 8, we have the following:*

$$\sum_{k=0}^{[\frac{n-1}{2}]} \begin{bmatrix} n \\ 2k+1 \end{bmatrix}_{p,q} (-1)^k p^{(2k+1)k} {}_C B_{n-(2k+1),p,q}(x,1) = \sum_{k=0}^{[\frac{n}{2}]} \begin{bmatrix} n \\ 2k \end{bmatrix}_{p,q} (-1)^k p^{(2k-1)k} {}_S B_{n-k,p,q}(x,1), \tag{62}$$

where $[x]$ is the greatest integer not exceeding x.

Corollary 9. *Setting $p = 1$ in Theorem 8, the following holds:*

$$\sum_{k=0}^{[\frac{n-1}{2}]} \begin{bmatrix} n \\ 2k+1 \end{bmatrix}_{q} (-1)^k y^{2k+1} {}_C B_{n-(2k+1),q}(x,y) = \sum_{k=0}^{[\frac{n}{2}]} \begin{bmatrix} n \\ 2k \end{bmatrix}_{q} (-1)^k y^{2k} {}_S B_{n-k,q}(x,y), \tag{63}$$

where ${}_C B_{n,q}(x,y)$ is the q-cosine Bernoulli polynomials and ${}_S B_{n,q}(x,y)$ is the q-sine Bernoulli polynomials.

Corollary 10. *Let $p = 1$ and $q \to 1$ in Theorem 8. Then one holds*

$$\sum_{k=0}^{[\frac{n-1}{2}]} \binom{n}{2k+1} (-1)^k y^{2k+1} {}_C B_{n-(2k+1)}(x,y) = \sum_{k=0}^{[\frac{n}{2}]} \binom{n}{2k} (-1)^k y^{2k} {}_S B_{n-k}(x,y), \tag{64}$$

where ${}_C B_n(x,y)$ is the cosine Bernoulli polynomials and ${}_S B_n(x,y)$ is the sine Bernoulli polynomials.

4. Several Symmetric Properties of the (p,q)-Cosine and (p,q)-Sine Bernoulli Polynomials

In this section, we point out several symmetric identities of the (p,q)-cosine and (p,q)-Bernoulli polynomials. Using various forms that are made by a and b, we obtain a few desired results regarding the (p,q)-cosine and (p,q)-sine Bernoulli polynomials. Moreover, we discover other relations of different Bernoulli polynomials by considering certain conditions in theorems. We also find the symmetric structure of the approximate roots based on the symmetric polynomials.

Theorem 9. *Let a and b be nonzero. Then, we obtain*

$$\begin{aligned}
(i)\quad & \sum_{k=0}^{n} \begin{bmatrix} n \\ k \end{bmatrix}_{p,q} a^{n-k-1} b^{k-1} {}_C B_{n-k,p,q}\left(\frac{x}{a},\frac{y}{a}\right) {}_C B_{k,p,q}\left(\frac{X}{b},\frac{Y}{b}\right) \\
&= \sum_{k=0}^{n} \begin{bmatrix} n \\ k \end{bmatrix}_{p,q} b^{n-k-1} a^{k-1} {}_C B_{n-k,p,q}\left(\frac{x}{b},\frac{y}{b}\right) {}_C B_{k,p,q}\left(\frac{X}{a},\frac{Y}{a}\right), \\
(ii)\quad & \sum_{k=0}^{n} \begin{bmatrix} n \\ k \end{bmatrix}_{p,q} a^{n-k-1} b^{k-1} {}_S B_{n-k,p,q}\left(\frac{x}{a},\frac{y}{a}\right) {}_S B_{k,p,q}\left(\frac{X}{b},\frac{Y}{b}\right) \\
&= \sum_{k=0}^{n} \begin{bmatrix} n \\ k \end{bmatrix}_{p,q} b^{n-k-1} a^{k-1} {}_S B_{n-k,p,q}\left(\frac{x}{b},\frac{y}{b}\right) {}_S B_{k,p,q}\left(\frac{X}{a},\frac{Y}{a}\right).
\end{aligned} \tag{65}$$

Proof.

(i) We consider form A as follows:

$$A := \frac{t^2 e_{p,q}(tx) e_{p,q}(tX) COS_{p,q}(ty) COS_{p,q}(tY)}{\left(e_{p,q}(at)-1\right)\left(e_{p,q}(bt)-1\right)} \tag{66}$$

From form A, we find

$$\begin{aligned} A &= \frac{t}{e_{p,q}(at)-1} e_{p,q}(tx) COS_{p,q}(ty) \frac{t}{e_{p,q}(bt)-1} e_{p,q}(tX) COS_{p,q}(tY) \\ &= \sum_{n=0}^{\infty} a^{n-1} {}_{C}B_{n,p,q}\left(\frac{x}{a}, \frac{y}{a}\right) \frac{t^n}{[n]_{p,q}!} \sum_{n=0}^{\infty} b^{n-1} {}_{C}B_{n,p,q}\left(\frac{X}{b}, \frac{Y}{b}\right) \frac{t^n}{[n]_{p,q}!} \\ &= \sum_{n=0}^{\infty} \left(\sum_{k=0}^{n} \begin{bmatrix} n \\ k \end{bmatrix}_{p,q} a^{n-k-1} b^{k-1} {}_{C}B_{n-k,p,q}\left(\frac{x}{a}, \frac{y}{a}\right) {}_{C}B_{k,p,q}\left(\frac{X}{b}, \frac{Y}{b}\right) \right) \frac{t^n}{[n]_{p,q}!}, \end{aligned} \tag{67}$$

and form A of Equation (66) can be transformed into the following:

$$\begin{aligned} A &= \frac{t}{e_{p,q}(bt)-1} e_{p,q}(tx) COS_{p,q}(ty) \frac{t}{e_{p,q}(at)-1} e_{p,q}(tX) COS_{p,q}(tY) \\ &= \sum_{n=0}^{\infty} \left(\sum_{k=0}^{n} \begin{bmatrix} n \\ k \end{bmatrix}_{p,q} b^{n-k-1} a^{k-1} {}_{C}B_{n-k,p,q}\left(\frac{x}{b}, \frac{y}{b}\right) {}_{C}B_{k,p,q}\left(\frac{X}{a}, \frac{Y}{a}\right) \right) \frac{t^n}{[n]_{p,q}!}. \end{aligned} \tag{68}$$

Using the comparison of coefficients in Equations (67) and (68), we find the desired result.

(ii) If we assume form B as follows:

$$B := \frac{t^2 e_{p,q}(tx) e_{p,q}(tX) SIN_{p,q}(ty) SIN_{p,q}(tY)}{\left(e_{p,q}(at)-1\right)\left(e_{p,q}(bt)-1\right)}, \tag{69}$$

then, we find Theorem 9 (ii) in the same manner.

□

Corollary 11. *Setting $a = 1$ in Theorem 9, the following holds:*

$$\begin{aligned} (i) \quad & \sum_{k=0}^{n} \begin{bmatrix} n \\ k \end{bmatrix}_{p,q} b^{k-1} {}_{C}B_{n-k,p,q}(x,y)\, {}_{C}B_{k,p,q}\left(\frac{X}{b}, \frac{Y}{b}\right) \\ & = \sum_{k=0}^{n} \begin{bmatrix} n \\ k \end{bmatrix}_{p,q} b^{n-k-1} {}_{C}B_{n-k,p,q}\left(\frac{x}{b}, \frac{y}{b}\right) {}_{C}B_{k,p,q}(X,Y), \\ (ii) \quad & \sum_{k=0}^{n} \begin{bmatrix} n \\ k \end{bmatrix}_{p,q} b^{k-1} {}_{S}B_{n-k,p,q}(x,y)\, {}_{S}B_{k,p,q}\left(\frac{X}{b}, \frac{Y}{b}\right) \\ & = \sum_{k=0}^{n} \begin{bmatrix} n \\ k \end{bmatrix}_{p,q} b^{n-k-1} {}_{S}B_{n-k,p,q}\left(\frac{x}{b}, \frac{y}{b}\right) {}_{S}B_{k,p,q}(X,Y). \end{aligned} \tag{70}$$

Corollary 12. *If $p = 1$ in Theorem 9, then we have*

$$\begin{aligned}
(i)\quad & \sum_{k=0}^{n} \begin{bmatrix} n \\ k \end{bmatrix}_q a^{n-k-1} b^{k-1} {}_{C}B_{n-k,q}\left(\frac{x}{a}, \frac{y}{a}\right) {}_{C}B_{k,q}\left(\frac{X}{b}, \frac{Y}{b}\right) \\
&= \sum_{k=0}^{n} \begin{bmatrix} n \\ k \end{bmatrix}_q b^{n-k-1} a^{k-1} {}_{C}B_{n-k,q}\left(\frac{x}{b}, \frac{y}{b}\right) {}_{C}B_{k,q}\left(\frac{X}{a}, \frac{Y}{a}\right), \\
(ii)\quad & \sum_{k=0}^{n} \begin{bmatrix} n \\ k \end{bmatrix}_q a^{n-k-1} b^{k-1} {}_{S}B_{n-k,q}\left(\frac{x}{a}, \frac{y}{a}\right) {}_{S}B_{k,q}\left(\frac{X}{b}, \frac{Y}{b}\right) \\
&= \sum_{k=0}^{n} \begin{bmatrix} n \\ k \end{bmatrix}_q b^{n-k-1} a^{k-1} {}_{S}B_{n-k,q}\left(\frac{x}{b}, \frac{y}{b}\right) {}_{S}B_{k,q}\left(\frac{X}{a}, \frac{Y}{a}\right),
\end{aligned} \tag{71}$$

where ${}_{C}B_{n,q}(x,y)$ denotes the q-cosine Bernoulli polynomials and ${}_{S}B_{n,q}(x,y)$ denotes the q-sine Bernoulli polynomials.

Corollary 13. *Putting $p = 1$ and $q \to 1$, one holds:*

$$\begin{aligned}
(i)\quad & \sum_{k=0}^{n} \binom{n}{k} a^{n-k-1} b^{k-1} {}_{C}B_{n-k}\left(\frac{x}{a}, \frac{y}{a}\right) {}_{C}B_{k}\left(\frac{X}{b}, \frac{Y}{b}\right) \\
&= \sum_{k=0}^{n} \binom{n}{k} b^{n-k-1} a^{k-1} {}_{C}B_{n-k}\left(\frac{x}{b}, \frac{y}{b}\right) {}_{C}B_{k}\left(\frac{X}{a}, \frac{Y}{a}\right), \\
(ii)\quad & \sum_{k=0}^{n} \binom{n}{k} a^{n-k-1} b^{k-1} {}_{S}B_{n-k}\left(\frac{x}{a}, \frac{y}{a}\right) {}_{S}B_{k}\left(\frac{X}{b}, \frac{Y}{b}\right) \\
&= \sum_{k=0}^{n} \binom{n}{k} b^{n-k-1} a^{k-1} {}_{S}B_{n-k}\left(\frac{x}{b}, \frac{y}{b}\right) {}_{S}B_{k}\left(\frac{X}{a}, \frac{Y}{a}\right),
\end{aligned} \tag{72}$$

where ${}_{C}B_n(x,y)$ is the cosine Bernoulli polynomials and ${}_{S}B_n(x,y)$ is the sine Bernoulli polynomials.

Theorem 9 is a basic symmetric property of (p,q)-cosine and (p,q)-sine Bernoulli polynomials. We aim to find several symmetric properties by mixing (p,q)-cosine and (p,q)-sine Bernoulli polynomials.

Theorem 10. *For nonzero integers a and b, we have*

$$\begin{aligned}
& \sum_{k=0}^{n} \begin{bmatrix} n \\ k \end{bmatrix}_{p,q} a^{n-k-1} b^{k-1} {}_{C}B_{n-k,p,q}\left(\frac{x}{a}, \frac{y}{a}\right) {}_{S}B_{k,p,q}\left(\frac{X}{b}, \frac{Y}{b}\right) \\
&= \sum_{k=0}^{n} \begin{bmatrix} n \\ k \end{bmatrix}_{p,q} b^{n-k-1} a^{k-1} {}_{C}B_{n-k,p,q}\left(\frac{x}{b}, \frac{y}{b}\right) {}_{S}B_{k,p,q}\left(\frac{X}{a}, \frac{Y}{a}\right).
\end{aligned} \tag{73}$$

Proof. We assume form C by mixing the (p,q)-cosine function with the (p,q)-sine function, such as the following:

$$C := \frac{t^2 e_{p,q}(tx) e_{p,q}(tX) COS_{p,q}(ty) SIN_{p,q}(tY)}{\left(e_{p,q}(at) - 1\right)\left(e_{p,q}(bt) - 1\right)}. \tag{74}$$

Form C of the above equation can be changed into

$$\begin{aligned} C &= \frac{t}{e_{p,q}(at)-1} e_{p,q}(tx)COS_{p,q}(ty) \frac{t}{e_{p,q}(bt)-1} e_{p,q}(tX)SIN_{p,q}(tY) \\ &= \sum_{n=0}^{\infty} a^{n-1} {}_{C}B_{n,p,q}\left(\frac{x}{a},\frac{y}{a}\right) \frac{t^n}{[n]_{p,q}!} \sum_{n=0}^{\infty} b^{n-1} {}_{S}B_{n,p,q}\left(\frac{X}{b},\frac{Y}{b}\right) \frac{t^n}{[n]_{p,q}!} \\ &= \sum_{n=0}^{\infty} \left(\sum_{k=0}^{n} \begin{bmatrix} n \\ k \end{bmatrix}_{p,q} a^{n-k-1} b^{k-1} {}_{C}B_{n-k,p,q}\left(\frac{x}{a},\frac{y}{a}\right) {}_{S}B_{k,p,q}\left(\frac{X}{b},\frac{Y}{b}\right) \right) \frac{t^n}{[n]_{p,q}!}, \end{aligned} \tag{75}$$

or, equivalently:

$$\begin{aligned} C &= \frac{t}{e_{p,q}(bt)-1} e_{p,q}(tx)COS_{p,q}(ty) \frac{t}{e_{p,q}(at)-1} e_{p,q}(tX)SIN_{p,q}(tY) \\ &= \sum_{n=0}^{\infty} \left(\sum_{k=0}^{n} \begin{bmatrix} n \\ k \end{bmatrix}_{p,q} b^{n-k-1} a^{k-1} {}_{C}B_{n-k,p,q}\left(\frac{x}{b},\frac{y}{b}\right) {}_{S}B_{k,p,q}\left(\frac{X}{a},\frac{Y}{a}\right) \right) \frac{t^n}{[n]_{p,q}!}. \end{aligned} \tag{76}$$

By comparing transformed Equations (75) and (76), we determine the result of Theorem 10. □

Corollary 14. *If $a = 1$ in Theorem 10, then we find*

$$\begin{aligned} &\sum_{k=0}^{n} \begin{bmatrix} n \\ k \end{bmatrix}_{p,q} b^{k-1} {}_{C}B_{n-k,p,q}(x,y)\, {}_{S}B_{k,p,q}\left(\frac{X}{b},\frac{Y}{b}\right) \\ &= \sum_{k=0}^{n} \begin{bmatrix} n \\ k \end{bmatrix}_{p,q} b^{n-k-1} {}_{C}B_{n-k,p,q}\left(\frac{x}{b},\frac{y}{b}\right) {}_{S}B_{k,p,q}(X,Y). \end{aligned} \tag{77}$$

Corollary 15. *Setting $p = 1$ in Theorem 10, one holds:*

$$\begin{aligned} &\sum_{k=0}^{n} \begin{bmatrix} n \\ k \end{bmatrix}_{q} a^{n-k-1} b^{k-1} {}_{C}B_{n-k,q}\left(\frac{x}{a},\frac{y}{a}\right) {}_{S}B_{k,q}\left(\frac{X}{b},\frac{Y}{b}\right) \\ &= \sum_{k=0}^{n} \begin{bmatrix} n \\ k \end{bmatrix}_{q} b^{n-k-1} a^{k-1} {}_{C}B_{n-k,q}\left(\frac{x}{b},\frac{y}{b}\right) {}_{S}B_{k,q}\left(\frac{X}{a},\frac{Y}{a}\right), \end{aligned} \tag{78}$$

where ${}_{C}B_{n,q}(x,y)$ is the q-cosine Bernoulli polynomials and ${}_{S}B_{n,q}(x,y)$ is the q-sine Bernoulli polynomials.

Corollary 16. *Assigning $p = 1$ and $q \to 1$ in Theorem 10, the following holds:*

$$\begin{aligned} &\sum_{k=0}^{n} \binom{n}{k} a^{n-k-1} b^{k-1} {}_{C}B_{n-k}\left(\frac{x}{a},\frac{y}{a}\right) {}_{S}B_{k}\left(\frac{X}{b},\frac{Y}{b}\right) \\ &= \sum_{k=0}^{n} \binom{n}{k} b^{n-k-1} a^{k-1} {}_{C}B_{n-k}\left(\frac{x}{b},\frac{y}{b}\right) {}_{S}B_{k}\left(\frac{X}{a},\frac{Y}{a}\right), \end{aligned} \tag{79}$$

where ${}_{C}B_n(x,y)$ is the cosine Bernoulli polynomials and ${}_{S}B_n(x,y)$ is the sine Bernoulli polynomials.

Theorem 11. *Let a and b be nonzero integers. Then, we derive*

$$\begin{aligned}&\sum_{k=0}^{n}\begin{bmatrix}n\\k\end{bmatrix}_{p,q} a^{n-k-1}b^{k-1}{}_{C}B_{n-k,p,q}\left(bx,\frac{y}{a}\right){}_{S}B_{k,p,q}\left(aX,\frac{Y}{b}\right)\\&=\sum_{k=0}^{n}\begin{bmatrix}n\\k\end{bmatrix}_{p,q} b^{n-k-1}a^{k-1}{}_{C}B_{n-k,p,q}\left(ax,\frac{y}{b}\right){}_{S}B_{k,p,q}\left(bX,\frac{Y}{a}\right).\end{aligned}\tag{80}$$

Proof. Let us consider form D containing a and b in the (p,q)-exponential functions as

$$D:=\frac{t^2 e_{p,q}(abtx)e_{p,q}(abtX)COS_{p,q}(ty)SIN_{p,q}(tY)}{\left(e_{p,q}(at)-1\right)\left(e_{p,q}(bt)-1\right)}.\tag{81}$$

From the above form D , we can obtain

$$\begin{aligned}D&=\frac{t}{e_{p,q}(at)-1}e_{p,q}(abtx)COS_{p,q}(ty)\frac{t}{e_{p,q}(bt)-1}e_{p,q}(abtX)SIN_{p,q}(tY)\\&=\sum_{n=0}^{\infty}a^{n-1}{}_{C}B_{n,p,q}\left(bx,\frac{y}{a}\right)\frac{t^n}{[n]_{p,q}!}\sum_{n=0}^{\infty}b^{n-1}{}_{S}B_{n,p,q}\left(aX,\frac{Y}{b}\right)\frac{t^n}{[n]_{p,q}!}\\&=\sum_{n=0}^{\infty}\left(\sum_{k=0}^{n}\begin{bmatrix}n\\k\end{bmatrix}_{p,q} a^{n-k-1}b^{k-1}{}_{C}B_{n-k,p,q}\left(bx,\frac{y}{a}\right){}_{S}B_{k,p,q}\left(aX,\frac{Y}{b}\right)\right)\frac{t^n}{[n]_{p,q}!},\end{aligned}\tag{82}$$

and

$$\begin{aligned}D&=\frac{t}{e_{p,q}(bt)-1}e_{p,q}(abtx)COS_{p,q}(ty)\frac{t}{e_{p,q}(at)-1}e_{p,q}(abtX)SIN_{p,q}(tY)\\&=\sum_{n=0}^{\infty}\left(\sum_{k=0}^{n}\begin{bmatrix}n\\k\end{bmatrix}_{p,q} b^{n-k-1}a^{k-1}{}_{C}B_{n-k,p,q}\left(ax,\frac{y}{b}\right){}_{S}B_{k,p,q}\left(bX,\frac{Y}{a}\right)\right)\frac{t^n}{[n]_{p,q}!}.\end{aligned}\tag{83}$$

By observing Equations (82) and (83) which are made by form D, we prove Theorem 11. □

Corollary 17. *Setting $a=1$ in Theorem 11, the following holds:*

$$\begin{aligned}&\sum_{k=0}^{n}\begin{bmatrix}n\\k\end{bmatrix}_{p,q} b^{k-1}{}_{C}B_{n-k,p,q}(bx,y){}_{S}B_{k,p,q}\left(X,\frac{Y}{b}\right)\\&=\sum_{k=0}^{n}\begin{bmatrix}n\\k\end{bmatrix}_{p,q} b^{n-k-1}{}_{C}B_{n-k,p,q}\left(x,\frac{y}{b}\right){}_{S}B_{k,p,q}(bX,Y).\end{aligned}\tag{84}$$

Corollary 18. *If $p=1$ in Theorem 11, then we obtain*

$$\begin{aligned}&\sum_{k=0}^{n}\begin{bmatrix}n\\k\end{bmatrix}_{q} a^{n-k-1}b^{k-1}{}_{C}B_{n-k,q}\left(bx,\frac{y}{a}\right){}_{S}B_{k,q}\left(aX,\frac{Y}{b}\right)\\&=\sum_{k=0}^{n}\begin{bmatrix}n\\k\end{bmatrix}_{q} b^{n-k-1}a^{k-1}{}_{C}B_{n-k,q}\left(ax,\frac{y}{b}\right){}_{S}B_{k,q}\left(bX,\frac{Y}{a}\right),\end{aligned}\tag{85}$$

where ${}_{C}B_{n,q}(x,y)$ is the q-cosine Bernoulli polynomials and ${}_{S}B_{n,q}(x,y)$ is the q-sine Bernoulli polynomials.

Corollary 19. *Let $p = 1$ and $q \to 1$ in Theorem 11. Then one holds*

$$\begin{aligned}&\sum_{k=0}^{n}\binom{n}{k}a^{n-k-1}b^{k-1}{}_{C}B_{n-k}\left(bx,\frac{y}{a}\right){}_{S}B_{k}\left(aX,\frac{Y}{b}\right)\\&=\sum_{k=0}^{n}\binom{n}{k}b^{n-k-1}a^{k-1}{}_{C}B_{n-k}\left(ax,\frac{y}{b}\right){}_{S}B_{k}\left(bX,\frac{Y}{a}\right),\end{aligned} \tag{86}$$

where ${}_{C}B_n(x,y)$ is the cosine Bernoulli polynomials and ${}_{S}B_n(x,y)$ is the sine Bernoulli polynomials.

Next, we investigate the structure of approximate roots in (p,q)-cosine and (p,q)-sine Bernoulli polynomials. Based on the theorems above, (p,q)-cosine and (p,q)-sine Bernoulli polynomials have symmetric properties. Thus, we assume that the approximate roots of (p,q)-cosine and (p,q)-sine Bernoulli polynomials also have symmetric properties as well. We aim to identify the stacking structure of the roots from the specific (p,q)-cosine and (p,q)-sine Bernoulli polynomials found in Section 3.

First, the structure of approximate roots in the (p,q)-cosine Bernoulli polynomials is illustrated in Figure 1 when $y = 5$, $q = 0.9$, and the value of p changes. Figure 1 reveals the pattern of the roots in the (p,q)-cosine Bernoulli polynomials when $p = 0.5$. In addition, the approximate roots appear when n changes from 1 to 30. The red points become closer together when n is 30 and n becomes smaller as illustrated by the blue points. Based on the graphs with real and imaginary axes, the (p,q)-cosine Bernoulli polynomials are symmetric.

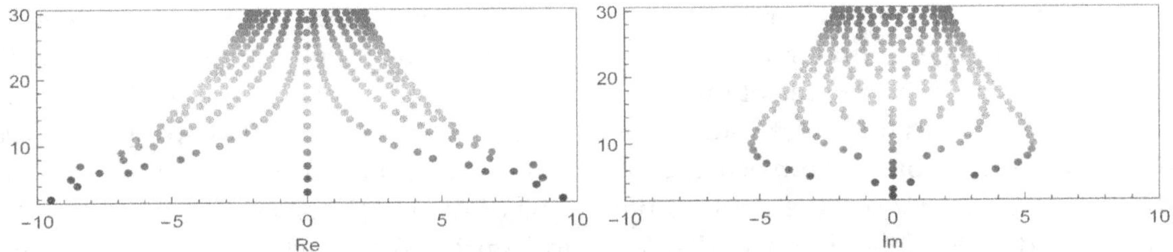

Figure 1. Stacking structure of approximate roots in the (p,q)-cosine Bernoulli polynomials when $p = 0.5, q = 0.9$, and $y = 5$.

Here, we aim to confirm that changes in the value of the (p,q)-cosine Bernoulli polynomials changes the structure of the approximate roots as the value changes. The structure of the approximate roots in polynomials when $p = 1$ and q changes, can be found in the q-cosine Bernoulli polynomials (see [11]).

Figure 2 below illustrates the stacking structure of the approximate roots of the (p,q)-cosine Bernoulli polynomials fixed at $p = 0.1, q = 0.5$ and $y = 5$ when $1 \leq n \leq 30$. Compared with Figure 1, Figure 2 displays a wider distribution of the approximate roots. The range of the left picture in Figure 1 is $-15 < \text{Re}\, x < 15$ and the range of the left picture in Figure 2 is $-50 < \text{Re}\, x < 50$. The structure of the approximate roots of $p = 0.1$ when $n = 30$ is wider on the real axis compared to when $p = 0.5$. The right-hand graphs in Figures 1 and 2 also reveal the same distribution. In addition, as n increases, the structure of the approximate roots appears symmetric.

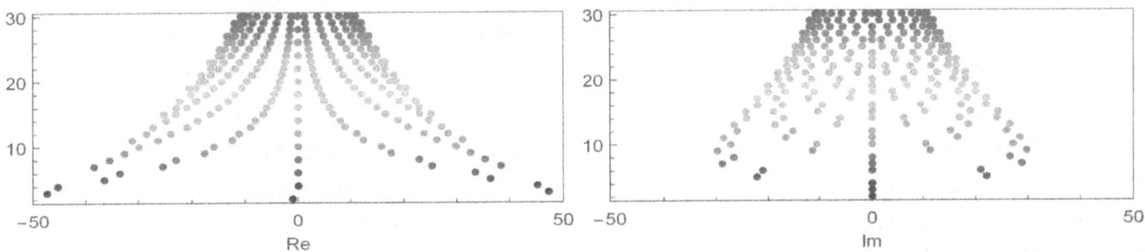

Figure 2. Stacking structure of approximate roots in the (p,q)-cosine Bernoulli polynomials when $p = 0.1, q = 0.9$, and $y = 5$.

Next, we examine the stacking structure of the approximate roots in the (p,q)-sine Bernoulli polynomials. The conditions are confirmed by equating them to the conditions of the (p,q)-cosine Bernoulli polynomials. The stacking structure of the approximate roots of the (p,q)-sine Bernoulli polynomials when $p = 0.5, q = 0.9$, and $y = 5$ can be checked in Figure 3. At $1 \leq n \leq 30$, the distribution range of the approximate roots appears wider in the values on the real axis than in the imaginary axis, as shown in the left picture in Figure 3. Figure 3 reveals that, as the value of n becomes larger, the approximate roots become more symmetric, and the approximate form approaches a circular shape, including the origin.

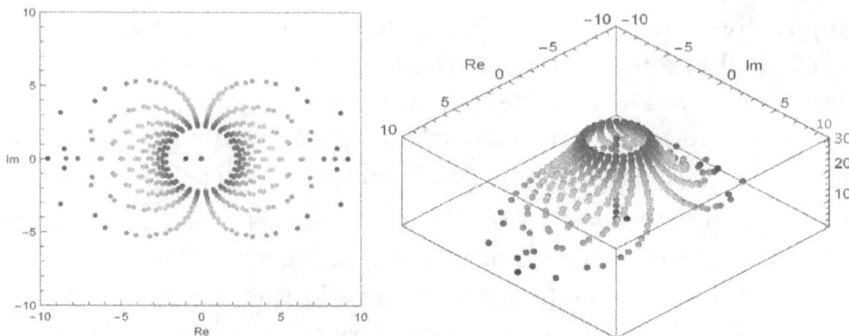

Figure 3. Stacking structure of approximate roots in the (p,q)-sine Bernoulli polynomials when $p = 0.5, q = 0.9$, and $y = 5$ in 3D.

When we change the value of p, the structure of the approximate roots of the (p,q)-sine Bernoulli polynomials when $p = 0.1$ under the same conditions as in Figure 3 is presented in Figure 4. In comparison with Figure 3, the area of the real and the imaginary axes in Figure 4 is greater, and the approximate roots have a wider distribution than observed in Figure 3. This property is common in the approximate roots of the (p,q)-cosine and (p,q)-sine Bernoulli polynomials. This indicates that, as the value of p decreases, the approximate roots of the (p,q)-cosine and (p,q)-sine Bernoulli polynomials spread wider. In addition, as displayed in Figure 4, the structure of the approximate roots of the (p,q)-sine Bernoulli polynomials is symmetric as the value of n increases.

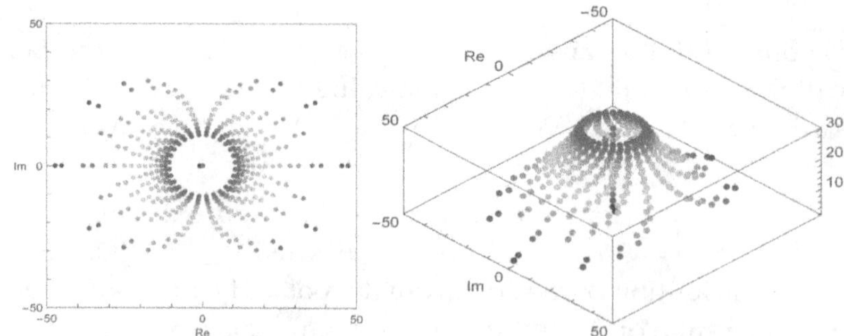

Figure 4. Stacking structure of approximate roots in the (p,q)-sine Bernoulli polynomials when $p = 0.1, q = 0.9$, and $y = 5$ in 3D.

5. Conclusions and Future Directions

In this paper, we explained about the (p,q)-cosine and (p,q)-sine Bernoulli polynomials, their basic properties, and various symmetric properties. Based on the above contents, we identified the structures of the approximate roots of the (p,q)-cosine and (p,q)-sine Bernoulli polynomials. As a result, we observed that the above polynomials obtain a structure of approximate roots, which has certain patterns and has a symmetric property under the given circumstances.

Further study is needed regarding whether the structure of approximate roots for the (p,q)-cosine and (p,q)-sine Bernoulli polynomials have symmetric properties under different circumstances. Furthermore, we think researching theories related to this topic is important to mathematicians.

Author Contributions: Conceptualization, J.Y.K.; Methodology, C.S.R.; Writing—original draft, J.Y.K. These autors contributed equally to this work. All authors have read and agreed to the published version of the manuscript.

Acknowledgments: The authors would like to thank the referees for their valuable comments, which improved the original manuscript in its present form.

References

1. Brodimas, G.; Jannussis, A.; Mignani, R. *Two-Parameter Quantum Groups*; Universita di Roma: Piazzale Aldo Moro, Roma, Italy, 1991; p. 820; View in KEK Scanned Document.
2. Chakrabarti, R.; Jagannathan, R. A (p,q)-oscillator realization of two-parameter quantum algebras. *J. Phys. A Math. Gen.* **1991**, *24*, L711–L718. [CrossRef]
3. Burban, I.M.; Klimyk, A.U. (p,q)-differentiation, (p,q)-integration and (p,q)-hypergeometric functions related to quantum groups. *Integral Transform. Spec. Funct.* **1994**, *2*, 15–36. [CrossRef]
4. Jagannathan, R. (P,Q)-special functions, special functions and differential equations. In *Proceedings of a Workshop*; The Institute of Mathematical Science: Matras, India, 1997; pp. 13–24.
5. Jagannathan, R.; Rao, K.S. Two-parameter quatum algebras, twin-basic numbers, and associated generalized hypergeometric series. In Proceedings of the Lnternational Conference on Number Theory and Mathematical Physics, Srinivasa Ramanujan Centre, Kumbakonam, India, 20–21 December 2005.
6. Corcino, R.B. On P,Q-Binomials coefficients. *Electron. J. Combin. Number Theory* **2008**, *8*, 1–16.
7. Sadjang, P.N. On the (p,q)-Gamma and the (p,q)-Beta functions. *arXiv* **2015**, arXiv:1506.07394v1.
8. Kupershmidt, B.A. Reflection Symmetries of q-Bernoulli Polynomials. *J. Nonlinear Math. Phys.* **2005**, *12*, 412–422. [CrossRef]
9. Kim, T.; Ryoo, C.S. Some identities for Euler and Bernoulli polynomials and their zeros. *Axioms* **2018**, *7*, 56. [CrossRef]
10. Kim, T. q-generalized Euler numbers and polynomials. *Russ. J. Math. Phys.* **2006**, *13*, 293–298. [CrossRef]
11. Kang, J.Y.; Ryoo, C.S. Various structures of the roots and explicit properties of q-cosine Bernoulli Polynomials and q-sine Bernoulli Polynomials. *Mathematics* **2020**, *8*, 463. [CrossRef]
12. Ryoo, C.S. A note on the zeros of the q-Bernoulli polynomials. *J. Appl. Math. Inform.* **2010**, *28*, 805–811.
13. Duran, U.; Acikgoz, M.; Araci, S. On some polynomials derived from (p,q)-calculus. *J. Comput. Theor. Nanosci.* **2016**, *13*, 7903–7908. [CrossRef]
14. Duran, U.; Acikgoz, M.; Araci, S. On (p,q)-Bernoulli, (p,q)-Euler and (p,q)-Genocchi polynomials. *J. Comput. Theor. Nanosci.* **2016**, *13*, 7833–7846. [CrossRef]
15. Acar, T.; Aral, A.; Mohiuddine, S.A. On Kantorovich modifivation of (p,q)-Baskakov operators. *J. Inequal. Appl.* **2016**, 98, 1–14.
16. Arik, M.; Demircan, E.; Turgut, T.; Ekinci, L.; Mungan, M. Fibonacci Oscillators. *Z. Phys. C-Part. Fiels* **1992**, *55*, 89–95. [CrossRef]
17. Duran, U.; Acikgoz, M.; Araci, S. A study on some new results arising from (p,q)-calculus. *TWMS J. Pure Appl. Math.* **2020**, *11*, 57–71.
18. Jackson, H.F. On q-definite integrals. *Q. J. Pure Appl. Math.* **1910**, *41*, 193–203.
19. lSadjang, P.N. On the fundamental theorem of (p,q)-calculus and some (p,q)-Taylor formulas. *arXiv* **2013**, arXiv:1309.3934.
20. Wachs, M.; White, D. (p,q)-Stirling numbers and set partition statistics. *J. Combin. Theorey Ser. A* **1991**, *56*, 27–46. [CrossRef]

4

Symmetric Identities of Hermite-Bernoulli Polynomials and Hermite-Bernoulli Numbers Attached to a Dirichlet Character χ

Serkan Araci [1,*], Waseem Ahmad Khan [2] and Kottakkaran Sooppy Nisar [3]

1 Department of Economics, Faculty of Economics, Administrative and Social Sciences, Hasan Kalyoncu University, TR-27410 Gaziantep, Turkey
2 Department of Mathematics, Faculty of Science, Integral University, Lucknow-226026, India; waseem08_khan@rediffmail.com
3 Department of Mathematics, College of Arts and Science-Wadi Aldawaser, Prince Sattam bin Abdulaziz University, 11991 Riyadh Region, Kingdom of Saudi Arabia; n.sooppy@psau.edu.sa or ksnisar1@gmail.com
* Correspondence: mtsrkn@hotmail.com

Abstract: We aim to introduce arbitrary complex order Hermite-Bernoulli polynomials and Hermite-Bernoulli numbers attached to a Dirichlet character χ and investigate certain symmetric identities involving the polynomials, by mainly using the theory of p-adic integral on $\mathbb{Z}_p$. The results presented here, being very general, are shown to reduce to yield symmetric identities for many relatively simple polynomials and numbers and some corresponding known symmetric identities.

Keywords: q-Volkenborn integral on $\mathbb{Z}_p$; Bernoulli numbers and polynomials; generalized Bernoulli polynomials and numbers of arbitrary complex order; generalized Bernoulli polynomials and numbers attached to a Dirichlet character χ

---

1. Introduction and Preliminaries

For a fixed prime number p, throughout this paper, let $\mathbb{Z}_p$, $\mathbb{Q}_p$, and $\mathbb{C}_p$ be the ring of p-adic integers, the field of p-adic rational numbers, and the completion of algebraic closure of $\mathbb{Q}_p$, respectively. In addition, let $\mathbb{C}$, $\mathbb{Z}$, and $\mathbb{N}$ be the field of complex numbers, the ring of rational integers and the set of positive integers, respectively, and let $\mathbb{N}_0 := \mathbb{N} \cup \{0\}$. Let $\mathrm{UD}(\mathbb{Z}_p)$ be the space of all uniformly differentiable functions on $\mathbb{Z}_p$. The notation $[z]_q$ is defined by

$$[z]_q := \frac{1-q^z}{1-q} \quad (z \in \mathbb{C};\, q \in \mathbb{C} \setminus \{1\};\, q^z \neq 1)\,.$$

Let ν_p be the normalized exponential valuation on $\mathbb{C}_p$ with $|p|_p = p^{\nu_p(p)} = p^{-1}$. For $f \in \mathrm{UD}(\mathbb{Z}_p)$ and $q \in \mathbb{C}_p$ with $|1-q|_p < 1$, q-Volkenborn integral on $\mathbb{Z}_p$ is defined by Kim [1]

$$I_q(f) = \int_{\mathbb{Z}_p} f(x)\, d\mu_q(x) = \lim_{N\to\infty} \frac{1}{[p^N]_q} \sum_{x=0}^{p^N-1} f(x)\, q^x. \tag{1}$$

For recent works including q-Volkenborn integration see References [1–10].

The ordinary p-adic invariant integral on $\mathbb{Z}_p$ is given by [7,8]

$$I_1(f) = \lim_{q\to 1} I_q(f) = \int_{\mathbb{Z}_p} f(x)\, dx. \tag{2}$$

It follows from Equation (2) that

$$I_1(f_1) = I_1(f) + f'(0), \tag{3}$$

where $f_n(x) := f(x+n)$ $(n \in \mathbb{N})$ and $f'(0)$ is the usual derivative. From Equation (3), one has

$$\int_{\mathbb{Z}_p} e^{xt}\,dx = \frac{t}{e^t - 1} = \sum_{n=0}^{\infty} B_n \frac{t^n}{n!}, \tag{4}$$

where B_n are the nth Bernoulli numbers (see References [11–14]; see also Reference [15] (Section 1.7)). From Equation (2) and (3), one gets

$$\begin{aligned} \frac{n\int_{\mathbb{Z}_p} e^{xt}dx}{\int_{\mathbb{Z}_p} e^{nxt}dx} &= \frac{1}{t}\left(\int_{\mathbb{Z}_p} e^{(x+n)t}dx - \int_{\mathbb{Z}_p} e^{xt}dx\right) \\ = \sum_{j=0}^{n-1} e^{jt} &= \sum_{k=0}^{\infty}\left(\sum_{j=0}^{n-1} j^k\right)\frac{t^k}{k!} = \sum_{k=0}^{\infty} S_k(n-1)\frac{t^k}{k!}, \end{aligned} \tag{5}$$

where

$$S_k(n) = 1^k + \cdots + n^k \quad (k \in \mathbb{N},\, n \in \mathbb{N}_0)\,. \tag{6}$$

From Equation (4), the generalized Bernoulli polynomials $B_n^{(\alpha)}(x)$ are defined by the following p-adic integral (see Reference [15] (Section 1.7))

$$\underbrace{\int_{\mathbb{Z}_p}\cdots\int_{\mathbb{Z}_p}}_{\alpha\text{ times}} e^{(x+y_1+y_2+\cdots+y_\alpha)t}dy_1dy_2\cdots dy_\alpha = \left(\frac{t}{e^t-1}\right)^{\alpha} e^{xt} = \sum_{n=0}^{\infty} B_n^{(\alpha)}(x)\frac{t^n}{n!} \tag{7}$$

in which $B_n^{(1)}(x) := B_n(x)$ are classical Bernoulli numbers (see, e.g., [1–10]).

Let $d,\ p \in \mathbb{N}$ be fixed with $(d,p) = 1$. For $N \in \mathbb{N}$, we set

$$\begin{aligned} X = X_d &= \varprojlim_{N} \left(\mathbb{Z}/dp^N\mathbb{Z}\right); \\ a + dp^N\mathbb{Z}_p &= \left\{x \in X \mid x \equiv a \left(\text{mod } dp^N\right)\right\} \\ &\quad \left(a \in \mathbb{Z} \text{ with } 0 \le a < dp^N\right); \\ X^* &= \bigcup_{\substack{0<a<dp \\ (a,p)=1}} (a + dp\mathbb{Z}_p)\,,\ X_1 = \mathbb{Z}_p. \end{aligned} \tag{8}$$

Let χ be a Dirichlet character with conductor $d \in \mathbb{N}$. The generalized Bernoulli polynomials attached to χ are defined by means of the generating function (see, e.g., [16])

$$\int_X \chi(y)e^{(x+y)t}dy = \frac{t\sum_{j=0}^{d-1}\chi(j)\,e^{jt}}{e^{dt}-1}e^{xt} = \sum_{n=0}^{\infty} B_{n,\chi}(x)\frac{t^n}{n!}. \tag{9}$$

Here $B_{n,\chi} := B_{n,\chi}(0)$ are the generalized Bernoulli numbers attached to χ. From Equation (9), we have (see, e.g., [16])

$$\int_X \chi(x)x^n dx = B_{n,\chi} \quad \text{and} \quad \int_X \chi(y)(x+y)^n\,dy = B_{n,\chi}(x). \tag{10}$$

Define the p-adic functional $T_k(\chi, n)$ by (see, e.g., [16])

$$T_k(\chi, n) = \sum_{\ell=0}^{n} \chi(\ell)\, \ell^k \quad (k \in \mathbb{N}). \tag{11}$$

Then one has (see, e.g., [16])

$$B_{k,\chi}(nd) - B_{k,\chi} = kT_{k-1}(\chi, nd-1) \quad (k,\, n,\, d \in \mathbb{N}). \tag{12}$$

Kim et al. [16] (Equation (2.14)) presented the following interesting identity

$$\frac{dn \int_X \chi(x)\, e^{xt}\, dx}{\int_X e^{dnxt}\, dx} = \sum_{\ell=0}^{nd-1} \chi(\ell)\, e^{\ell t} = \sum_{k=0}^{\infty} T_k(\chi, nd-1)\, \frac{t^k}{k!} \quad (n \in \mathbb{N}). \tag{13}$$

Very recently, Khan [17] (Equation (2.1)) (see also Reference [11]) introduced and investigated λ-Hermite-Bernoulli polynomials of the second kind ${}_HB_n(x, y|\lambda)$ defined by the following generating function

$$\begin{aligned} &\int_{\mathbb{Z}_p} (1+\lambda t)^{\frac{x+u}{\lambda}} (1+\lambda t^2)^{\frac{y}{\lambda}}\, \mathrm{d}\mu_0(u) \\ &\quad = \frac{\log(1+\lambda t)^{\frac{1}{\lambda}}}{(1+\lambda t)^{\frac{1}{\lambda}} - 1} (1+\lambda t)^{\frac{x}{\lambda}} (1+\lambda t^2)^{\frac{y}{\lambda}} = \sum_{m=0}^{\infty} {}_HB_m(x, y|\lambda) \frac{t^m}{m!} \\ &\qquad \left(\lambda,\, t \in \mathbb{C}_p \text{ with } \lambda \neq 0,\; |\lambda t| < p^{-\frac{1}{p-1}}\right). \end{aligned} \tag{14}$$

Hermite-Bernoulli polynomials ${}_HB_k^{(\alpha)}(x, y)$ of order α are defined by the following generating function

$$\left(\frac{t}{e^t - 1}\right)^{\alpha} e^{xt+yt^2} = \sum_{k=0}^{\infty} {}_HB_k^{(\alpha)}(x, y) \frac{t^k}{k!} \quad (\alpha,\, x,\, y \in \mathbb{C};\; |t| < 2\pi) \tag{15}$$

where ${}_HB_k^{(1)}(x, y) := {}_HB_k(x, y)$ are Hermite-Bernoulli polynomials, cf. [18,19]. For more information related to systematic works of some special functions and polynomials, see References [20–29].

We aim to introduce arbitrary complex order Hermite-Bernoulli polynomials attached to a Dirichlet character χ and investigate certain symmetric identities involving the polynomials (15) and (31), by mainly using the theory of p-adic integral on $\mathbb{Z}_p$. The results presented here, being very general, are shown to reduce to yield symmetric identities for many relatively simple polynomials and numbers and some corresponding known symmetric identities.

2. Symmetry Identities of Hermite-Bernoulli Polynomials of Arbitrary Complex Number Order

Here, by mainly using Kim's method in References [30,31], we establish certain symmetry identities of Hermite-Bernoulli polynomials of arbitrary complex number order.

Theorem 1. *Let α, x, y, $z \in \mathbb{C}$, η_1, $\eta_2 \in \mathbb{N}$, and $n \in \mathbb{N}_0$. Then,*

$$\begin{aligned} &\sum_{m=0}^{n} \sum_{\ell=0}^{m} \binom{n}{m} \binom{m}{\ell}\, {}_HB_{n-m}^{(\alpha)}(\eta_2 x, \eta_2^2 z)\, S_{m-\ell}(\eta_1 - 1)\, B_{\ell}^{(\alpha-1)}(\eta_1 y)\, \eta_1^{n-m-1}\, \eta_2^{m} \\ &= \sum_{m=0}^{n} \sum_{\ell=0}^{m} \binom{n}{m} \binom{m}{\ell}\, {}_HB_{n-m}^{(\alpha)}(\eta_1 x, \eta_1^2 z)\, S_{m-\ell}(\eta_2 - 1)\, B_{\ell}^{(\alpha-1)}(\eta_2 y)\, \eta_2^{n-m-1}\, \eta_1^{m} \end{aligned} \tag{16}$$

and

$$\sum_{m=0}^{n}\sum_{j=0}^{\eta_1-1}\binom{n}{m}\eta_1^{m-1}\eta_2^{n-m}B_{n-m}^{(\alpha-1)}(\eta_1 y)\,{}_HB_m^{(\alpha)}\left(\eta_2 x+\frac{\eta_2}{\eta_1}j,\eta_2^2 z\right)$$
$$=\sum_{m=0}^{n}\sum_{j=0}^{\eta_1-1}\binom{n}{m}\eta_2^{m-1}\eta_1^{n-m}B_{n-m}^{(\alpha-1)}(\eta_2 y)\,{}_HB_m^{(\alpha)}\left(\eta_1 x+\frac{\eta_1}{\eta_2}j,\eta_1^2 z\right). \tag{17}$$

Proof. Let

$$F(\alpha;\eta_1,\eta_2)(t):=\frac{e^{\eta_1\eta_2 t}-1}{\eta_1\eta_2 t}\left(\frac{\eta_1 t}{e^{\eta_1 t}-1}\right)^{\alpha}e^{\eta_1\eta_2 xt+\eta_1^2\eta_2^2 zt^2}\left(\frac{\eta_2 t}{e^{\eta_2 t}-1}\right)^{\alpha}e^{\eta_1\eta_2 yt} \tag{18}$$
$$(\alpha,\,x,\,y,\,z\in\mathbb{C};\;t\in\mathbb{C}\setminus\{0\};\;\eta_1,\,\eta_2\in\mathbb{N};\;1^{\alpha}:=1).$$

Since $\lim_{t\to 0}\eta t/(e^{\eta t}-1)=1=\lim_{t\to 0}(e^{\eta t}-1)/(\eta t)$ $(\eta\in\mathbb{N})$, $F(\alpha;\eta_1,\eta_2)(t)$ may be assumed to be analytic in $|t|<2\pi/(\eta_1\eta_2)$. Obviously $F(\alpha;\eta_1,\eta_2)(t)$ is symmetric with respect to the parameters η_1 and η_2.

Using Equation (4), we have

$$F(\alpha;\eta_1,\eta_2)(t):=\left(\frac{\eta_1 t}{e^{\eta_1 t}-1}\right)^{\alpha}e^{\eta_1\eta_2 xt+\eta_1^2\eta_2^2 zt^2}\frac{\int_{\mathbb{Z}_p}e^{\eta_2 t u}du}{\int_{\mathbb{Z}_p}e^{\eta_1\eta_2 t u}du}\left(\frac{\eta_2 t}{e^{\eta_2 t}-1}\right)^{\alpha-1}e^{\eta_1\eta_2 yt}. \tag{19}$$

Using Equations (5) and (15), we find

$$F(\alpha;\eta_1,\eta_2)(t)=\sum_{n=0}^{\infty}{}_HB_n^{(\alpha)}(\eta_2 x,\eta_2^2 z)\frac{(\eta_1 t)^n}{n!}\cdot\frac{1}{\eta_1}\sum_{m=0}^{\infty}S_m(\eta_1-1)\frac{(\eta_2 t)^m}{m!}$$
$$\cdot\sum_{\ell=0}^{\infty}B_\ell^{(\alpha-1)}(\eta_1 y)\frac{(\eta_2 t)^\ell}{\ell!}. \tag{20}$$

Employing a formal manipulation of double series (see, e.g., [32] (Equation (1.1)))

$$\sum_{n=0}^{\infty}\sum_{k=0}^{\infty}A_{k,n}=\sum_{n=0}^{\infty}\sum_{k=0}^{[n/p]}A_{k,n-pk}\quad(p\in\mathbb{N}) \tag{21}$$

with $p=1$ in the last two series in Equation (20), and again, the resulting series and the first series in Equation (20), we obtain

$$F(\alpha;\eta_1,\eta_2)(t)=\sum_{n=0}^{\infty}\sum_{m=0}^{n}\sum_{\ell=0}^{m}\frac{{}_HB_{n-m}^{(\alpha)}(\eta_2 x,\eta_2^2 z)\,S_{m-\ell}(\eta_1-1)\,B_\ell^{(\alpha-1)}(\eta_1 y)}{(n-m)!\,(m-\ell)!\,\ell!}$$
$$\times\eta_1^{n-m-1}\eta_2^{m}t^n. \tag{22}$$

Noting the symmetry of $F(\alpha;\eta_1,\eta_2)(t)$ with respect to the parameters η_1 and η_2, we also get

$$F(\alpha;\eta_1,\eta_2)(t)=\sum_{n=0}^{\infty}\sum_{m=0}^{n}\sum_{\ell=0}^{m}\frac{{}_HB_{n-m}^{(\alpha)}(\eta_1 x,\eta_1^2 z)\,S_{m-\ell}(\eta_2-1)\,B_\ell^{(\alpha-1)}(\eta_2 y)}{(n-m)!\,(m-\ell)!\,\ell!}$$
$$\times\eta_2^{n-m-1}\eta_1^{m}t^n. \tag{23}$$

Equating the coefficients of t^n in the right sides of Equations (22) and (23), we obtain the first equality of Equation (16).

For (17), we write

$$F(\alpha;\eta_1,\eta_2)(t) = \frac{1}{\eta_1}\left(\frac{\eta_1 t}{e^{\eta_1 t}-1}\right)^{\alpha} e^{\eta_1\eta_2 xt+\eta_1^2\eta_2^2 zt^2}\,\frac{e^{\eta_1\eta_2 t}-1}{e^{\eta_2 t}-1}\left(\frac{\eta_2 t}{e^{\eta_2 t}-1}\right)^{\alpha-1} e^{\eta_1\eta_2 yt}. \tag{24}$$

Noting

$$\frac{e^{\eta_1\eta_2 t}-1}{e^{\eta_2 t}-1} = \sum_{j=0}^{\eta_1-1} e^{\eta_2 jt} = \sum_{j=0}^{\eta_1-1} e^{\eta_1 \frac{\eta_2}{\eta_1} jt},$$

we have

$$F(\alpha;\eta_1,\eta_2)(t) = \frac{1}{\eta_1}\sum_{j=0}^{\eta_1-1}\left(\frac{\eta_1 t}{e^{\eta_1 t}-1}\right)^{\alpha} e^{\eta_1\left(\eta_2 x+\frac{\eta_2}{\eta_1}j\right)t+\eta_1^2\eta_2^2 zt^2}\left(\frac{\eta_2 t}{e^{\eta_2 t}-1}\right)^{\alpha-1} e^{\eta_1\eta_2 yt}. \tag{25}$$

Using Equation (15), we obtain

$$\begin{aligned} F(\alpha;\eta_1,\eta_2)(t) &= \frac{1}{\eta_1}\sum_{n=0}^{\infty} B_n^{(\alpha-1)}(\eta_1 y)\frac{(\eta_2 t)^n}{n!} \\ &\times \sum_{m=0}^{\infty}\sum_{j=0}^{\eta_1-1} {}_H B_m^{(\alpha)}\left(\eta_2 x+\frac{\eta_2}{\eta_1}j,\eta_2^2 z\right)\frac{(\eta_1 t)^m}{m!}. \end{aligned} \tag{26}$$

Applying Equation (21) with $p=1$ to the right side of Equation (26), we get

$$\begin{aligned} F(\alpha;\eta_1,\eta_2)(t) &= \sum_{n=0}^{\infty}\sum_{m=0}^{n}\sum_{j=0}^{\eta_1-1} B_{n-m}^{(\alpha-1)}(\eta_1 y) \\ &\times {}_H B_m^{(\alpha)}\left(\eta_2 x+\frac{\eta_2}{\eta_1}j,\eta_2^2 z\right)\frac{\eta_1^{m-1}\eta_2^{n-m}}{m!(n-m)!}t^n. \end{aligned} \tag{27}$$

In view of symmetry of $F(\alpha;\eta_1,\eta_2)(t)$ with respect to the parameters η_1 and η_2, we also obtain

$$\begin{aligned} F(\alpha;\eta_1,\eta_2)(t) &= \sum_{n=0}^{\infty}\sum_{m=0}^{n}\sum_{j=0}^{\eta_1-1} B_{n-m}^{(\alpha-1)}(\eta_2 y) \\ &\times {}_H B_m^{(\alpha)}\left(\eta_1 x+\frac{\eta_1}{\eta_2}j,\eta_1^2 z\right)\frac{\eta_2^{m-1}\eta_1^{n-m}}{m!(n-m)!}t^n. \end{aligned} \tag{28}$$

Equating the coefficients of t^n in the right sides of Equation (27) and Equation (28), we have Equation (17). □

Corollary 1. *By substituting $\alpha=1$ in Theorem 1, we have*

$$\begin{aligned} &\sum_{m=0}^{n}\sum_{\ell=0}^{m}\binom{n}{m}\binom{m}{\ell}\,{}_H B_{n-m}(\eta_2 x,\eta_2^2 z)\,S_{m-\ell}(\eta_1-1)\,(\eta_1 y)^{\ell}\,\eta_1^{n-m-1}\,\eta_2^{m} \\ &= \sum_{m=0}^{n}\sum_{\ell=0}^{m}\binom{n}{m}\binom{m}{\ell}\,B_{n-m}(\eta_1 x,\eta_1^2 z)\,S_{m-\ell}(\eta_2-1)\,(\eta_2 y)^{\ell}\,\eta_2^{n-m-1}\,\eta_1^{m} \end{aligned}$$

and

$$\sum_{m=0}^{n}\sum_{j=0}^{\eta_1-1}\binom{n}{m}\eta_1^{m-1}\eta_2^{n-m}(\eta_1 y)^{n-m}\,{}_HB_m\left(\eta_2 x+\frac{\eta_2}{\eta_1}j,\eta_2^2 z\right)$$
$$=\sum_{m=0}^{n}\sum_{j=0}^{\eta_1-1}\binom{n}{m}\eta_2^{m-1}\eta_1^{n-m}(\eta_2 y)^{n-m}\,{}_HB_m\left(\eta_1 x+\frac{\eta_1}{\eta_2}j,\eta_1^2 z\right). \tag{29}$$

Corollary 2. *Taking* $\alpha=1$ *and* $z=0$ *in Theorem 1, we have*

$$\sum_{m=0}^{n}\sum_{\ell=0}^{m}\binom{n}{m}\binom{m}{\ell}B_{n-m}(\eta_2 x)\,S_{m-\ell}(\eta_1-1)\,(\eta_1 y)^{\ell}\,\eta_1^{n-m-1}\,\eta_2^{m}$$
$$=\sum_{m=0}^{n}\sum_{\ell=0}^{m}\binom{n}{m}\binom{m}{\ell}B_{n-m}(\eta_1 x)\,S_{m-\ell}(\eta_2-1)\,(\eta_2 y)^{\ell}\,\eta_2^{n-m-1}\,\eta_1^{m}$$

and

$$\sum_{m=0}^{n}\sum_{j=0}^{\eta_1-1}\binom{n}{m}\eta_1^{m-1}\eta_2^{n-m}(\eta_1 y)^{n-m}B_m\left(\eta_2 x+\frac{\eta_2}{\eta_1}j\right)$$
$$=\sum_{m=0}^{n}\sum_{j=0}^{\eta_1-1}\binom{n}{m}\eta_2^{m-1}\eta_1^{n-m}(\eta_2 y)^{n-m}B_m\left(\eta_1 x+\frac{\eta_1}{\eta_2}j\right). \tag{30}$$

3. Symmetry Identities of Arbitrary Order Hermite-Bernoulli Polynomials Attached to a Dirichlet Character χ

We begin by introducing generalized Hermite-Bernoulli polynomials attached to a Dirichlet character χ of order $\alpha\in\mathbb{C}$ defined by means of the following generating function:

$$\left(\frac{t\sum_{j=0}^{d-1}\chi(j)\,e^{jt}}{e^{dt}-1}\right)^{\alpha}e^{xt+yt^2}=\sum_{n=0}^{\infty}{}_HB_{n,\chi}^{(\alpha)}(x,y)\frac{t^n}{n!} \tag{31}$$
$$(\alpha,\,x,\,y\in\mathbb{C}),$$

where χ is a Dirichlet character with conductor d.

Here, $B_{n,\chi}^{(\alpha)}(x) := {}_HB_{n,\chi}^{(\alpha)}(x,0)$, $B_{n,\chi}^{(\alpha)} := {}_HB_{n,\chi}^{(\alpha)}(0,0)$, and $B_{n,\chi} := {}_HB_{n,\chi}^{(1)}(0,0)$ are called the generalized Hermite-Bernoulli polynomials and numbers attached to χ of order α and Hermite-Bernoulli numbers attached to χ, respectively.

Remark 1. *Taking* $y=0$ *in Equation (31) gives* ${}_HB_{n,\chi}^{(\alpha)}(x,0) :={}_H B_{n,\chi}^{(\alpha)}(x)$, *cf. [33].*

Remark 2. *Equation (15) is obtained when* $\chi := 1$ *in Equation (31).*

Remark 3. *The Hermite-Bernoulli polynomials* ${}_HB_n(x,y)$ *are obtained when* $\chi := 1$ *and* $\alpha = 1$ *in Equation (31).*

Remark 4. *The generalized Bernoulli polynomials* $B_n^{(\alpha)}(x)$ *is obtained when* $\chi := 1$ *and* $y=0$ *in Equation (31).*

Remark 5. *The classical Bernoulli polynomials attached to* χ *is obtained when* $\alpha = 1$ *and* $y = 0$ *in Equation (31).*

Theorem 2. *Let α, x, y, $z \in \mathbb{C}$, η_1, $\eta_2 \in \mathbb{N}$, and $n \in \mathbb{N}_0$. Then,*

$$\begin{aligned}&\sum_{m=0}^{n}\sum_{\ell=0}^{m}\binom{n}{m}\binom{m}{\ell}\eta_1^{n-m-1}\eta_2^{m}\,{}_HB_{n-m,\chi}^{(\alpha)}\left(\eta_2 x,\eta_2^2 z\right)\,B_{m-\ell,\chi}^{(\alpha-1)}(\eta_1 y)\,T_\ell(\chi,d\eta_1-1)\\&=\sum_{m=0}^{n}\sum_{\ell=0}^{m}\binom{n}{m}\binom{m}{\ell}\eta_2^{n-m-1}\eta_1^{m}\,{}_HB_{n-m,\chi}^{(\alpha)}\left(\eta_1 x,\eta_1^2 z\right)\,B_{m-\ell,\chi}^{(\alpha-1)}(\eta_2 y)\,T_\ell(\chi,d\eta_2-1)\end{aligned}\tag{32}$$

and

$$\begin{aligned}&\sum_{m=0}^{n}\sum_{\ell=0}^{d\eta_1-1}\chi(\ell)\binom{n}{m}\eta_1^{n-m-1}\eta_2^{m}\,{}_HB_{n-m,\chi}^{(\alpha)}\left(\eta_2 x+\frac{\ell\eta_2}{\eta_1},\eta_2^2 z\right)B_{m,\chi}^{(\alpha-1)}(\eta_1 y)\\&=\sum_{m=0}^{n}\sum_{\ell=0}^{d\eta_2-1}\chi(\ell)\binom{n}{m}\eta_2^{n-m-1}\eta_1^{m}\,{}_HB_{n-m,\chi}^{(\alpha)}\left(\eta_1 x+\frac{\ell\eta_1}{\eta_2},\eta_1^2 z\right)B_{m,\chi}^{(\alpha-1)}(\eta_2 y),\end{aligned}\tag{33}$$

where χ is a Dirichlet character with conductor d.

Proof. Let

$$\begin{aligned}G(\alpha;\eta_1,\eta_2)(t):=&\frac{d}{\int_X e^{d\eta_1\eta_2 ut}du}\left(\frac{\eta_1 t\sum_{j=0}^{d-1}\chi(j)\,e^{j\eta_1 t}}{e^{d\eta_1 t}-1}\right)^{\alpha}e^{\eta_1\eta_2 xt+\eta_1^2\eta_2^2 zt^2}\\&\times\left(\frac{\eta_2 t\sum_{j=0}^{d-1}\chi(j)\,e^{j\eta_2 t}}{e^{d\eta_2 t}-1}\right)^{\alpha}e^{\eta_1\eta_2 yt}\end{aligned}\tag{34}$$

$$\left(\alpha,\ x,\ y,\ z\in\mathbb{C};\ t\in\mathbb{C}\setminus\{0\};\ \eta_1,\ \eta_2\in\mathbb{N};\ 1^\alpha:=1\right).$$

Obviously $G(\alpha;\eta_1,\eta_2)(t)$ is symmetric with respect to the parameters η_1 and η_2. As in the function $F(\alpha;\eta_1,\eta_2)(t)$ in Equation (18), $G(\alpha;\eta_1,\eta_2)(t)$ can be considered to be analytic in a neighborhood of $t=0$. Using Equation (9), we have

$$\begin{aligned}G(\alpha;\eta_1,\eta_2)(t)=&\frac{d\int_X\chi(u)e^{\eta_2 ut}du}{\int_X e^{d\eta_1\eta_2 ut}du}\left(\frac{\eta_1 t\sum_{j=0}^{d-1}\chi(j)\,e^{j\eta_1 t}}{e^{d\eta_1 t}-1}\right)^{\alpha}e^{\eta_1\eta_2 xt+\eta_1^2\eta_2^2 zt^2}\\&\times\left(\frac{\eta_2 t\sum_{j=0}^{d-1}\chi(j)\,e^{j\eta_2 t}}{e^{d\eta_2 t}-1}\right)^{\alpha-1}e^{\eta_1\eta_2 yt}.\end{aligned}\tag{35}$$

Applying Equations (13) and (31) to Equation (35), we obtain

$$\begin{aligned}G(\alpha;\eta_1,\eta_2)(t):=&\frac{1}{\eta_1}\sum_{n=0}^{\infty}{}_HB_{n,\chi}^{(\alpha)}\left(\eta_2 x,\eta_2^2 z\right)\frac{(\eta_1 t)^n}{n!}\sum_{m=0}^{\infty}B_{m,\chi}^{(\alpha-1)}(\eta_1 y)\frac{(\eta_2 t)^m}{m!}\\&\times\sum_{\ell=0}^{\infty}T_\ell(\chi,d\eta_1-1)\frac{(\eta_2 t)^\ell}{\ell!}.\end{aligned}\tag{36}$$

Similarly as in the proof of Theorem 1, we find

$$\begin{aligned}G(\alpha;\eta_1,\eta_2)(t)=&\sum_{n=0}^{\infty}\sum_{m=0}^{n}\sum_{\ell=0}^{m}\frac{\eta_1^{n-m-1}\eta_2^{m}}{(n-m)!(m-\ell)!\ell!}\\&\times{}_HB_{n-m,\chi}^{(\alpha)}\left(\eta_2 x,\eta_2^2 z\right)B_{m-\ell,\chi}^{(\alpha-1)}(\eta_1 y)\,T_\ell(\chi,d\eta_1-1)\,t^n.\end{aligned}\tag{37}$$

In view of the symmetry of $G(\alpha;\eta_1,\eta_2)(t)$ with respect to the parameters η_1 and η_2, we also get

$$\begin{aligned} G(\alpha;\eta_1,\eta_2)(t) = \sum_{n=0}^{\infty}\sum_{m=0}^{n}\sum_{\ell=0}^{m} \frac{\eta_2^{n-m-1}\eta_1^{m}}{(n-m)!(m-\ell)!\ell!} \\ \times {}_HB_{n-m,\chi}^{(\alpha)}\left(\eta_1 x,\eta_1^2 z\right) B_{m-\ell,\chi}^{(\alpha-1)}(\eta_2 y)\, T_\ell(\chi, d\eta_2-1)\, t^n. \end{aligned} \tag{38}$$

Equating the coefficients of t^n of the right sides of Equations (37) and (38), we obtain Equation (32).

From Equation (13), we have

$$\frac{d\int_X \chi(u)e^{\eta_2 ut}du}{\int_X e^{d\eta_1\eta_2 ut}du} = \frac{1}{\eta_1}\sum_{\ell=0}^{d\eta_1-1}\chi(\ell)\, e^{\ell\eta_2 t}. \tag{39}$$

Using Equation (39) in Equation (35), we get

$$\begin{aligned} G(\alpha;\eta_1,\eta_2)(t) = \frac{1}{\eta_1}\sum_{\ell=0}^{d\eta_1-1}\chi(\ell)\left(\frac{\eta_1 t\sum_{j=0}^{d-1}\chi(j)\, e^{j\eta_1 t}}{e^{d\eta_1 t}-1}\right)^{\alpha} e^{\left(\eta_2 x+\frac{\ell\eta_2}{\eta_1}\right)\eta_1 t+\eta_1^2\eta_2^2 z t^2} \\ \times\left(\frac{\eta_2 t\sum_{j=0}^{d-1}\chi(j)\, e^{j\eta_2 t}}{e^{d\eta_2 t}-1}\right)^{\alpha-1} e^{\eta_1\eta_2 yt}. \end{aligned} \tag{40}$$

Using Equation (31), similarly as above, we obtain

$$\begin{aligned} G(\alpha;\eta_1,\eta_2)(t) = \sum_{n=0}^{\infty}\sum_{m=0}^{n}\sum_{\ell=0}^{d\eta_1-1}\chi(\ell)\, {}_HB_{n-m,\chi}^{(\alpha)}\left(\eta_2 x+\frac{\ell\eta_2}{\eta_1},\eta_2^2 z\right) \\ \times B_{m,\chi}^{(\alpha-1)}(\eta_1 y)\frac{\eta_1^{n-m-1}\eta_2^{m}}{(n-m)!m!}t^n. \end{aligned} \tag{41}$$

Since $G(\alpha;\eta_1,\eta_2)(t)$ is symmetric with respect to the parameters η_1 and η_2, we also have

$$\begin{aligned} G(\alpha;\eta_1,\eta_2)(t) = \sum_{n=0}^{\infty}\sum_{m=0}^{n}\sum_{\ell=0}^{d\eta_2-1}\chi(\ell)\, {}_HB_{n-m,\chi}^{(\alpha)}\left(\eta_1 x+\frac{\ell\eta_1}{\eta_2},\eta_1^2 z\right) \\ \times B_{m,\chi}^{(\alpha-1)}(\eta_2 y)\frac{\eta_2^{n-m-1}\eta_1^{m}}{(n-m)!m!}t^n. \end{aligned} \tag{42}$$

Equating the coefficients of t^n of the right sides in Equation (41) and Equation (42), we get Equation (33). □

4. Conclusions

The results in Theorems 1 and 2, being very general, can reduce to yield many symmetry identities associated with relatively simple polynomials and numbers using Remarks 1–5. Setting $z = 0$ and $\in \mathbb{N}$ in the results in Theorem 1 and Theorem 2 yields the corresponding known identities in References [33,34], respectively.

Author Contributions: All authors contributed equally.

References

1. Kim, T. q-Volkenborn integration. *Russ. J. Math. Phys.* **2002**, *9*, 288–299.
2. Cenkci, M. The p-adic generalized twisted h, q-Euler-l-function and its applications. *Adv. Stud. Contem. Math.* **2007**, *15*, 37–47.
3. Cenkci, M.; Simsek, Y.; Kurt, V. Multiple two-variable p-adic q-L-function and its behavior at $s = 0$. *Russ. J. Math. Phys.* **2008**, *15*, 447–459. [CrossRef]
4. Kim, T. On a q-analogue of the p-adic log gamma functions and related integrals. *J. Numb. Theor.* **1999**, *76*, 320–329. [CrossRef]
5. Kim, T. A note on q-Volkenborn integration. *Proc. Jangeon Math. Soc.* **2005**, *8*, 13–17.
6. Kim, T. q-Euler numbers and polynomials associated with p-adic q-integrals. *J. Nonlinear Math. Phys.* **2007**, *14*, 15–27. [CrossRef]
7. Kim, T. A note on p-adic q-integral on $\mathbb{Z}_p$ associated with q-Euler numbers. *Adv. Stud. Contem. Math.* **2007**, *15*, 133–137.
8. Kim, T. On p-adic q-l-functions and sums of powers. *J. Math. Anal. Appl.* **2007**, *329*, 1472–1481. [CrossRef]
9. Kim, T.; Choi, J.Y.; Sug, J.Y. Extended q-Euler numbers and polynomials associated with fermionic p-adic q-integral on $\mathbb{Z}_p$. *Russ. J. Math. Phy.* **2007**, *14*, 160–163. [CrossRef]
10. Simsek, Y. On p-adic twisted q-L-functions related to generalized twisted Bernoulli numbers. *Russ. J. Math. Phy.* **2006**, *13*, 340–348. [CrossRef]
11. Haroon, H.; Khan, W.A. Degenerate Bernoulli numbers and polynomials associated with degenerate Hermite polynomials. *Commun. Korean Math. Soc.* **2017**, in press.
12. Khan, N.; Usman, T.; Choi, J. A new generalization of Apostol-type Laguerre-Genocchi polynomials. *C. R. Acad. Sci. Paris Ser. I* **2017**, *355*, 607–617. [CrossRef]
13. Pathan, M.A.; Khan, W.A. Some implicit summation formulas and symmetric identities for the generalized Hermite-Bernoulli polynomials. *Mediterr. J. Math.* **2015**, *12*, 679–695. [CrossRef]
14. Pathan, M.A.; Khan, W.A. A new class of generalized polynomials associated with Hermite and Euler polynomials. *Mediterr. J. Math.* **2016**, *13*, 913–928. [CrossRef]
15. Srivastava, H.M.; Choi, J. *Zeta and q-Zeta Functions and Associated Series and Integrals*; Elsevier Science Publishers: Amsterdam, The Netherlands; London, UK; New York, NY, USA, 2012.
16. Kim, T.; Rim, S.H.; Lee, B. Some identities of symmetry for the generalized Bernoulli numbers and polynomials. *Abs. Appl. Anal.* **2009**, *2009*, 848943. [CrossRef]
17. Khan, W.A. Degenerate Hermite-Bernoulli numbers and polynomials of the second kind. *Prespacetime J.* **2016**, *7*, 1297–1305.
18. Cesarano, C. Operational Methods and New Identities for Hermite Polynomials. *Math. Model. Nat. Phenom.* **2017**, *12*, 44–50. [CrossRef]
19. Dattoli, G.; Lorenzutta, S.; Cesarano, C. Finite sums and generalized forms of Bernoulli polynomials. *Rend. Mat.* **1999**, *19* , 385–391.
20. Bell, E.T. Exponential polynomials. *Ann. Math.* **1934**, *35*, 258–277. [CrossRef]
21. Andrews, L.C. *Special Functions for Engineers and Applied Mathematicians*; Macmillan Publishing Company: New York, NY, USA, 1985.
22. Jang, L.C.; Kim, S.D.; Park, D.W.; Ro, Y.S. A note on Euler number and polynomials. *J. Inequ. Appl.* **2006**, *2006*, 34602. [CrossRef]
23. Kim, T. On the q-extension of Euler and Genocchi numbers, *J. Math. Anal. Appl.* **2007**, *326*, 1458–1465. [CrossRef]
24. Kim, T. q-Bernoulli numbers and polynomials associated with Gaussian binomial coefficients. *Russ. J. Math. Phys.* **2008**, *15*, 51–57. [CrossRef]
25. Kim, T. On the multiple q-Genocchi and Euler numbers. *Russ. J. Math. Phy.***2008**, **15**, 481–486. [CrossRef]
26. Kim, T. New approach to q-Euler, Genocchi numbers and their interpolation functions. *Adv. Stud. Contem. Math.* **2009**, *18*, 105–112.
27. Kim, T. Sums of products of q-Euler numbers. *J. Comput. Anal. Appl.* **2010**, *12*, 185–190.
28. Kim, Y.H.; Kim, W.; Jang, L.C. On the q-extension of Apostol-Euler numbers and polynomials. *Abs. Appl. Anal.* **2008**, *2008*, 296159.

29. Simsek, Y. Complete sum of products of (h, q)-extension of the Euler polynomials and numbers. *J. Differ. Eqn. Appl.* **2010**, *16*, 1331–1348. [CrossRef]
30. Kim, T.; Kim, D.S. An identity of symmetry for the degenerate Frobenius-Euler polynomials. *Math. Slovaca* **2018**, *68*, 239–243. [CrossRef]
31. Kim, T. Symmetry p-adic invariant integral on $\mathbb{Z}_p$ for Bernoulli and Euler polynomials. *J. Differ. Equ. Appl.* **2008**, *14*, 1267–1277. [CrossRef]
32. Choi, J. Notes on formal manipulations of double series. *Commun. Korean Math. Soc.* **2003**, *18*, 781–789. [CrossRef]
33. Kim, T.; Jang, L.C.; Kim, Y.H.; Hwang, K.W. On the identities of symmetry for the generalized Bernoulli polynomials attached to χ of higher order. *J. Inequ. Appl.* **2009**, *2009*, 640152. [CrossRef]
34. Kim, T.; Hwang, K.W.; Kim, Y.H. Symmetry properties of higher order Bernoulli polynomials. *Adv. Differ. Equ.* **2009**, *2009*, 318639. [CrossRef]

5

Note on Type 2 Degenerate q-Bernoulli Polynomials

Dae San Kim [1], Dmitry V. Dolgy [2], Jongkyum Kwon [3,*] and Taekyun Kim [4]

1 Department of Mathematics, Sogang University, Seoul 121-742, Korea
2 Kwangwoon Institute for Advanced Studies, Kwangwoon University, Seoul 139-701, Korea
3 Department of Mathematics Education and ERI, Gyeongsang National University, Jinju, Gyeongsangnamdo 52828, Korea
4 Department of Mathematics, Kwangwoon University, Seoul 139-701, Korea
* Correspondence: mathkjk26@gnu.ac.kr

Abstract: The purpose of this paper is to introduce and study type 2 degenerate q-Bernoulli polynomials and numbers by virtue of the bosonic p-adic q-integrals. The obtained results are, among other things, several expressions for those polynomials, identities involving those numbers, identities regarding Carlitz's q-Bernoulli numbers, identities concerning degenerate q-Bernoulli numbers, and the representations of the fully degenerate type 2 Bernoulli numbers in terms of moments of certain random variables, created from random variables with Laplace distributions. It is expected that, as was done in the case of type 2 degenerate Bernoulli polynomials and numbers, we will be able to find some identities of symmetry for those polynomials and numbers.

Keywords: type 2 degenerate q-Bernoulli polynomials; p-adic q-integral

MSC: 11S80; 11B83

---

1. Introduction

There are various ways of studying special polynomials and numbers, including generating functions, combinatorial methods, umbral calculus techniques, matrix theory, probability theory, p-adic analysis, differential equations, and so on.

In [1], it was shown that odd integer power sums (alternating odd integer power sums) can be represented in terms of some values of the type 2 Bernoulli polynomials (the type 2 Euler polynomials). In addition, some identities of symmetry, involving the type 2 Bernoulli polynomials, odd integer power sums, the type 2 Euler polynomials, and alternating odd integer power sums, were obtained by introducing appropriate quotients of bosonic and fermionic p-adic integrals on $\mathbb{Z}_p$. Furthermore, in [1], it was shown that the moments of two random variables, constructed from random variables with Laplace distributions, are closely connected with the type 2 Bernoulli numbers and the type 2 Euler numbers.

In recent years, studying degenerate versions of various special polynomials and numbers, which began with the paper by Carlitz in [2], has attracted the interest of many mathematicians. For example, in [3], the degenerate type 2 Bernoulli and Euler polynomials, and their corresponding numbers were introduced and some properties of them, which include distribution relations, Witt type formulas, and analogues for the interpretation of integer power sums in terms of Bernoulli polynomials, were investigated by means of both types of p-adic integrals.

As a q-analogue of the Volkenborn integrals for uniformly differentiable functions, the bosonic p-adic q-integrals were introduced in [4] by Kim. These integrals, together with the fermionic p-adic integrals and the fermionic p-adic q-integrals, have proven to be very useful tools in studying many problems arising from number theory and combinatorics. For instance, in [5], the type 2 q-Bernoulli

(q-Euler) polynomials were introduced by virtue of the bosonic (fermionic) p-adic q-integrals. Then, it was noted, among other things, that the odd q-integer (alternating odd q-integer) power sums are expressed in terms of the type 2 q-Bernoulli (q-Euler) polynomials.

In this short paper, we would like to introduce the type 2 degenerate q-Bernoulli polynomials and the corresponding numbers by making use of the bosonic p-adic q-integrals, as a degenerate version of and also as a q-analogue of the type 2 Bernoulli polynomials, and derive several basic results for them. The obtained results are several expressions for those polynomials, identities involving those numbers, identities regarding Carlitz's q-Bernoulli numbers, identities concerning degenerate q-Bernoulli numbers, and the representations of the fully degenerate type 2 Bernoulli numbers ($q = 1$ and $x = 1$ cases of the type 2 degenerate q-Bernoulli polynomials) in terms of moments of certain random variables, created from random variables with Laplace distributions.

The motivation for introducing the type 2 degenerate q-Bernoulli polynomials and numbers is to study their number-theoretic and combinatorial properties, and their applications in mathematics and other sciences in general. One novelty of this paper is that they arise naturally by means of the bosonic p-adic q-integrals so that it is possible to easily find some identities of symmetry for those polynomials and numbers, as it was done, for example, in [1]. In the rest of this section, we recall what is needed in the latter part of the paper.

Throughout this paper, p is a fixed odd prime number. We use the standard notations $\mathbb{Z}_p$, $\mathbb{Q}_p$, and $\mathbb{C}_p$ to denote the ring of p-adic integers, the field of p-adic rational numbers, and the completion of the algebraic closure of $\mathbb{Q}_p$, respectively. The p-adic norm on $\mathbb{C}_p$ is normalized as $|p|_p = \frac{1}{p}$.

As is well known, the Bernoulli numbers are given by the recurrence relation

$$B_0 = 1, \quad (B+1)^n = \begin{cases} 1, & \text{if } n = 1, \\ 0, & \text{if } n > 1, \end{cases} \tag{1}$$

where, as usual, B^n are to be replaced by B_n (see [2,6,7]).

Additionally, the Bernoulli polynomials of degree n are given by

$$B_n(x) = \sum_{l=0}^{n} \binom{n}{l} B_l x^{n-l}, \tag{2}$$

(see [3,8,9]).

Let q be an indeterminate in $\mathbb{C}_p$. For $q \in \mathbb{C}_p$, we assume that $|1-q|_p < p^{-\frac{1}{p-1}}$.

In [7], Carlitz considered the q-Bernoulli numbers which are given by the recurrence relation:

$$\beta_{0,q} = 1, \quad q(q\beta_q + 1)^n - \beta_{n,q} = \begin{cases} 1, & \text{if } n = 1, \\ 0, & \text{if } n > 1, \end{cases} \tag{3}$$

where β_q^n are to be replaced by $\beta_{n,q}$, as usual.

In addition, he defined the q-Bernoulli polynomials as

$$\beta_{n,q}(x) = \sum_{l=0}^{n} \binom{n}{l} [x]_q^{n-l} q^{lx} \beta_{l,q}, \quad (n \geq 0), \tag{4}$$

where $[x]_q = \frac{1-q^x}{1-q}$, (see [7]).

Recently, the type 2 Bernoulli polynomials have been defined as

$$\frac{t}{e^t - e^{-t}} e^{xt} = \sum_{n=0}^{\infty} b_n(x) \frac{t^n}{n!}, \tag{5}$$

(see [1,3,8]).

When $x = 0$, $b_n = b_n(0)$ are called the type 2 Bernoulli numbers.

From (5), we note that

$$\sum_{l=0}^{n-1}(2l+1)^k = \frac{1}{k+1}(b_{k+1}(2n) - b_{k+1}),\ \ (k \geq 0). \tag{6}$$

Let f be a uniformly differentiable function on $\mathbb{Z}_p$. Then, Kim defined the p-adic q-integral of f on $\mathbb{Z}_p$ as

$$\begin{aligned} I_q(f) &= \int_{\mathbb{Z}_p} f(x)d\mu_q(x) \\ &= \lim_{N\to\infty} \frac{1}{[p^N]_q} \sum_{x=0}^{p^N-1} f(x)q^x \\ &= \lim_{N\to\infty} \sum_{x=0}^{p^N-1} f(x)\mu_q(x+p^N\mathbb{Z}_p), \end{aligned} \tag{7}$$

(see [4]). Here, we note that $\mu_q(x+p^N\mathbb{Z}_p) = \frac{q^x}{[p^N]_q}$ is a distribution but not a measure. The details on the existence of the p-adic q-integrals for uniformly differentiable functions f on $\mathbb{Z}_p$ can be found in [4,10].

From (7), we note that

$$qI_q(f_1) = I_q(f) + (q-1)f(0) + \frac{q-1}{\log q}f'(0), \tag{8}$$

where $f_1(x) = f(x+1)$.

By virtue of (8) and induction, we get

$$q^n I_q(f_n) = I_q(f) + (q-1)\sum_{l=0}^{n-1} q^l f(l) + \frac{q-1}{\log q}\sum_{l=0}^{n-1} q^l f'(l), \tag{9}$$

where $f_n(x) = f(x+n),\ \ (n \geq 1)$.

The degenerate exponential function is defined by

$$e_\lambda^x(t) = (1+\lambda t)^{\frac{x}{\lambda}}, \tag{10}$$

(see [11]), where $\lambda \in \mathbb{C}_p$ with $|\lambda|_p < p^{-\frac{1}{p-1}}$.
For brevity, we also set

$$e_\lambda(t) = e_\lambda^1(t) = (1+\lambda t)^{\frac{1}{\lambda}}. \tag{11}$$

Carlitz defined the degenerate Bernoulli polynomials as

$$\frac{t}{e_\lambda(t)-1}e_\lambda^x(t) = \sum_{n=0}^{\infty} B_{n,\lambda}(x)\frac{t^n}{n!}, \tag{12}$$

where $B_{n,\lambda} = B_{n,\lambda}(0)$ are called the degenerate Bernoulli numbers.

From (12), we note that

$$\sum_{l=0}^{n-1}(l)_{k,\lambda} = \frac{1}{k+1}(B_{k+1,\lambda}(n) - B_{k+1,\lambda}),\ \ (n \geq 0), \tag{13}$$

where $(l)_{0,\lambda} = 1$, $(l)_{k,\lambda} = l(l-\lambda)\cdots(l-(k-1)\lambda),\ \ (k \geq 1)$.

In the special case of $\lambda = 1$, the falling factorial sequence (also called the Pochammer symbol) is given by

$$(l)_0 = 1, (l)_k = l(l-1)\cdots(l-(k-1)), \quad (k \geq 1). \tag{14}$$

In this paper, we study type 2 degenerate q-Bernoulli polynomials and investigate some identities and properties for these polynomials.

2. Type 2 Degenerate q-Bernoulli Polynomials

Throughout this section, we assume that $q \in \mathbb{C}_p$ with $|1-q|_p < p^{-\frac{1}{p-1}}$ and $\lambda \in \mathbb{C}_p$. Now, we define the *type 2 degenerate q-Bernoulli polynomials* by

$$\sum_{n=0}^{\infty} b_{n,q}(x \mid \lambda)\frac{t^n}{n!} = \frac{1}{2}\int_{\mathbb{Z}_p} e_\lambda^{[x+2y]_q}(t)d\mu_q(y). \tag{15}$$

By (15), we get

$$\frac{1}{2}\int_{\mathbb{Z}_p}\Big([x+2y]_q\Big)_{n,\lambda} d\mu_q(y) = b_{n,q}(x \mid \lambda), \quad (n \geq 0). \tag{16}$$

When $x = 1$, $b_{n,q}(\lambda) = b_{n,q}(1 \mid \lambda)$ are called the type 2 degenerate q-Bernoulli numbers. We observe here that

$$\begin{aligned}&\lim_{q\to 1}\lim_{\lambda\to 0}\frac{1}{2}\int_{\mathbb{Z}_p}\Big([x+2y]_q\Big)_{n,\lambda} d\mu_q(y)\\ &=\lim_{q\to 1}\lim_{\lambda\to 0}\frac{1}{2}b_{n,q}(x \mid \lambda) = b_n(x-1), \quad (n \geq 0).\end{aligned} \tag{17}$$

The degenerate Stirling numbers of the first kind appear as the coefficients in the expansion

$$(x)_{n,\lambda} = \sum_{l=0}^{n} S_{1,\lambda}(n,l)x^l, \quad (n \geq 0), \tag{18}$$

(see [12]).

Thus, by (18), we get

$$\begin{aligned}&\frac{1}{2}\int_{\mathbb{Z}_p}\Big([x+2y]_q\Big)_{n,\lambda} d\mu_q(y)\\ &=\frac{1}{2}\sum_{l=0}^{n} S_{1,\lambda}(n,l)\int_{\mathbb{Z}_p}[x+2y]_q^l d\mu_q(y)\\ &=\sum_{l=0}^{n} S_{1,\lambda}(n,l)b_{l,q}(x),\end{aligned} \tag{19}$$

where $b_{l,q}(x)$ is the type 2 q-Bernoulli polynomials given by $\frac{1}{2}\int_{\mathbb{Z}_p}[x+2y]_q^n d\mu_q(y) = b_{n,q}(x)$, $(n \geq 0)$, (see [5]).

Therefore, by (16) and (19), we obtain the following theorem.

Theorem 1. *For $n \geq 0$, we have*

$$b_{n,q}(x \mid \lambda) = \sum_{l=0}^{n} S_{1,\lambda}(n,l)b_{l,q}(x). \tag{20}$$

Now, we observe that

$$\begin{aligned}
&\frac{1}{2}\int_{\mathbb{Z}_p}\Big([x+2y]_q\Big)_{n,\lambda}d\mu_q(y)\\
=&\frac{1}{2}\sum_{l=0}^{n}S_{1,\lambda}(n,l)\int_{\mathbb{Z}_p}[x+2y]_q^l d\mu_q(y)\\
=&\frac{1}{2}\sum_{l=0}^{n}S_{1,\lambda}(n,l)\frac{1}{(1-q)^l}\sum_{m=0}^{l}\binom{l}{m}(-q^x)^m\frac{2m+1}{[2m+1]_q}.
\end{aligned} \tag{21}$$

Therefore, by (21), we obtain the following theorem.

Theorem 2. *For $n \geq 0$, we have*

$$b_{n,q}(x \mid \lambda)=\frac{1}{2}\sum_{l=0}^{n}\frac{S_{1,\lambda}(n,l)}{(1-q)^l}\sum_{m=0}^{l}\binom{l}{m}(-q^x)^m\frac{2m+1}{[2m+1]_q}. \tag{22}$$

From (16), we note that

$$\begin{aligned}
\sum_{n=0}^{\infty}b_{n,q}(x \mid \lambda)\frac{t^n}{n!}&=\frac{1}{2}\sum_{n=0}^{\infty}\int_{\mathbb{Z}_p}\Big([x+2y]_q\Big)_{n,\lambda}d\mu_q(y)\frac{t^n}{n!}\\
&=\frac{1}{2}\int_{\mathbb{Z}_p}\Big(1+\lambda t\Big)^{\frac{[x+2y]_q}{\lambda}}d\mu_q(y)\\
&=\sum_{n=0}^{\infty}\left(\sum_{k=0}^{n}\sum_{m=0}^{k}\binom{k}{m}q^{mx}[x]_q^{k-m}S_1(n,k)\lambda^{n-k}b_{m,q}\right)\frac{t^n}{n!},
\end{aligned} \tag{23}$$

where $S_1(n,k)$ are the Stirling numbers of the first kind and $b_{n,q}$ are the type 2 q-Bernoulli numbers.
Therefore, by (23), we get the following theorem.

Theorem 3. *For $n \geq 0$, we have*

$$b_{n,q}(x \mid \lambda)=\sum_{k=0}^{n}\sum_{m=0}^{k}\binom{k}{m}q^{mx}[x]_q^{k-m}S_1(n,k)\lambda^{n-k}b_{m,q}. \tag{24}$$

In particular,

$$b_{n,q}(\lambda)=\sum_{k=0}^{n}q^{kx}S_1(n,k)\lambda^{n-k}b_{k,q}.$$

In [4], Kim expressed Carlitz's q-Bernoulli polynomials in terms of the following p-adic q-integrals on $\mathbb{Z}_p$:

$$\int_{\mathbb{Z}_p}[x+y]_q^n d\mu_q(y)=\beta_{n,q}(x),\ \ (n\geq 0). \tag{25}$$

From (9) and (25), we have

$$\begin{aligned} q^n\beta_{m,q}(n) &= \int_{\mathbb{Z}_p} q^n[x+n]_q^m d\mu_q(x) \\ &= \int_{\mathbb{Z}_p}[x]_q^m d\mu_q(x) + (q-1)\sum_{l=0}^{n-1} q^l[l]_q^m + m\sum_{l=0}^{n-1}[l]_q^{m-1}q^{2l} \\ &= \beta_{m,q} + (q-1)\sum_{l=0}^{n-1} q^l[l]_q^m + m\sum_{l=0}^{n-1}[l]_q^{m-1}q^{2l} \\ &= \beta_{m,q} + (m+1)\sum_{l=0}^{n-1} q^{2l}[l]_q^{m-1} - \sum_{l=0}^{n-1} q^l[l]_q^{m-1}, \end{aligned} \tag{26}$$

where n is a positive integer.

Therefore, we obtain the following theorem.

Theorem 4. *For $n \geq 0$, we have*

$$q^n\beta_{m,q}(n) - \beta_{m,q} = (m+1)\sum_{l=0}^{n-1} q^{2l}[l]_q^{m-1} - \sum_{l=0}^{n-1} q^l[l]_q^{m-1}. \tag{27}$$

Let us take $f(x) = \big([x]_q\big)_{m,\lambda}$, $(m \geq 1)$. From (9), we have

$$\begin{aligned} & q^n\int_{\mathbb{Z}_p}\big([x+n]_q\big)_{m,\lambda} d\mu_q(x) \\ =& \int_{\mathbb{Z}_p}\big([x]_q\big)_{m,\lambda} d\mu_q(x) + (q-1)\sum_{l=0}^{n-1} q^l\big([l]_q\big)_{m,\lambda} \\ &+ \sum_{l=0}^{n-1}\left(\sum_{k=0}^{m-1}\frac{1}{[l]_q - k\lambda}\right)\big([l]_q\big)_{m,\lambda} q^{2l}. \end{aligned} \tag{28}$$

In [13], the degenerate q-Bernoulli polynomials are defined by Kim as

$$\int_{\mathbb{Z}_p} e_\lambda^{[x+y]_q}(t)d\mu_q(y) = \sum_{n=0}^{\infty}\beta_{n,\lambda,q}(x)\frac{t^n}{n!}. \tag{29}$$

In particular, the degenerate q-Bernoulli numbers are given by $\beta_{n,\lambda,q} = \beta_{n,\lambda,q}(0)$.

From (29), we have

$$\int_{\mathbb{Z}_p}\big([x+y]_q\big)_{n,\lambda} d\mu_q(y) = \beta_{n,\lambda,q}(x), \ \ (n \geq 0). \tag{30}$$

By (28) and (30), this completes the proof for the next theorem.

Theorem 5. *For $m, n \in \mathbb{N}$, we have*

$$\begin{aligned} & q^n\beta_{m,\lambda,q}(n) - \beta_{m,\lambda,q} \\ =& (q-1)\sum_{l=0}^{n-1} q^l\big([l]_q\big)_{m,\lambda} + \sum_{l=0}^{n-1}\left(\sum_{k=0}^{m-1}\frac{1}{[l]_q - k\lambda}\right)\big([l]_q\big)_{m,\lambda} q^{2l}. \end{aligned} \tag{31}$$

Let us take $f(x) = \Big([2x+1]_q\Big)_{m,\lambda}$, $(m \geq 1)$. From (9), we have

$$\begin{aligned} & q^n \int_{\mathbb{Z}_p} \Big([2x+2n+1]_q\Big)_{m,\lambda} d\mu_q(x) \\ = & \int_{\mathbb{Z}_p} \Big([2x+1]_q\Big)_{m,\lambda} d\mu_q(x) + (q-1)\sum_{l=0}^{n-1} q^l \Big([2l+1]_q\Big)_{m,\lambda} \\ & + 2\sum_{l=0}^{n-1}\Big(\sum_{k=0}^{m-1}\frac{1}{[2l+1]_q - k\lambda}\Big)\Big([2l+1]_q\Big)_{m,\lambda} q^{3l+1}. \end{aligned} \tag{32}$$

From (16) and (32), we have

$$\begin{aligned} & q^n b_{m,q}(2n+1 \mid \lambda) - b_{m,q}(\lambda) \\ = & \frac{q-1}{2}\sum_{l=0}^{n-1} q^l \Big([2l+1]_q\Big)_{m,\lambda} + \sum_{l=0}^{n-1}\Big(\sum_{k=0}^{m-1}\frac{1}{[2l+1]_q - k\lambda}\Big)\Big([2l+1]_q\Big)_{m,\lambda} q^{3l+1}. \end{aligned} \tag{33}$$

Therefore, by (33), we obtain the following theorem.

Theorem 6. *For $m, n \in \mathbb{N}$, we have*

$$\begin{aligned} & \frac{q-1}{2}\sum_{l=0}^{n-1} q^l \Big([2l+1]_q\Big)_{m,\lambda} + \sum_{l=0}^{n-1}\Big(\sum_{k=0}^{m-1}\frac{1}{[2l+1]_q - k\lambda}\Big)\Big([2l+1]_q\Big)_{m,\lambda} q^{3l+1} \\ = & q^n b_{m,q}(2n+1 \mid \lambda) - b_{m,q}(\lambda). \end{aligned} \tag{34}$$

From (7), we can derive the following integral equation:

$$\begin{aligned} \int_{\mathbb{Z}_p} f(x) d\mu_q(x) &= \lim_{N\to\infty} \frac{1}{[p^N]_q} \sum_{x=0}^{p^N-1} f(x) q^x \\ &= \lim_{N\to\infty} \frac{1}{[dp^N]_q} \sum_{x=0}^{dp^N-1} f(x) q^x \\ &= \lim_{N\to\infty} \frac{1}{[dp^N]_q} \sum_{a=0}^{d-1}\sum_{x=0}^{p^N-1} f(a+dx) q^{a+dx} \\ &= \sum_{a=0}^{d-1} q^a \frac{1}{[d]_q} \lim_{N\to\infty} \frac{1}{[p^N]_{q^d}} \sum_{x=0}^{p^N-1} f(a+dx) q^{dx} \\ &= \frac{1}{[d]_q} \sum_{a=0}^{d-1} q^a \int_{\mathbb{Z}_p} f(a+dx) d\mu_{q^d}(x), \end{aligned} \tag{35}$$

where d is a positive integer.

Lemma 1. *For $d \in \mathbb{N}$, we have*

$$\int_{\mathbb{Z}_p} f(x) d\mu_q(x) = \frac{1}{[d]_q} \sum_{a=0}^{d-1} q^a \int_{\mathbb{Z}_p} f(a+dx) d\mu_{q^d}(x). \tag{36}$$

We obtain the following theorem from Lemma 1.

Theorem 7. *For $n, d \in \mathbb{N}$, we have*

$$b_{n,q}(\lambda) = [d]_q^{n-1} \sum_{a=0}^{d-1} q^a b_{n,q^d}\left(\frac{2a+1}{d} \mid \frac{\lambda}{[d]_q}\right). \tag{37}$$

Proof. Let us apply Lemma 1 with $f(x) = \left([2x+1]_q\right)_{n,\lambda}$, $(n \in \mathbb{N})$. Then, by virtue of (16), we have

$$\begin{aligned}
&\int_{\mathbb{Z}_p} \left([2x+1]_q\right)_{n,\lambda} d\mu_q(x) = \frac{1}{[d]_q} \sum_{a=0}^{d-1} q^a \int_{\mathbb{Z}_p} \left([2(a+dx)+1]_q\right)_{n,\lambda} d\mu_{q^d}(x) \\
=& \frac{1}{[d]_q} \sum_{a=0}^{d-1} q^a [d]_q^n \int_{\mathbb{Z}_p} \left(\left[\frac{2a+1}{d} + 2x\right]_{q^d}\right)_{n,\frac{\lambda}{[d]_q}} d\mu_{q^d}(x) \\
=& [d]_q^{n-1} \sum_{a=0}^{d-1} q^a \int_{\mathbb{Z}_p} \left(\left[\frac{2a+1}{d} + 2x\right]_{q^d}\right)_{n,\frac{\lambda}{[d]_q}} d\mu_{q^d}(x) \\
=& 2[d]_q^{n-1} \sum_{a=0}^{d-1} q^a b_{n,q^d}\left(\frac{2a+1}{d} \mid \frac{\lambda}{[d]_q}\right).
\end{aligned}$$

□

3. Further Remarks

Assume that $X_1, X_2, X_3, \cdots$ are independent random variables, each of which has the Laplace distribution with parameters 0 and 1. Namely, each of them has the probability density function given by $\frac{1}{2}\exp(-|x|)$.

Let Z be the random variable given by $Z = \sum_{k=1}^{\infty} \frac{X_k}{2k\pi}$. In addition, let b_n be the type 2 Bernoulli numbers defined by

$$\frac{t}{e^t - e^{-t}} = \sum_{n=0}^{\infty} b_n \frac{t^n}{n!}. \tag{38}$$

Then, it was shown in [1] that

$$\begin{aligned}
\sum_{n=0}^{\infty} E[Z^n] \frac{(it)^n}{n!} &= \frac{t}{e^{\frac{t}{2}} - e^{-\frac{t}{2}}} \\
&= \sum_{n=0}^{\infty} \left(\frac{1}{2}\right)^{n-1} b_n \frac{t^n}{n!}.
\end{aligned} \tag{39}$$

Thereby, it was obtained that

$$i^n E[Z^n] = \left(\frac{1}{2}\right)^{n-1} b_n. \tag{40}$$

Before proceeding further, we recall that the Volkenborn integral (also called the p-adic invariant integral) for a uniformly differentiable function f on $\mathbb{Z}_p$ is given by

$$\int_{\mathbb{Z}_p} f(y) d\mu(y) = \lim_{N \to \infty} \frac{1}{p^N} \sum_{y=0}^{p^N - 1} f(y). \tag{41}$$

Then, it is well known (see [14]) that this integral satisfies the following integral equation:

$$\int_{\mathbb{Z}_p} f(y+1) d\mu(y) = \int_{\mathbb{Z}_p} f(y) d\mu(y) + f'(0). \tag{42}$$

When $q = 1$ and $x = 1$, by virtue of (41), (15) becomes

$$\begin{aligned}\sum_{n=0}^{\infty} b_n(\lambda)\frac{t^n}{n!} &= \frac{1}{2}\int_{\mathbb{Z}_p} e_\lambda^{1+2y}(t)d\mu(y)\\ &= \frac{\frac{1}{\lambda}\log(1+\lambda t)}{e_\lambda(t) - e_\lambda^{-1}(t)}.\end{aligned} \tag{43}$$

Here, $b_n(\lambda)$ may be called the fully degenerate type 2 Bernoulli numbers, even though they were defined slightly differently in [3]. Replacing t with $\frac{2}{\lambda}\log(1+\lambda t)$ in (39) and by making use of (43), we have

$$\begin{aligned}&\sum_{m=0}^{\infty} E[Z^m](2i)^m \frac{1}{m!}\left(\frac{\log(1+\lambda t)}{\lambda}\right)^m\\ &= \sum_{n=0}^{\infty}\left(\sum_{m=0}^{n} S_{1,\lambda}(n,m)(2i)^m E[Z^m]\right)\frac{t^n}{n!}\\ &= 2\sum_{n=0}^{\infty} b_n(\lambda)\frac{t^n}{n!}.\end{aligned} \tag{44}$$

Here, $S_{1,\lambda}(n,k)$ are the degenerate Stirling numbers of the first kind (see [12]) either given by

$$\frac{1}{m}\left(\frac{\log(1+\lambda t)}{\lambda}\right)^m = \sum_{n=m}^{\infty} S_{1,\lambda}(n,m)\frac{t^n}{n!}, \tag{45}$$

or given by

$$(x)_{n,\lambda} = \sum_{m=0}^{n} S_{1,\lambda}(n,m)x^m = \sum_{m=0}^{n} S_1(n,m)\lambda^{n-m}x^m. \tag{46}$$

Thus, by (44), we have shown that

$$2b_n(\lambda) = \sum_{m=0}^{n} S_{1,\lambda}(n,m)(2i)^m E[Z^m].$$

Here, we remark that we only considered $q = 1$ and $x = 1$ cases of (15), namely the fully degenerate type 2 Bernoulli numbers. This is because we do not see how to express type 2 degenerate q-Bernoulli polynomials or type 2 degenerate q-Bernoulli numbers in terms of the moments of some suitable random variables, constructed from random variables with Laplace distributions. We leave this as an open problem to the interested reader.

4. Conclusions

Studies on various special polynomials and numbers have been preformed using several different methods, such as generating functions, combinatorial methods, umbral calculus techniques, matrix theory, probability theory, p-adic analysis, differential equations, and so on.

One way of introducing new special polynomials and numbers is to study various degenerate versions of some known special polynomials and numbers, which began with Carlitz's paper in [2]. Actually, degenerate versions were investigated not only for some polynomials but also for a transcendental function, namely the gamma function. For this, we refer the reader to [11]. Another way of introducing new special polynomials and numbers is to study various q-analogues of some known special polynomials and numbers. The bosonic p-adic q-integrals, together with the fermionic p-adic q-integrals, turned out to be very powerful and fruitful tools for naturally constructing such q-analogues. They were introduced by Kim in [4] and have been widely used ever since their invention.

In this paper, the type 2 degenerate q-Bernoulli polynomials and the corresponding numbers were introduced and investigated as a degenerate version of and also as a q-analogue of type 2 Bernoulli polynomials by making use of the bosonic p-adic q-integrals [1,3,5]. Here, as an introductory paper on the subject, only very basic results were obtained. The obtained results are several expressions for those polynomials, identities involving those numbers, identities regarding Carlitz's q-Bernoulli numbers, identities concerning degenerate q-Bernoulli numbers, and the representations of the fully degenerate type 2 Bernoulli numbers ($q = 1$ and $x = 1$ cases of the type 2 degenerate q-Bernoulli polynomials) in terms of moments of certain random variables, created from random variables with Laplace distributions. We are planning to study more detailed results relating to these polynomials and numbers in a forthcoming paper.

Author Contributions: All authors contributed equally to the manuscript and typed, read, and approved the final manuscript.

Acknowledgments: The authors would like to thank the referees for their valuable comments and suggestions.

References

1. Kim, D.S.; Kim, H.Y.; Kim, D.; Kim, T. Identities of symmetry for type 2 Bernoulli and Euler polynomials. *Symmetry* **2019**, *11*, 613. [CrossRef]
2. Carlitz, L. Degenerate Stirling, Bernoulli and Eulerian numbers. *Utilitas Math.* **1979**, *15*, 51–88.
3. Kim, D.S.; Kim, H.Y.; Pyo, S.-S.; Kim, T. Some identities of special numbers and polynomials arising from p-adic integrals on $\mathbb{Z}_p$. *Adv. Differ. Equ.* **2019**, *2019*, 190. [CrossRef]
4. Kim, T. q-Volkenborn integration. *Russ. J. Math. Phys.* **2002**, *9*, 288–299.
5. Kim, D.S.; Kim, T.; Kim, H.Y.; Kwon, J. A note on type 2 q-Bernoulli and type 2 q-Euler polynomials. *J. Inequal. Appl.* **2019**, accepted. [CrossRef]
6. Araci, S.; Acikgoz, M. A note on the Frobenius-Euler numbers and polynomials associated with Bernstein polynomials. *Adv. Stud. Contemp. Math.* **2012**, *22*, 399–406.
7. Carlitz, L. Expansions of q-Bernoulli numbers. *Duke Math. J.* **1958**, *25*, 355–364. [CrossRef]
8. Jang, G.-W.; Kim, T. A note on type 2 degenerate Euler and Bernoulli polynomials. *Adv. Stud. Contemp. Math.* **2019**, *29*, 147–159.
9. Kim, D.S.; Dolgy, D.V.; Kim, D.; Kim, T. Some identities on r-central factorial numbers and r-central Bell polynomials. *Adv. Differ. Equ.* **2019**, accepted. [CrossRef]
10. Diarra, B. The p-adic q-distributions, Advances in ultrametric analysis. *Contemp. Math.* **2013**, *596*, 45–62.
11. Kim, T.; Kim, D.S. Degenerate Laplace transform and degenerate gamma function. *Russ. J. Math. Phys.* **2017**, *24*, 241–248. [CrossRef]
12. Kim, D.S.; Kim, T.; Jang, G.-W. A note on degenerate Stirling numbers of the first kind. *Proc. Jangjeon Math. Soc.* **2018**, *21*, 393–404.
13. Kim, T. On degenerate q-Bernoulli polynomials. *Bull. Korean Math. Soc.* **2016**, *53*, 1149–1156. [CrossRef]
14. Schikhof, W.H. *Ultrametric Calculus. An Introduction to p-Adic Analysis*; Cambridge Studies in Advanced Mathematics, 4; Cambridge University Press: Cambridge, UK, 1984.

6

Connection Problem for Sums of Finite Products of Chebyshev Polynomials of the Third and Fourth Kinds

Dmitry Victorovich Dolgy [1], Dae San Kim [2], Taekyun Kim [3,4] and Jongkyum Kwon [5,*]

1 Institute of National Sciences, Far Eastern Federal University, 690950 Vladivostok, Russia; d_dol@mail.ru
2 Department of Mathematics, Sogang University, Seoul 04107, Korea; dskim@sogang.ac.kr
3 Department of Mathematics, College of Science, Tianjin Polytechnic University, Tianjin 300160, China; tkkim@kw.ac.kr
4 Department of Mathematics, Kwangwoon University, Seoul 01897, Korea
5 Department of Mathematics Education and ERI, Gyeongsang National University, Jinju 52828, Gyeongsangnamdo, Korea
* Correspondence: mathkjk26@gnu.ac.kr.

Abstract: This paper treats the connection problem of expressing sums of finite products of Chebyshev polynomials of the third and fourth kinds in terms of five classical orthogonal polynomials. In fact, by carrying out explicit computations each of them are expressed as linear combinations of Hermite, generalized Laguerre, Legendre, Gegenbauer, and Jacobi polynomials which involve some terminating hypergeometric functions ${}_2F_0$, ${}_2F_1$, and ${}_3F_2$.

Keywords: sums of finite products of Chebyshev polynomials of the third and fourth kinds; Hermite; generalized Laguerre; Legendre; Gegenbauer; Jacobi

MSC: 11B83; 33C05; 33C20; 33C45

---

1. Introduction and Preliminaries

In this section, we will recall some basic facts about relevant orthogonal polynomials that will be needed throughout this paper. For this, we will first fix some notations. For any nonnegative integer n, the falling factorial polynomials $(x)_n$ and the rising factorial polynomials $<x>_n$ are respectively given by

$$(x)_n = x(x-1)\cdots(x-n+1),\ (n \geq 1),\ (x)_0 = 1, \tag{1}$$

$$<x>_n = x(x+1)\cdots(x+n-1),\ (n \geq 1),\ <x>_0 = 1. \tag{2}$$

The two factorial polynomials are evidently related by

$$(-1)^n(x)_n = <-x>_n,\ (-1)^n <x>_n = (-x)_n. \tag{3}$$

$$\frac{(2n-2s)!}{(n-s)!} = \frac{2^{2n-2s}(-1)^s <\frac{1}{2}>_n}{<\frac{1}{2}-n>_s},\ (n \geq s \geq 0). \tag{4}$$

$$B(x,y) = \int_0^1 t^{x-1}(1-t)^{y-1}dt = \frac{\Gamma(x)\Gamma(y)}{\Gamma(x+y)},\ (Re\, x, Re\, y > 0). \tag{5}$$

$$\Gamma(n+\frac{1}{2}) = \frac{(2n)!\sqrt{\pi}}{2^{2n}n!},\ (n \geq 0). \tag{6}$$

$$\frac{\Gamma(x+1)}{\Gamma(x+1-n)} = (x)_n, \frac{\Gamma(x+n)}{\Gamma(x)} =< x >_n, \ (n \geq 0), \tag{7}$$

where $\Gamma(x)$ and $B(x,y)$ are the gamma and beta functions respectively.

The hypergeometric function is defined by

$$\begin{aligned} &{}_pF_q(a_1, \cdots, a_p; b_1, \cdots, b_q; x) \\ &= \sum_{n=0}^{\infty} \frac{< a_1 >_n \cdots < a_p >_n}{< b_1 >_n \cdots < b_q >_n} \frac{x^n}{n!}. \end{aligned} \tag{8}$$

We are now ready to state some basic facts about Chebyshev polynomials of the third kind $V_n(x)$, those of the fourth kind $W_n(x)$, Hermite polynomials $H_n(x)$, generalized (extended) Laguerre polynomials $L_n^{\alpha}(x)$, Legendre polynomials $P_n(x)$, Gegenbauer polynomials $C_n^{(\lambda)}(x)$, and Jacobi polynomials $P_n^{(\alpha,\beta)}$. Chebyshev polynomials are diversely used in approximation theory and numerical analysis, Hermite polynomials appear as the eigenfunctions of the harmonic oscillator in quantum mechanics, Laguerre polynomials have important applications to the solution of Schrödinger's equation for the hydrogen atom, Legendre polynomials can be used to write the Coulomb potential as a series, Gegenbauer polynomials play an important role in the constructive theory of spherical functions and Jacobi polynomials occur in the solution to the equations of motion of the symmetric top. All the necessary facts on those special polynomials can be found in [1–9]. For the full accounts of this fascinating area of orthogonal polynomials, the reader may refer to [10–13].

The above special polynomials are given in terms of generating functions by

$$F(t,x) = \frac{1-t}{1-2xt+t^2} = \sum_{n=0}^{\infty} V_n(x)t^n, \tag{9}$$

$$G(t,x) = \frac{1+t}{1-2xt+t^2} = \sum_{n=0}^{\infty} W_n(x)t^n, \tag{10}$$

$$e^{2xt-t^2} = \sum_{n=0}^{\infty} H_n(x)\frac{t^n}{n!}, \tag{11}$$

$$(1-t)^{-\alpha-1}\exp(-\tfrac{xt}{1-t}) = \sum_{n=0}^{\infty} L_n^{\alpha}(x)t^n, \ (\alpha > -1), \tag{12}$$

$$(1-2xt+t^2)^{-\frac{1}{2}} = \sum_{n=0}^{\infty} P_n(x)t^n, \tag{13}$$

$$\frac{1}{(1-2xt+t^2)^{\lambda}} = \sum_{n=0}^{\infty} C_n^{(\lambda)}(x)t^n, \ (\lambda > -\frac{1}{2}, \lambda \neq 0, |t| < 1, |x| \leq 1), \tag{14}$$

$$\frac{2^{\alpha+\beta}}{R(1-t+R)^{\alpha}(1+t+R)^{\beta}} = \sum_{n=0}^{\infty} P_n^{(\alpha,\beta)}(x)t^n, \tag{15}$$

$$(R = \sqrt{1-2xt+t^2}, \ \alpha, \beta > -1).$$

Explicit expressions for the above special polynomials are as in the following:

$$\begin{aligned} V_n(x) &= {}_2F_1(-n, n+1; \tfrac{1}{2}; \tfrac{1-x}{2}) \\ &= \sum_{l=0}^{n} \binom{2n-l}{l} 2^{n-l}(x-1)^{n-l}, \end{aligned} \tag{16}$$

$$\begin{aligned} W_n(x) &= (2n+1){}_2F_1(-n, n+1; \tfrac{3}{2}; \tfrac{1-x}{2}) \\ &= (2n+1)\sum_{l=0}^{n} \frac{2^{n-l}}{2n-2l+1}\binom{2n-l}{l}(x-1)^{n-l}, \end{aligned} \tag{17}$$

$$H_n(x) = n!\sum_{l=0}^{[\frac{n}{2}]} \frac{(-1)^l}{l!(n-2l)!}(2x)^{n-2l}, \tag{18}$$

$$\begin{aligned} L_n^{\alpha}(x) &= \frac{<\alpha+1>_n}{n!}{}_1F_1(-n, \alpha+1; x) \\ &= \sum_{l=0}^{n} \frac{(-1)^l\binom{n+\alpha}{n-l}}{l!}x^l, \end{aligned} \tag{19}$$

$$\begin{aligned} P_n(x) &= {}_2F_1(-n, n+1; 1; \tfrac{1-x}{2}) \\ &= \frac{1}{2^n}\sum_{l=0}^{[\frac{n}{2}]}(-1)^l\binom{n}{l}\binom{2n-2l}{n}x^{n-2l}, \end{aligned} \tag{20}$$

$$\begin{aligned} C_n^{(\lambda)}(x) &= \binom{n+2\lambda-1}{n}{}_2F_1(-n, n+2\lambda; \lambda+\tfrac{1}{2}; \tfrac{1-x}{2}) \\ &= \sum_{k=0}^{[\frac{n}{2}]}(-1)^k\frac{\Gamma(n-k+\lambda)}{\Gamma(\lambda)k!(n-2k)!}(2x)^{n-2k}, \end{aligned} \tag{21}$$

$$\begin{aligned} P_n^{(\alpha,\beta)}(x) &= \frac{<\alpha+1>_n}{n!}{}_2F_1(-n, 1+\alpha+\beta+n; \alpha+1; \tfrac{1-x}{2}) \\ &= \sum_{k=0}^{n}\binom{n+\alpha}{n-k}\binom{n+\beta}{k}(\tfrac{x-1}{2})^k(\tfrac{x+1}{2})^{n-k}. \end{aligned} \tag{22}$$

Next, we state Rodrigues-type formulas for Hermite and generalized Laguerre polynomials and Rodrigues' formulas for Legendre, Gegenbauer and Jacobi polynomials.

$$H_n(x) = (-1)^n e^{x^2}\frac{d^n}{dx^n}e^{-x^2}, \tag{23}$$

$$L_n^{\alpha}(x) = \frac{1}{n!}x^{-\alpha}e^x\frac{d^n}{dx^n}(e^{-x}x^{n+\alpha}), \tag{24}$$

$$P_n(x) = \frac{1}{2^n n!}\frac{d^n}{dx^n}(x^2-1)^n, \tag{25}$$

$$(1-x^2)^{\lambda-\frac{1}{2}}C_n^{(\lambda)}(x) = \frac{(-2)^n}{n!}\frac{<\lambda>_n}{<n+2\lambda>_n}\frac{d^n}{dx^n}(1-x^2)^{n+\lambda-\frac{1}{2}}, \tag{26}$$

$$(1-x)^{\alpha}(1+x)^{\beta}P_n^{(\alpha,\beta)}(x) = \frac{(-1)^n}{2^n n!}\frac{d^n}{dx^n}(1-x)^{n+\alpha}(1+x)^{n+\beta}. \tag{27}$$

The last thing we want to mention is the orthogonalities with respect to various weight functions enjoyed by Hermite, generalized Laguerre, Legendre, Gegenbauer and Jacobi polynomials.

$$\int_{-\infty}^{\infty} e^{-x^2}H_n(x)H_m(x)\,dx = 2^n n!\sqrt{\pi}\delta_{n,m}, \tag{28}$$

$$\int_0^{\infty} x^{\alpha}e^{-x}L_n^{\alpha}(x)L_m^{\alpha}(x)\,dx = \frac{1}{n!}\Gamma(\alpha+n+1)\delta_{n,m}, \tag{29}$$

$$\int_{-1}^{1} P_n(x)P_m(x)\,dx = \frac{2}{2n+1}\delta_{n,m}, \tag{30}$$

$$\int_{-1}^{1}(1-x^2)^{\lambda-\frac{1}{2}}C_n^{(\lambda)}(x)C_m^{(\lambda)}(x)\,dx = \frac{\pi 2^{1-2\lambda}\Gamma(n+2\lambda)}{n!(n+\lambda)\Gamma(\lambda)^2}\delta_{n,m}, \tag{31}$$

$$\int_{-1}^{1}(1-x)^{\alpha}(1+x)^{\beta}P_n^{(\alpha,\beta)}(x)P_m^{(\alpha,\beta)}(x)\,dx = \frac{2^{\alpha+\beta+1}\Gamma(n+\alpha+1)\Gamma(n+\beta+1)}{(2n+\alpha+\beta+1)\Gamma(n+\alpha+\beta+1)\Gamma(n+1)}\delta_{n,m}. \tag{32}$$

For convenience, let us put

$$\gamma_{n,r}(x) = \sum_{l=0}^{n}\sum_{i_1+i_2+\cdots+i_{r+1}=l}\binom{r-1+n-l}{r-1}V_{i_1}(x)V_{i_2}(x)\cdots V_{i_{r+1}}(x),\ (n\geq 0, r\geq 1), \tag{33}$$

$$\mathcal{E}_{n,r}(x) = \sum_{l=0}^{n}\sum_{i_1+i_2+\cdots+i_{r+1}=l}(-1)^{n-l}\binom{r-1+n-l}{r-1}W_{i_1}(x)W_{i_2}(x)\cdots W_{i_{r+1}}(x), (n\geq 0, r\geq 1). \tag{34}$$

We observe here that both $\gamma_{n,r}(x)$ and $\mathcal{E}_{n,r}(x)$ have degree n.

In this paper, we will consider the connection problem of expressing the sums of finite products in (33) and (34) as linear combinations of $H_n(x), L_n^{\alpha}(x), P_n(x), C_n^{(\lambda)}(x)$, and $P_n^{(\alpha,\beta)}(x)$. These will be done by performing explicit computations based on Proposition 1. We observe here that the formulas in Proposition 1 follow from their orthogonalities, Rodrigues' and Rodrigues-type formulas and integration by parts.

Our main results are the following Theorems 1 and 2.

Theorem 1. *Let n, r be any integers with $n\geq 0, r\geq 1$. Then we have the following.*

$$\begin{aligned}
&\sum_{l=0}^{n}\sum_{i_1+i_2+\cdots+i_{r+1}=l}\binom{r-1+n-l}{r-1}V_{i_1}(x)V_{i_2}(x)\cdots V_{i_{r+1}}(x)\\
&= \frac{(2n+2r)!}{r!4^{n+r}(n+r-\frac{1}{2})_{n+r}}\sum_{k=0}^{n}\frac{(-2)^k}{(n-k)!}\\
&\quad\times\sum_{j=0}^{[\frac{k}{2}]}\frac{{}_2F_1(2j-k,\frac{1}{2}-n-r;-2n-2r;2)}{j!4^j(k-2j)!}H_{n-k}(x) &(35)\\
&= \frac{1}{r!}\sum_{k=0}^{n}\sum_{l=0}^{k}\frac{(-2)^{n-l}(2n+2r-l)!(n+r-l)!}{l!(2n+2r-2l)!(k-l)!}\\
&\quad\times {}_2F_0(l-k,n-k+\alpha+1;-;1)L_{n-k}^{\alpha}(x) &(36)\\
&= \frac{(-1)^n n!(2n+2r)!}{r!4^r(n+r-\frac{1}{2})_{n+r}}\sum_{k=0}^{n}\frac{(-1)^k(2k+1)}{(n-k)!(n+k+1)!}\\
&\quad\times {}_3F_2(k-n,\frac{1}{2}-n-r,-n-k-1;-2n-2r,-n;1)P_k(x) &(37)\\
&= \frac{(-1)^n(2n+2r)!4^{\lambda-r}\Gamma(\lambda)\Gamma(n+\lambda+\frac{1}{2})}{\sqrt{\pi}r!(n+r-\frac{1}{2})_{n+r}}\sum_{k=0}^{n}\frac{(-1)^k(k+\lambda)}{\Gamma(n+k+2\lambda+1)(n-k)!}\\
&\quad\times {}_3F_2(k-n,\frac{1}{2}-n-r,-n-k-2\lambda;-2n-2r,-n-\lambda+\frac{1}{2};1)C_k^{(\lambda)}(x) &(38)\\
&= \frac{(-1)^n(2n+2r)!\Gamma(n+\alpha+1)}{r!4^r(n+r-\frac{1}{2})_{n+r}}\sum_{k=0}^{n}\frac{(-1)^k(2k+\alpha+\beta+1)\Gamma(k+\alpha+\beta+1)}{(n-k)!\Gamma(\alpha+k+1)\Gamma(n+k+\alpha+\beta+2)}\\
&\quad\times {}_3F_2(k-n,\frac{1}{2}-n-r,-n-k-\alpha-\beta-1;-2n-2r,-n-\alpha;1)P_k^{(\alpha,\beta)}(x). &(39)
\end{aligned}$$

Theorem 2. *Let n, r be any integers with $n \geq 0$, $r \geq 1$. Then we have the following.*

$$\begin{aligned}
&\sum_{l=0}^{n} \sum_{i_1+i_2+\cdots+i_{r+1}=l} (-1)^{n-l} \binom{r-1+n-l}{r-1} W_{i_1}(x) W_{i_2}(x) \cdots W_{i_{r+1}}(x) \\
&= \frac{(2n+1)(2n+2r)!}{r! 2^{2n+2r+1} (n+r+\frac{1}{2})_{n+r+1}} \sum_{k=0}^{n} \frac{(-2)^k}{(n-k)!} \\
&\times \sum_{j=0}^{[\frac{k}{2}]} \frac{{}_2F_1(2j-k, -n-r-\frac{1}{2}; -2n-2r; 2)}{j! 4^j (k-2j)!} H_{n-k}(x) \qquad (40) \\
&= \frac{(2n+1)}{r!} \sum_{k=0}^{n} \sum_{l=0}^{k} \frac{(-2)^{n-l} (2n+2r-l)! (n+r-l)!}{l! (2n+2r-2l+1)! (k-l)!} \\
&\times {}_2F_0(l-k, n-k+\alpha+1; -; 1) L_{n-k}^{\alpha}(x) \qquad (41)
\end{aligned}$$

$$\begin{aligned}
&= \frac{(-1)^n n! (2n+1)(2n+2r)!}{r! 2^{2r+1} (n+r+\frac{1}{2})_{n+r+1}} \sum_{k=0}^{n} \frac{(-1)^k (2k+1)}{(n-k)! (n+k+1)!} \\
&\times {}_3F_2(k-n, -n-r-\frac{1}{2}, -n-k-1; -2n-2r, -n; 1) P_k(x) \qquad (42) \\
&= \frac{(-1)^n (2n+2r)! 2^{2\lambda-2r-1} (2n+1) \Gamma(\lambda) \Gamma(n+\lambda+\frac{1}{2})}{\sqrt{\pi} r! (n+r+\frac{1}{2})_{n+r+1}} \\
&\times \sum_{k=0}^{n} \frac{(-1)^k (k+\lambda)}{\Gamma(n+k+2\lambda+1)(n-k)!} \\
&\times {}_3F_2(k-n, -n-r-\frac{1}{2}, -n-k-2\lambda; -2n-2r, -n-\lambda+\frac{1}{2}; 1) C_k^{(\lambda)}(x) \qquad (43) \\
&= \frac{(-1)^n (2n+2r)! (2n+1) \Gamma(n+\alpha+1)}{r! 2^{2r+1} (n+r+\frac{1}{2})_{n+r+1}} \\
&\times \sum_{k=0}^{n} \frac{(-1)^k (2k+\alpha+\beta+1) \Gamma(k+\alpha+\beta+1)}{(n-k)! \Gamma(\alpha+k+1) \Gamma(n+k+\alpha+\beta+2)} \\
&\times {}_3F_2(k-n, -n-r-\frac{1}{2}, -n-k-\alpha-\beta-1; -2n-2r, -n-\alpha; 1) P_k^{(\alpha,\beta)}(x). \qquad (44)
\end{aligned}$$

Before closing the section, we are going to mention some of previous results on the related connection problems. The papers [14–16] treat the connection problem of expressing sums of finite products of Bernoulli, Euler and Genocchi polynomials in terms of Bernoulli polynomials. In fact, they were carried out by deriving Fourier series expansions for the functions closely related to those sums of finite products. Moreover, the same was done for the sums of finite products of Chebyshev polynomials of the second kind and of Fibonacci polynomials in [17].

Along the same line as the present paper, sums of finite products of Chebyshev polynomials of the second kind and Fibonacci polynomials were expressed in [18] as linear combinations of the orthogonal polynomials $H_n(x), L_n^{\alpha}(x), P_n(x), C_n^{(\lambda)}(x)$, and $P_n^{(\alpha,\beta)}(x)$. Also, the connection problem of expressing in terms of all kinds of Chebyshev polynomials were done for sums of finite products of Chebyshev polynomials of the second, third and fourth kinds and of Fibonacci, Legendre and Laguerre polynomials in [19–21].

Finally, we let the reader refer to [22,23] for some applications of Chebyshev polynomials.

2. Proof of Theorem 1

First, we will state Propositions 1 and 2 that will be needed in showing Theorems 1 and 2.

The results in (a), (b), (c), (d) and (e) in Proposition 1 follow respectively from (3.7) of [4], (2.3) of [4] (see also (2.4) of [2]), (2.3) of [5], (2.3) of [3] and (2.7) of [7]. They can be derived from their orthogonalities in (28)–(32), Rodrigues-type and Rodrigues' formulas in (23)–(27) and integration by parts.

Proposition 1. *Let $q(x) \in \mathbb{R}[x]$ be a polynomial of degree n. Then we have the following.*

$$
\begin{aligned}
(a)\quad & q(x)=\sum_{k=0}^{n} C_{k,1}H_k(x), \text{ where} \\
& C_{k,1}=\frac{(-1)^k}{2^k k!\sqrt{\pi}}\int_{-\infty}^{\infty} q(x)\frac{d^k}{dx^k}e^{-x^2}dx, \\
(b)\quad & q(x)=\sum_{k=0}^{n} C_{k,2}L_k^{\alpha}(x), \text{ where} \\
& C_{k,2}=\frac{1}{\Gamma(\alpha+k+1)}\int_{0}^{\infty} q(x)\frac{d^k}{dx^k}(e^{-x}x^{k+\alpha})\,dx, \\
(c)\quad & q(x)=\sum_{k=0}^{n} C_{k,3}P_k(x), \text{ where} \\
& C_{k,3}=\frac{2k+1}{2^{k+1}k!}\int_{-1}^{1} q(x)\frac{d^k}{dx^k}(x^2-1)^k dx, \\
(d)\quad & q(x)=\sum_{k=0}^{n} C_{k,4}C_k^{(\lambda)}(x), \text{ where} \\
& C_{k,4}=\frac{(k+\lambda)\Gamma(\lambda)}{(-2)^k\sqrt{\pi}\Gamma(k+\lambda+\frac{1}{2})}\int_{-1}^{1} q(x)\frac{d^k}{dx^k}(1-x^2)^{k+\lambda-\frac{1}{2}}dx, \\
(e)\quad & q(x)=\sum_{k=0}^{n} C_{k,5}P_n^{(\alpha,\beta)}(x), \text{ where} \\
& C_{k,5}=\frac{(-1)^k(2k+\alpha+\beta+1)\Gamma(k+\alpha+\beta+1)}{2^{\alpha+\beta+k+1}\Gamma(\alpha+k+1)\Gamma(\beta+k+1)} \\
& \quad\times\int_{-1}^{1} q(x)\frac{d^k}{dx^k}(1-x)^{k+\alpha}(1+x)^{k+\beta}dx.
\end{aligned}
$$

The following proposition will be used in showing Theorems 1 and 2. In fact, (a) is needed for (35) and (40), (b) for (39) and (44), and (b) or (c) for (37), (38), (42) and (43).

Proposition 2. *The following holds true.*

(a) *For any nonnegative integer m,*

$$
\int_{-\infty}^{\infty} x^m e^{-x^2}dx=\begin{cases} 0, & \text{if } m\equiv 1 \pmod 2, \\ \frac{m!\sqrt{\pi}}{(\frac{m}{2})!2^m}, & \text{if } m\equiv 0 \pmod 2, \end{cases}
$$

(b) *For any real numbers $r, s > -1$, we have*

$$
\int_{-1}^{1}(1-x)^r(1+x)^s dx = 2^{r+s+1}\frac{\Gamma(r+1)\Gamma(s+1)}{\Gamma(r+s+2)},
$$

(c) *For any real numbers r, s with $r+s > -1, s > -1$, we have*

$$
\int_{-1}^{1}(1-x)^r(1-x^2)^s dx = 2^{r+2s+1}\frac{\Gamma(r+s+1)\Gamma(s+1)}{\Gamma(r+2s+2)}.
$$

Proof.
(a) This is an easy exercise.
(c) This follows from (b) with r replaced by $r+s$.
(b) This follows from the change of variable $1+x=2y$ and (5). □

The following lemma can be obtained by differentiating (9), as was shown in [24].

Lemma 1. *Let n, r be integers with $n \geq 0, r \geq 1$. Then we have the following identity.*

$$\sum_{l=0}^{n} \sum_{i_1+i_2+\cdots+i_{r+1}=l} \binom{r-1+n-l}{r-1} V_{i_1}(x)V_{i_2}(x)\cdots V_{i_{r+1}}(x) = \frac{1}{2^r r!} V_{n+r}^{(r)}(x), \tag{45}$$

where the inner sum runs over all nonnegative integers $i_1, i_2, \cdots i_{r+1}$, with $i_1+i_2+\cdots+i_{r+1}=l$.

From (16), we see that the rth derivative of $V_n(x)$ is given by

$$V_n^{(r)}(x) = \sum_{l=0}^{n-r} \binom{2n-l}{l} 2^{n-l}(n-l)_r(x-1)^{n-l-r}. \tag{46}$$

Especially, we have

$$V_{n+r}^{(r+k)}(x) = \sum_{l=0}^{n-k} \binom{2n+2r-l}{l} 2^{n+r-l}(n+r-l)_{r+k}(x-1)^{n-k-l}. \tag{47}$$

Now, we are ready to prove Theorem 1. As (38) and (39) can be shown similarly to (43) and (44) in the next section, we will show only (35), (36) and (37). With $\gamma_{n,r}(x)$ as in (33), we let

$$\gamma_{n,r}(x) = \sum_{k=0}^{n} C_{k,1}H_k(x). \tag{48}$$

Then, from (a) of Proposition 1, (45), (47), and integration by parts k times, we obtain

$$\begin{aligned} C_{k,1} &= \frac{(-1)^k}{2^k k!\sqrt{\pi}} \int_{-\infty}^{\infty} \gamma_{n,r}(x) \frac{d^k}{dx^k} e^{-x^2} dx \\ &= \frac{(-1)^k}{2^{k+r} k! r! \sqrt{\pi}} \int_{-\infty}^{\infty} V_{n+r}^{(r)}(x) \frac{d^k}{dx^k} e^{-x^2} dx \\ &= \frac{1}{2^{k+r} k! r! \sqrt{\pi}} \int_{-\infty}^{\infty} V_{n+r}^{(r+k)}(x) e^{-x^2} dx \\ &= \frac{1}{2^{k+r} k! r! \sqrt{\pi}} \sum_{l=0}^{n-k} \binom{2n+2r-l}{l} 2^{n+r-l}(n+r-l)_{r+k} \\ &\times \int_{-\infty}^{\infty} (x-1)^{n-k-l} e^{-x^2} dx. \end{aligned} \tag{49}$$

Before proceeding further, by making use of (a) in Proposition 2, we note that

$$\begin{aligned}
&\int_{-\infty}^{\infty}(x-1)^m e^{-x^2}dx \\
&=\sum_{s=0}^{m}\binom{m}{s}(-1)^{m-s}\int_{-\infty}^{\infty}x^s e^{-x^2}dx \\
&=\sum_{\substack{0\le s\le m\\ s\equiv 0 \pmod 2}}\binom{m}{s}(-1)^{m-s}\frac{s!\sqrt{\pi}}{(\frac{s}{2})!2^s} \\
&=(-1)^m\sqrt{\pi}\sum_{j=0}^{[\frac{m}{2}]}\binom{m}{2j}\frac{(2j)!}{j!2^{2j}},\ (m\ge 0).
\end{aligned} \tag{50}$$

From (48)–(50), and after simplifications, we have

$$\begin{aligned}
\gamma_{n,r}(x) &= \frac{1}{r!}\sum_{k=0}^{n}\frac{(-2)^k}{(n-k)!}\sum_{l=0}^{k}\sum_{j=0}^{[\frac{k-l}{2}]}\frac{(-\frac{1}{2})^l(2n+2r-l)!(n+r-l)!}{l!(2n+2r-2l)!(k-l-2j)!j!4^j}H_{n-k}(x) \\
&= \frac{1}{r!}\sum_{k=0}^{n}\frac{(-2)^k}{(n-k)!}\sum_{j=0}^{[\frac{k}{2}]}\frac{1}{j!4^j}\sum_{l=0}^{k-2j}\frac{(-\frac{1}{2})^l(2n+2r-l)!(n+r-l)!}{l!(2n+2r-2l)!(k-l-2j)!}H_{n-k}(x) \\
&= \frac{(2n+2r)!}{r!4^{n+r}<\frac{1}{2}>_{n+r}}\sum_{k=0}^{n}\frac{(-2)^k}{(n-k)!}\sum_{j=0}^{[\frac{k}{2}]}\frac{1}{j!4^j(k-2j)!} \\
&\quad\times\sum_{l=0}^{k-2j}\frac{2^l<2j-k>_l<\frac{1}{2}-n-r>_l}{l!<-2n-2r>_l}H_{n-k}(x) \\
&= \frac{(2n+2r)!}{r!4^{n+r}(n+r-\frac{1}{2})_{n+r}}\sum_{k=0}^{n}\frac{(-2)^k}{(n-k)!} \\
&\quad\times\sum_{j=0}^{[\frac{k}{2}]}\frac{{}_2F_1(2j-k,\frac{1}{2}-n-r;-2n-2r;2)}{j!4^j(k-2j)!}H_{n-k}(x).
\end{aligned} \tag{51}$$

This shows (35) of Theorem 1.
Next, we let

$$\gamma_{n,r}(x)=\sum_{k=0}^{n}C_{k,2}L_k^{\alpha}(x). \tag{52}$$

Then, from (b) of Proposition 1, (45), (47) and integration by parts k times, we get

$$
\begin{aligned}
C_{k,2} &= \frac{1}{2^r r!\Gamma(\alpha+k+1)} \int_0^\infty V_{n+r}^{(r)}(x) \frac{d^k}{dx^k}(e^{-x}x^{k+\alpha})dx \\
&= \frac{(-1)^k}{2^r r!\Gamma(\alpha+k+1)} \int_0^\infty V_{n+r}^{(r+k)}(x) e^{-x}x^{k+\alpha}dx \\
&= \frac{(-1)^k}{2^r r!\Gamma(\alpha+k+1)} \sum_{l=0}^{n-k} \binom{2n+2r-l}{l} 2^{n+r-l}(n+r-l)_{r+k} \\
&\times \int_0^\infty (x-1)^{n-k-l} e^{-x}x^{k+\alpha}dx \\
&= \frac{(-1)^k}{2^r r!\Gamma(\alpha+k+1)} \sum_{l=0}^{n-k} \binom{2n+2r-l}{l} 2^{n+r-l}(n+r-l)_{r+k} \\
&\times \sum_{s=0}^{n-k-l} \binom{n-k-l}{s} (-1)^{n-k-l-s}\Gamma(s+k+\alpha+1) \\
&= \frac{(-1)^k}{2^r r!} \sum_{l=0}^{n-k} \binom{2n+2r-l}{l} 2^{n+r-l}(n+r-l)_{r+k} \\
&\times \sum_{s=0}^{n-k-l} \binom{n-k-l}{s} (-1)^{n-k-l-s} < k+\alpha+1 >_s \\
&= \frac{1}{r!} \sum_{l=0}^{n-k} \frac{(2n+2r-l)!(-2)^{n-l}(n+r-l)!}{l!(2n+2r-2l)!(n-k-l)!} \\
&\times \sum_{s=0}^{n-k-l} \frac{1}{s!} < k+l-n >_s < k+\alpha+1 >_s \\
&= \frac{1}{r!} \sum_{l=0}^{n-k} \frac{(2n+2r-l)!(-2)^{n-l}(n+r-l)!}{l!(2n+2r-2l)!(n-k-l)!} \\
&\times {}_2F_0(k+l-n, k+\alpha+1; -; 1).
\end{aligned} \tag{53}
$$

Combining (52)–(53), we finally have

$$
\begin{aligned}
\gamma_{n,r}(x) = \frac{1}{r!} \sum_{k=0}^{n} \sum_{l=0}^{k} & \frac{(2n+2r-l)!(-2)^{n-l}(n+r-l)!}{l!(2n+2r-2l)!(k-l)!} \\
& \times {}_2F_0(l-k, n-k+\alpha+1; -; 1) L_{n-k}^{\alpha}(x).
\end{aligned} \tag{54}
$$

This completes the proof for (36) of Theorem 1.
Finally, let us put

$$
\gamma_{n,r}(x) = \sum_{k=0}^{n} C_{k,3} P_k(x). \tag{55}
$$

Then, from (c) of Proposition 1, (45), (47) and integration by parts k times, we have

$$
\begin{aligned}
C_{k,3} &= \frac{2k+1(-1)^k}{2^{k+r+1}k!r!} \int_{-1}^{1} V_{n+r}^{(r+k)}(x)(x^2-1)^k dx \\
&= \frac{(2k+1)(-1)^k}{2^{k+r+1}k!r!} \sum_{l=0}^{n-k} \binom{2n+2r-l}{l} 2^{n+r-l}(n+r-l)_{r+k} \\
&\times \int_{-1}^{1} (x-1)^{n-k-l}(x^2-1)^k dx.
\end{aligned} \tag{56}
$$

By making use of (c) in Proposition 2 and after simplifications, from (56) we obtain

$$\begin{aligned} C_{k,3} &= \frac{(-1)^{n+k}(2k+1)}{r!}\sum_{l=0}^{n-k} \\ &\times \frac{(-1)^l 4^{n-l}(2n+2r-l)!(n+r-l)!(n-l)!}{l!(2n+2r-2l)!(n-k-l)!(n+k-l+1)!} \\ &= \frac{(-1)^n(2n+2r)!n!}{r!4^r(n+r-\frac{1}{2})_{n+r}}\frac{(-1)^k(2k+1)}{(n-k)!(n+k+1)!} \\ &\times \sum_{l=0}^{n-k}\frac{<k-n>_l<\frac{1}{2}-n-r>_l<-n-k-1>_l}{l!<-2n-2r>_l<-n>_l} \\ &= \frac{(-1)^n(2n+2r)!n!}{r!4^r(n+r-\frac{1}{2})_{n+r}}\frac{(-1)^k(2k+1)}{(n-k)!(n+k+1)!} \\ &\times {}_3F_2(k-n,\frac{1}{2}-n-r,-n-k-1;-2n-2r,-n;1). \end{aligned} \tag{57}$$

From (55) and (57), we get

$$\begin{aligned} \gamma_{n,r}(x) = &\frac{(-1)^n(2n+2r)!n!}{r!4^r(n+r-\frac{1}{2})_{n+r}}\sum_{k=0}^{n}\frac{(-1)^k(2k+1)}{(n-k)!(n+k+1)!} \\ &\times {}_3F_2(k-n,\frac{1}{2}-n-r,-n-k-1;-2n-2r,-n;1)P_k(x). \end{aligned} \tag{58}$$

This proves (37) of Theorem 1.

3. Proof of Theorem 2

Here we will show only (43) and (44) in Theorem 2, as (40)–(42) can be shown analogously to the proofs for (35)–(37), respectively. The following can be derived by differentiating the Equation (10) and is stated in [24].

Lemma 2. *Let n, r be integers with $n \geq 0$, $r \geq 1$. Then we have the following identity.*

$$\sum_{l=0}^{n}\sum_{i_1+i_2+\cdots+i_{r+1}=l}(-1)^{n-l}\binom{r-1+n-l}{r-1}W_{i_1}(x)W_{i_2}(x)\cdots W_{i_{r+1}}(x) = \frac{1}{2^r r!}W_{n+r}^{(r)}(x), \tag{59}$$

where the inner sum runs over all nonnegative integers $i_1, i_2, \cdots, i_{r+1}$, with $i_1 + i_2 + \cdots + i_{r+1} = l$.

From (17), the rth derivative of $W_n(x)$ is given by

$$W_n^{(r)}(x) = (2n+1)\sum_{l=0}^{n-r}\frac{2^{n-l}}{2n+1-2l}\binom{2n-l}{l}(n-l)_r(x-1)^{n-l-r}. \tag{60}$$

In particular, we have

$$W_{n+r}^{(r+k)}(x) = (2n+1)\sum_{l=0}^{n-k}\frac{2^{n+r-l}}{2n+2r+1-2l}\binom{2n+2r-l}{l}(n+r-l)_{r+k}(x-1)^{n-k-l}. \tag{61}$$

With $\mathcal{E}_{n,r}(x)$ as in (34), we let

$$\mathcal{E}_{n,r}(x) = \sum_{k=0}^{n} C_{k,4}C_k^{(\alpha)}(x). \tag{62}$$

Then, from (d) of Proposition 1, (59), (61) and integration by parts k times, we get

$$\begin{aligned}
C_{k,4} &= \frac{(k+\lambda)\Gamma(\lambda)}{2^{k+r}r!\sqrt{\pi}\Gamma(k+\lambda+\frac{1}{2})} \\
&\times \int_{-1}^{1} W_{n+r}^{(r+k)}(x)(1-x^2)^{k+\lambda-\frac{1}{2}}dx \\
&= \frac{(k+\lambda)\Gamma(\lambda)(2n+1)}{2^{k+r}r!\sqrt{\pi}\Gamma(k+\lambda+\frac{1}{2})} \\
&\times \sum_{l=0}^{n-k} \frac{2^{n+r-l}}{2n+2r+1-2l}\binom{2n+2r-l}{l}(n+r-l)_{r+k} \\
&\times \int_{-1}^{1}(x-1)^{n-k-l}(1-x^2)^{k+\lambda-\frac{1}{2}}dx \\
&= \frac{(k+\lambda)\Gamma(\lambda)(2n+1)(-2)^{n-k}}{r!\sqrt{\pi}\Gamma(k+\lambda+\frac{1}{2})} \\
&\times \sum_{l=0}^{n-k} \frac{(-\frac{1}{2})^l(2n+2r-l)!(n+r-l)!}{(2n+2r-2l+1)l!(2n+2r-2l)!(n-k-l)!} \\
&\times \int_{-1}^{1}(1-x)^{n-k-l}(1-x^2)^{k+\lambda-\frac{1}{2}}dx.
\end{aligned} \tag{63}$$

Invoking (c) of Proposition 2 and after simplifications, from (63) we obtain

$$\begin{aligned}
C_{k,4} &= \frac{(-1)^{n-k}(k+\lambda)\Gamma(\lambda)(2n+1)2^{2n+2\lambda+1}\Gamma(n+\lambda+\frac{1}{2})(2n+2r)!}{\Gamma(n+k+2\lambda+1)(n-k)!r!\sqrt{\pi}} \\
&\times \sum_{l=0}^{n-k} \frac{(-\frac{1}{4})^l(2n+2r-l)!(n+r-l+1)!(n-k)!(n+k+2\lambda)_l}{l!(2n+2r)!(2n+2r-2l+2)!(n-k-l)!(n+\lambda-\frac{1}{2})_l} \\
&= \frac{(-1)^k(k+\lambda)\Gamma(\lambda)(2n+1)2^{2\lambda-2r-1}(-1)^n\Gamma(n+\lambda+\frac{1}{2})(2n+2r)!}{\Gamma(n+k+2\lambda+1)(n-k)!r!\sqrt{\pi}(n+r+\frac{1}{2})_{n+r+1}} \\
&\times \sum_{l=0}^{n-k} \frac{<k-n>_l<-n-r-\frac{1}{2}>_l<-n-k-2\lambda>_l}{l!<-2n-2r>_l<-n-\lambda+\frac{1}{2}>_l} \\
&= \frac{(-1)^k(k+\lambda)\Gamma(\lambda)(2n+1)2^{2\lambda-2r-1}(-1)^n\Gamma(n+\lambda+\frac{1}{2})(2n+2r)!}{\Gamma(n+k+2\lambda+1)(n-k)!r!\sqrt{\pi}(n+r+\frac{1}{2})_{n+r+1}} \\
&\times {}_3F_2(k-n,-n-r-\frac{1}{2},-n-k-2\lambda;-2n-2r,-n-\lambda+\frac{1}{2};1).
\end{aligned} \tag{64}$$

From (62) and (64), we have

$$\begin{aligned}
\mathcal{E}_{n,r}(x) &= \frac{\Gamma(\lambda)(2n+1)2^{2\lambda-2r-1}(-1)^n\Gamma(n+\lambda+\frac{1}{2})(2n+2r)!}{r!\sqrt{\pi}(n+r+\frac{1}{2})_{n+r+1}} \\
&\times \sum_{k=0}^{n} \frac{(-1)^k(k+\lambda)}{\Gamma(n+k+2\lambda+1)(n-k)!} \\
&\times {}_3F_2(k-n,-n-r-\frac{1}{2},-n-k-2\lambda;-2n-2r,-n-\lambda+\frac{1}{2};1)C_k^{(\alpha)}(x).
\end{aligned} \tag{65}$$

This shows (43) of Theorem 2.

Next, we let

$$\mathcal{E}_{n,r}(x) = \sum_{k=0}^{n} C_{k,5}P_n^{(\alpha,\beta)}(x). \tag{66}$$

Then, from (e) of Proposition 1, and (59), (61), and integrating by parts k times, we obtain

$$\begin{aligned}C_{k,5} &= \frac{(2k+\alpha+\beta+1)\Gamma(k+\alpha+\beta+1)}{2^{\alpha+\beta+k+r+1}r!\Gamma(\alpha+k+1)\Gamma(\beta+k+1)} \\ &\times \int_{-1}^{1} W_{n+r}^{(r+k)}(x)(1-x)^{k+\alpha}(1+x)^{k+\beta}dx \\ &= \frac{(2k+\alpha+\beta+1)\Gamma(k+\alpha+\beta+1)(2n+1)}{2^{\alpha+\beta+k+r+1}r!\Gamma(\alpha+k+1)\Gamma(\beta+k+1)} \\ &\times \sum_{l=0}^{n-k} \frac{2^{n+r-l}}{2n+2r-2l+1}\binom{2n+2r-l}{l}(n+r-l)_{r+k}(-1)^{n-k-l} \\ &\times \int_{-1}^{1}(1-x)^{n+\alpha-l}(1+x)^{k+\beta}dx.\end{aligned} \tag{67}$$

By exploiting (b) in Proposition 2 and after simplifications, from (67) we get

$$\begin{aligned}C_{k,5} &= \frac{(-1)^{n-k}(2k+\alpha+\beta+1)\Gamma(k+\alpha+\beta+1)2^{2n+1}(2n+1)\Gamma(n+\alpha+1)}{\Gamma(\alpha+k+1)\Gamma(n+k+\alpha+\beta+2)r!} \\ &\times \sum_{l=0}^{n-k} \frac{(-\frac{1}{4})^l(2n+2r-l)!(n+r-l+1)!(n+k+\alpha+\beta+1)_l}{l!(2n+2r-2l+2)!(n-k-l)!(n+\alpha)_l} \\ &= \frac{(-1)^{n-k}(2k+\alpha+\beta+1)\Gamma(k+\alpha+\beta+1)(2n+1)\Gamma(n+\alpha+1)(2n+2r)!}{\Gamma(\alpha+k+1)\Gamma(n+k+\alpha+\beta+2)(n-k)!r!2^{2r+1}(n+r+\frac{1}{2})_{n+r+1}} \\ &\times \sum_{l=0}^{n-k} \frac{<k-n>_l<-n-r-\frac{1}{2}>_l<-n-k-\alpha-\beta-1>_l}{l!<-2n-2r>_l<-n-\alpha>_l} \\ &= \frac{(-1)^{n-k}(2k+\alpha+\beta+1)\Gamma(k+\alpha+\beta+1)(2n+1)\Gamma(n+\alpha+1)(2n+2r)!}{\Gamma(\alpha+k+1)\Gamma(n+k+\alpha+\beta+2)(n-k)!r!2^{2r+1}(n+r+\frac{1}{2})_{n+r+1}} \\ &\times {}_3F_2(k-n,-n-r-\frac{1}{2},-n-k-\alpha-\beta-1;-2n-2r,-n-\alpha;1).\end{aligned} \tag{68}$$

Thus, from (66) and (68), we have

$$\begin{aligned}\mathcal{E}_{n,r}(x) &= \frac{(-1)^n(2n+1)\Gamma(n+\alpha+1)(2n+2r)!}{r!2^{2r+1}(n+r+\frac{1}{2})_{n+r+1}} \\ &\times \sum_{k=0}^{n} \frac{(-1)^k(2k+\alpha+\beta+1)\Gamma(k+\alpha+\beta+1)}{\Gamma(\alpha+k+1)\Gamma(n+k+\alpha+\beta+2)(n-k)!} \\ &\times {}_3F_2(k-n,-n-r-\frac{1}{2},-n-k-\alpha-\beta-1;-2n-2r,-n-\alpha;1)P_n^{(\alpha,\beta)}(x).\end{aligned}$$

4. Conclusions

In this paper, we considered sums of finite products of Chebyshev polynomials of the third and fourth kinds and expressed each of them in terms of five orthogonal polynomials. Indeed, by explicit computations we expressed each of them as linear combinations of Hermite, generalized Laguerre, Legendre, Gegenbauer and Jacobi polynomials which involve some terminating hypergeometric functions. This can be viewed as a generalization of the classical linearization problem. In general, the linearization problem deals with determining the coefficients in the expansion of the products of two polynomials $a_m(x)$ and $b_n(x)$ in terms of an arbitrary polynomial sequence $\{p_k(x)\}_{k\geq 0}$:

$$a_m(x)b_n(x) = \sum_{k=0}^{m+n} c_{mn}(k)p_k(x).$$

Those sums of finite products were also represented by all kinds of Chebyshev polynomials in [20]. In addition, the same had been done for sums of finite products of Chebyshev polynomials of the first and second kinds, Fibonacci polynomials, Legendre polynomials, Laguerre polynomials and Lucas polynomials.

Author Contributions: T.K. and D.S.K. conceived the framework and structured the whole paper; T.K. wrote the paper; J.K. typed; D.V.D. checked for typos; D.S.K.completed the revision of the article.

Acknowledgments: We would like to thank the referees for their valuable comments and suggestions.

References

1. Kim, D.S.; Dolgy, D.V.; Kim, T.; Rim, S.H. Identities involving Bernoulli and Euler polynomials arising from Chebyshev polynomials. *Proc. Jangjeon Math. Soc.* **2012**, *15*, 361–370.
2. Kim, D.S.; Kim, T.; Dolgy, D.V. Some identities on Laguerre polynomials in connection with Bernoulli and Euler numbers. *Discret. Dyn. Nat. Soc.* **2012**, *2012*, 619197. [CrossRef]
3. Kim, D.S.; Kim, T.; Rim, S.-H. Some identities involving Gegenbauer polynomials. *Adv. Differ. Equ.* **2012**, *2012*, 219. [CrossRef]
4. Kim, D.S.; Kim, T.; Rim, S.-H.; Lee, S.H. Hermite polynomials and their applications associated with Bernoulli and Euler numbers. *Discret. Dyn. Nat. Soc.* **2012**, *2012*, 974632. [CrossRef]
5. Kim, D.S.; Rim, S.-H.; Kim, T. Some identities on Bernoulli and Euler polynomials arising from orthogonality of Legendre polynomials. *J. Inequal. Appl.* **2012**, *2012*, 227. [CrossRef]
6. Kim, T.; Kim, D.S. Extended Laguerre polynomials associated with Hermite, Bernoulli, and Euler numbers and polynomials. *Abstr. Appl. Anal.* **2012**, *2012*, 957350. [CrossRef]
7. Kim, T.; Kim, D.S.; Dolgy, D.V. Some identities on Bernoulli and Hermite polynomials associated with Jacobi polynomials. *Discret. Dyn. Nat. Soc.* **2012**, *2012*, 584643. [CrossRef]
8. Kim, T.; Kim, D.S. Identities for degenerate Bernoulli polynomials and Korobov polynomials of the first kind. *Sci. China Math.* **2018**. [CrossRef]
9. Kim, T.; Kim, D.S.; Kwon, J.; Gang, -W. J. Sums of finite products of Legendre and Laguerre polynomials by Chebyshev polynomials. *Adv. Stud. Contemp. Math. (Kyungshang)* **2018**, *28*, 551–565.
10. Andrews, G.E.; Askey, R.; Roy, R. *Special Functions*; Encyclopedia of Mathematics and its Applications 71; Cambridge University Press: Cambridge, UK, 1999.
11. Beals, R.; Wong, R. *Special Functions and Orthogonal Polynomials*; Cambridge Studies in Advanced Mathematics 153; Cambridge University Press: Cambridge, UK, 2016.
12. Wang, Z.X.; Guo, D.R. *Special Functions*; Translated by Guo, D.R., Xia, X.J.; World Scientific Publishing Co., Inc.: Teaneck, NJ, USA, 1989.
13. Mason, J.C.; Handscomb, D.C. *Chebyshev Polynomials*; Chapman & Hall/CRC: Boca Raton, FL, USA, 2003.
14. Agarwal, R.P.; San Kim, D.; Kim, T.; Kwon, J. Sums of finite products of Bernoulli functions. *Adv. Differ. Equ.* **2017**, *2017*, 237. [CrossRef]
15. Kim, T.; San Kim, D.; Jang, G.W.; Kwon, J. Sums of finite products of Euler functions. In *Advances in Real and Complex Analysis with Applications*; Trends in Mathematics; Birkhäuser: Singapore, 2017; pp. 243–260.
16. Kim, T.; Kim, D.S.; Jang, L.C.; Jang, G.-W. Sums of finite products of Genocchi functions. *Adv. Differ. Equ.* **2017**, *2017*, 268. [CrossRef]
17. Kim, T.; San Kim, D.; Dolgy, D.V.; Park, J.W. Sums of finite products of Chebyshev polynomials of the second kind and Fibonacci polynomials. *J. Inequal. Appl.* **2018**, *2018*, 148. [CrossRef] [PubMed]
18. Kim, T.; Kim, D.S.; Kwon, J.; Dolgy, D.V. Expressing sums of finite products of Chebyshev polynomials of the second kind and Fibonacci polynomials by several orthogonal polynomials. *Mathematics* **2018**, *6*, 210. [CrossRef]
19. Kim, T.; Dolgy, D.V.; Kim, D.S. Representing sums of finite products of Chebyshev polynomials of the second kind and Fibonacci polynomials in terms of Chebyshev polynomials. *Adv. Stud. Contemp. Math. (Kyungshang)* **2018**, *28*, 321–335.

20. Kim, T.; Kim, D.S.; Dolgy, D.V.; Ryoo, C.-S. Representing sums of finite products of Chebyshev polynomials of the third and fourth kinds by Chebyshev polynomials. *Symmetry* **2018**, *10*, 258. [CrossRef]
21. Kim, T.; Kim, D.; Victorovich, D.; Ryoo, C. Representing sums of finite products of Legendre and Laguerre polynomials by Chebyshev polynomials. *Adv. Stud. Contemp. Math. (Kyungshang)* **2018**, *28*, in press.
22. Doha, E.H.; Abd-Elhameed, W.M.; Alsuyuti, M.M. On using third and fourth kinds Chebyshev polynomials for solving the integrated forms of high odd-order linear boundary value problems. *J. Egyptian Math. Soc.* **2015**, *23*, 397–405. [CrossRef]
23. Mason, J.C. Chebyshev polynomials of the second, third and fourth kinds in approximation, indefinite integration, and integral transforms. *J. Comput. Appl. Math.* **1993**, *49*, 169–178. [CrossRef]
24. Kim, T.; San Kim, D.; Dolgy, D.V.; Kwon, J. Sums of finite products of Chebyshev polynomials of the third and fourth kinds. *Adv. Differ. Equ.* **2018**, *2018*, 283. [CrossRef]

7

On p-Adic Fermionic Integrals of q-Bernstein Polynomials Associated with q-Euler Numbers and Polynomials

Lee-Chae Jang [1], Taekyun Kim [2,*], Dae San Kim [3] and Dmitry Victorovich Dolgy [4,5]

1 Graduate School of Education, Konkuk University, Seoul 143-701, Korea; Lcjang@konkuk.ac.kr
2 Department of Mathematics, Kwangwoon University, Seoul 139-701, Korea
3 Department of Mathematics, Sogang University, Seoul 121-742, Korea; dskim@sogang.ac.kr
4 Kwangwoon Institute for Advanced Studies, Kwangwoon University, Seoul 139-701, Korea; d_dol@mail.ru
5 Institute of National Sciences, Far Eastern Federal University, 690950 Vladivostok, Russia
* Correspondence: tkkim@kw.ac.kr
† 2010 *Mathematics Subject Classication.* 11B83; 11S80.

Abstract: We study a q-analogue of Euler numbers and polynomials naturally arising from the p-adic fermionic integrals on $\mathbb{Z}_p$ and investigate some properties for these numbers and polynomials. Then we will consider p-adic fermionic integrals on $\mathbb{Z}_p$ of the two variable q-Bernstein polynomials, recently introduced by Kim, and demonstrate that they can be written in terms of the q-analogues of Euler numbers. Further, from such p-adic integrals we will derive some identities for the q-analogues of Euler numbers.

Keywords: two variable q-Berstein polynomial; two variable q-Berstein operator; q-Euler number; q-Euler polynomial

---

1. Introduction

As is well known, the classical Bernstein polynomial of order n for $f \in C[0,1]$ is defined by (see [1–3]),

$$\mathbb{B}_n(f|x) = \sum_{k=0}^{n} f\left(\frac{k}{n}\right) B_{k,n}(x), \quad 0 \le x \le 1, \tag{1}$$

where $\mathbb{B}_n$ is called the Bernstein operater of order n, and (see [4–30]),

$$B_{k,n}(x) = \binom{n}{k} x^k (1-x)^{n-k}, \quad n,k \ge 0, \tag{2}$$

are called the Bernstein basis polynomials (or Bernstein polynomials of degree n).

The Weierstrass approximation theorem states that every continuous function defined on $[0,1]$ can be uniformly approximated as closely as desired by a polynomial function. In 1912, S. N. Bernstein explicitly constructed a sequence of polynomials that uniformly approximates any given continuous function f on $[0,1]$. Namely, he showed that $\mathbb{B}_n(f|x)$ tends uniformly to $f(x)$ as $n \to \infty$ on $[0,1]$ (see [3]). For $q \in \mathbb{C}$, with $0 < |q| < 1$, and $n,k \in \mathbb{Z}_{\ge 0}$, with $n \ge k$, the q-Bernstein polynomials of degree n are defined by Kim as (see [8])

$$B_{k,n}(x,q) = \binom{n}{k} [x]_q^k [1-x]_{\frac{1}{q}}^{n-k}, \tag{3}$$

where $[x]_q = \frac{1-q^x}{1-q}$. For any $f \in C[0,1]$, the q-Bernstein operator of order n is defined as

$$\mathbb{B}_{n,q}(f|x) = \sum_{k=0}^{n} f\left(\frac{k}{n}\right) B_{k,n}(x,q) = \sum_{k=0}^{n} f\left(\frac{k}{n}\right) \binom{n}{k} [x]_q^k [1-x]_{\frac{1}{q}}^{n-k}, \tag{4}$$

where $0 \leq x \leq 1$, and $n \in \mathbb{Z}_{\geq 0}$, (see [8,13]).

Here we note that a different version of q-Bernstein polynomials from Kim's was introduced earlier in 1997 by Phillips (see [22]). His q-Bernstein polynomial of order n for f is defined by

$$\mathbb{B}_n(f,q;x) = \sum_{k=0}^{n} f(\frac{[k]_q}{[n]_q}) \begin{bmatrix} n \\ k \end{bmatrix}_q x^k \prod_{s=0}^{n-1-k} (1-q^s x),$$

where f is a function defined on $[0,1]$, q is any positive real number, and

$$\begin{bmatrix} n \\ k \end{bmatrix}_q = \frac{[n]_q!}{[k]_q![n-k]_q!}, \quad [n]_q! = [1]_q[2]_q \ldots [n]_q, (n \geq 1), \quad [0]_q! = 1.$$

The properties of Phillips' q-Bernstein polynomilas for $q \in (0,1)$ were treated for example in [6,15,16,22–24], while those for $q > 1$ were developed for instance in [17–20].

A Bernoulli trial is an experiment where only two outcomes, whether a particular event A occurs or not, are possible. Flipping of coin is an example of Bernoulli trial, where only two outcomes, namely head and tail, are possible. Conventionally, it is said that the outcome of Bernoulli trial is a "success" if A occurs and a "failure" otherwise. Let $P_n(k)$ denote the probability of k successes in n independent Bernoulli trials with the probability of success r. Then it is given by the binomial probability law

$$P_n(k) = \binom{n}{k} r^k (1-r)^{n-k}, \text{ for } k = 0,1,2,\cdots,n. \tag{5}$$

We remark here that the Bernstein basis is the probability mass function of the binomial distribution from the definition of Bernstein polynomials. Let p be a fixed odd prime number. Throughout this paper, we will use the notations $\mathbb{Z}_p, \mathbb{Q}_p$, and $\mathbb{C}_p$ to denote respectively the ring of p-adic integers, the field of p-adic rational numbers and the completion of the algebraic closure of $\mathbb{Q}_p$. The p-adic norm in $\mathbb{C}_p$ is normalized in such a way that $|p|_p = \frac{1}{p}$. It is known that in terms of the recurrence relation the Euler numbers are given as follows (see [10,11]):

$$E_0 = 1, \ (E+1)^n + E_n = 2\delta_{0,n}, \tag{6}$$

where $\delta_{n,k}$ is the Kronecker's symbol. Then the Euler polynomials can be given as (see [10])

$$E_n(x) = \sum_{l=0}^{n} \binom{n}{l} E_l x^{n-l}, \tag{7}$$

The q-Euler polynomials, considered by L. Carlitz, are given by

$$\mathcal{E}_{0,q} = 1, \ q(q\mathcal{E}_q + 1)^n + \mathcal{E}_{n,q} = \begin{cases} [2]_q, & \text{if } n = 0, \\ 0, & \text{if } n > 0, \end{cases} \tag{8}$$

with the understanding that $\mathcal{E}_q^n$ is to be replaced by $\mathcal{E}_{n,q}$ (see [5]). Note that $\lim_{q\to 1} \mathcal{E}_{n,q} = E_n, \ (n \geq 0)$.

Let $f(x)$ be a continuous function on $\mathbb{Z}_p$. Then the p-adic fermionic integral on $\mathbb{Z}_p$ is defined by Kim as (see [12])

$$I_{-1}(f) = \int_{\mathbb{Z}_p} f(x) d\mu_{-1}(x) = \lim_{N\to\infty} \sum_{x=0}^{p^N-1} f(x)(-1)^x, \tag{9}$$

where we notice that $\mu_{-1}(x + p^N\mathbb{Z}_p) = (-1)^x$ is a measure.

From (9), we note that (see [12])

$$I_{-1}(f_1) + I_{-1}(f) = 2f(0), \tag{10}$$

where $f_1(x) = f(x+1)$. By (10), we easily get (see [25])

$$\int_{\mathbb{Z}_p} (x+y)^n d\mu_{-1}(y) = E_n(x), \ (n \geq 0), \tag{11}$$

When $x = 0$, we note that $\int_{\mathbb{Z}_p} x^n d\mu_{-1}(x) = E_n, \ (n \geq 0)$. Let q be an indeterminate in $\mathbb{C}_p$ with $|1-q|_p < p^{-\frac{1}{p-1}}$. Taking (11) into consideration, we may investigate a q-analogue of Euler polynomials which are given by (see [12,26])

$$\int_{\mathbb{Z}_p} [x+y]_q^n d\mu_{-1}(y) = E_{n,q}(x), \ (n \geq 0), \tag{12}$$

When $x = 0$, $E_{n,q} = E_{n,q}(0)$, $(n \geq 0)$ are said to be the q-Euler numbers. Using (9), we can easily see that

$$\begin{aligned} \int_{\mathbb{Z}_p} [x]_q^n d\mu_{-1}(x) &= \frac{2}{(1-q)^n} \sum_{l=0}^{n} \binom{n}{l} (-1)^l \frac{1}{1+q^l} \\ &= 2 \sum_{m=0}^{\infty} (-1)^m [m]_q^n, \ (n \geq 0). \end{aligned} \tag{13}$$

Thus, by (13), we get

$$E_{n,q} = 2 \sum_{m=0}^{\infty} (-1)^m [m]_q^n = \frac{2}{(1-q)^n} \sum_{l=0}^{n} \binom{n}{l} (-1)^l \frac{1}{1+q^l}. \tag{14}$$

For $n, k \geq 0$, with $n \geq k$, and $q \in \mathbb{C}_p$, with $|1-q|_p < p^{-\frac{1}{p-1}}$, we define the p-adic q-Bernstein polynomials as follows:

$$B_{k,n}(x,q) = \binom{n}{k} [x]_q^k [1-x]_{\frac{1}{q}}^{n-k}. \tag{15}$$

Then we consider the p-adic q-Bernstein operator defined for continuous functions f on $\mathbb{Z}_p$ and given by

$$\mathbb{B}_{n,q}(f|x) = \sum_{k=0}^{n} f\left(\frac{k}{n}\right) B_{k,n}(x,q), \ (x \in \mathbb{Z}_p). \tag{16}$$

We study a q-analogue of Euler numbers and polynomials naturally arising from the p-adic fermionic integrals on $\mathbb{Z}_p$ and investigate some properties for these numbers and polynomials. Then we will consider p-adic fermionic integrals on $\mathbb{Z}_p$ of the two variable q-Bernstein polynomials, recently introduced by Kim in [8], and demonstrate that they can be written in terms of the q-analogues of Euler numbers. Further, from such p-adic integrals we will derive some identities for the q-analogues of Euler numbers.

2. q-Bernstein Polynomials Associated with q-Euler Numbers and Polynomials

We assume that $q \in \mathbb{C}_p$, with $|1-q|_p < p^{-\frac{1}{p-1}}$, throughout this section. From (12), we notice that

$$\sum_{n=0}^{\infty} E_{n,q}(x)\frac{t^n}{n!} = \sum_{m=0}^{\infty}(-1)^m e^{[m+x]_q t}. \tag{17}$$

By (10), we get

$$\int_{\mathbb{Z}_p}[x+1]_q^n d\mu_{-1}(x) + \int_{\mathbb{Z}_p}[x]_q^n d\mu_{-1}(x) = 2\delta_{0,n}, \ (n \geq 0). \tag{18}$$

Thus, from (12), we have

$$E_{n,q}(1) + E_{n,q} = \begin{cases} 2, & \text{if } n = 0, \\ 0, & \text{if } n > 0. \end{cases} \tag{19}$$

On the other hand,

$$\begin{aligned} E_{n,q}(x) &= \int_{\mathbb{Z}_p}[x+y]_q^n d\mu_{-1}(y) \\ &= \sum_{l=0}^{n}\binom{n}{l}[x]_q^{n-l}q^{lx}\int_{\mathbb{Z}_p}[y]_q^l d\mu_{-1}(y) \\ &= \sum_{l=0}^{n}\binom{n}{l}q^{lx}E_{l,q}[x]_q^{n-l} = \left(q^x E_q + [x]_q\right)^n, \end{aligned} \tag{20}$$

with the understanding that E_q^n is to be replaced by $E_{n,q}$. From (19) and (20), we note that

$$E_{0,q} = 1, \ (qE_q+1)^n + E_{n,q} = \begin{cases} 2, & \text{if } n = 0, \\ 0, & \text{if } n > 0. \end{cases} \tag{21}$$

Now, we observe that

$$\begin{aligned} E_{n,q}(2) &= (q^2E_q + 1 + q)^n = (q(qE_q+1)+1)^n \\ &= \sum_{l=0}^{n} q^l (qE_q+1)^l \binom{n}{l} = 2 - E_{0,q} - \sum_{l=1}^{n} q^l E_{l,q}\binom{n}{l} \\ &= 2 - \sum_{l=0}^{n} q^l E_{l,q}\binom{n}{l} = 2 - (qE_q+1)^n. \end{aligned} \tag{22}$$

Now, by combining (21) with (22), we have the following theorem.

Theorem 1. *For any $n \geq 0$, we have*

$$E_{n,q}(2) = 2 + E_{n,q}, \ (n > 0), \ E_{0,q}(2) = 1. \tag{23}$$

Invoking (9), we can derive the following equation

$$\int_{\mathbb{Z}_p}[1-x+y]_{q^{-1}}^n d\mu_{-1}(y) = (-1)^n q^n \int_{\mathbb{Z}_p}[x+y]_q^n d\mu_{-1}(y), \tag{24}$$

where n is a nonnegative integer. By (12) and (24), we get

$$E_{n,q^{-1}}(1-x) = (-1)^n q^n E_{n,q}(x), \ (n > 0). \tag{25}$$

On the other hand, we have

$$\begin{aligned}\int_{\mathbb{Z}_p}[1-x]^n_{q^{-1}}d\mu_{-1}(x) &= (-1)^n q^n \int_{\mathbb{Z}_p}[x-1]^n_q d\mu_{-1}(x)\\ &= (-1)^n q^n E_{n,q}(-1),\end{aligned} \tag{26}$$

as $[-x]_{q^{-1}} = -q[x]_q$. By (25) and (26), we get

$$\int_{\mathbb{Z}_p}[1-x]^n_{q^{-1}}d\mu_{-1}(x) = (-1)^n q^n E_{n,q}(-1) = E_{n,q^{-1}}(2). \tag{27}$$

Therefore, by (23) and (27), we have

Theorem 2. *For any $n > 0$, we have*

$$\int_{\mathbb{Z}_p}[1-x]^n_{q^{-1}}d\mu_{-1}(x) = 2 + E_{n,q^{-1}}. \tag{28}$$

For $q \in \mathbb{C}_p$, with $|1-q|_p < p^{-\frac{1}{p-1}}$, and $x_1, x_1 \in \mathbb{Z}_p$, the two variable q-Bernstein polynomials are defined by

$$B_{k,n}(x_1, x_2|q) = \begin{cases} \binom{n}{k}[x_1]^k_q[1-x_2]^{n-k}_{q^{-1}}, & \text{if } n \geq k,\\ 0, & \text{if } n < k,\end{cases} \tag{29}$$

where $n, k \geq 0$. From (29), we note that

$$B_{n-k,n}(1-x_2, 1-x_1|q^{-1}) = B_{k,n}(x_1, x_2|q), B_{k,n}(x, x|q) = B_{k,n}(x, q), \tag{30}$$

where $n, k \geq 0$ and $x_1, x_2 \in \mathbb{Z}_p$. For continuous functions f on $\mathbb{Z}_p$, the two variable q-Bernstein operator of order n is defined by

$$\begin{aligned}\mathbb{B}_{n,q}(f|x_1, x_2) &= \sum_{k=0}^{n} f\left(\frac{k}{n}\right)\binom{n}{k}[x_1]^k_q[1-x_2]^{n-k}_{q^{-1}}\\ &= \sum_{k=0}^{n} f\left(\frac{k}{n}\right) B_{k,n}(x_1, x_2|q),\end{aligned} \tag{31}$$

where $n, k \in \mathbb{Z}_{\geq 0}$, and $x_1, x_2 \in \mathbb{Z}_p$. In particular, if $f = 1$, then we have

$$\begin{aligned}\mathbb{B}_{n,q}(1|x_1, x_2) &= \sum_{k=0}^{n}\binom{n}{k}[x_1]^k_q[1-x_2]^{n-k}_{q^{-1}}\\ &= (1+[x_1]_q - [x_2]_q)^n,\end{aligned} \tag{32}$$

where we used the fact

$$[1-x]_{q^{-1}} = 1 - [x]_q. \tag{33}$$

Taking the double p-adic fermionic integral on $\mathbb{Z}_p$ as in the following, we have

$$\begin{aligned}
&\int_{\mathbb{Z}_p}\int_{\mathbb{Z}_p} B_{k,n}(x_1,x_2|q)d\mu_{-1}(x_1)d\mu_{-1}(x_2)\\
&=\binom{n}{k}\int_{\mathbb{Z}_p}[x_1]_q^k d\mu_{-1}(x_1)\int_{\mathbb{Z}_p}[1-x_2]_{q^{-1}}^{n-k}d\mu_{-1}(x_2)\\
&=\begin{cases}\binom{n}{k}E_{k,q}(2+E_{n-k,q^{-1}}), & \text{if } n>k,\\ E_{k,q}, & \text{if } n=k.\end{cases}
\end{aligned} \tag{34}$$

Therefore, from (34) we obtain the next theorem.

Theorem 3. *For any $n,k\in\mathbb{Z}_{\geq 0}$, with $n\geq k$, we have*

$$\begin{aligned}
&\int_{\mathbb{Z}_p}\int_{\mathbb{Z}_p} B_{k,n}(x_1,x_2|q)d\mu_{-1}(x_1)d\mu_{-1}(x_2)\\
&=\begin{cases}\binom{n}{k}E_{n,q}(2+E_{n,q^{-1}}), & \text{if } n>k,\\ E_{k,q}, & \text{if } n=k.\end{cases}
\end{aligned} \tag{35}$$

Making the use of the definition of the two variable q-Bernstein polynomials and from (33), we notice that

$$\begin{aligned}
&\int_{\mathbb{Z}_p}\int_{\mathbb{Z}_p} B_{k,n}(x_1,x_2|q)d\mu_{-1}(x_1)d\mu_{-1}(x_2)\\
&=\sum_{l=0}^{k}\binom{n}{n-k}\binom{k}{l}(-1)^{k+l}\int_{\mathbb{Z}_p}\int_{\mathbb{Z}_p}[1-x_1]_{q^{-1}}^{k-l}[1-x_2]_{q^{-1}}^{n-k}d\mu_{-1}(x_1)d\mu_{-1}(x_2)\\
&=\binom{n}{k}\int_{\mathbb{Z}_p}[1-x_2]_{q^{-1}}^{n-k}d\mu_{-1}(x_2)\sum_{l=0}^{k}\binom{k}{l}(-1)^{k-l}\int_{\mathbb{Z}_p}[1-x_1]_{q^{-1}}^{k-l}d\mu_{-1}(x_1)\\
&=\binom{n}{k}\int_{\mathbb{Z}_p}[1-x_2]_{q^{-1}}^{n-k}d\mu_{-1}(x_2)\left\{1+\sum_{l=0}^{k-1}\binom{k}{l}(2+E_{k-l},q^{-1})\right\}.
\end{aligned} \tag{36}$$

Therefore, from (34) and (36) we deduce the following theorem.

Theorem 4. *For any $k\geq 0$, we have*

$$E_{k,q}=2(2^k-1)+\sum_{l=0}^{k}\binom{k}{l}E_{k-l,q^{-1}}. \tag{37}$$

For $m,n,k\in\mathbb{Z}_{\geq 0}$, we observe that

$$\begin{aligned}
&\int_{\mathbb{Z}_p}\int_{\mathbb{Z}_p} B_{k,n}(x_1,x_2|q)B_{k,m}(x_1,x_2|q)d\mu_{-1}(x_1)d\mu_{-1}(x_2)\\
&=\binom{n}{k}\binom{m}{k}\int_{\mathbb{Z}_p}[x_1]_q^{2k}d\mu_{-1}(x_1)\int_{\mathbb{Z}_p}[1-x_2]_{q^{-1}}^{n+m-2k}d\mu_{-1}(x_2)\\
&=\binom{n}{k}\binom{m}{k}E_{2k,q}\int_{\mathbb{Z}_p}[1-x_2]_{q^{-1}}^{n+m-2k}d\mu_{-1}(x_2).
\end{aligned} \tag{38}$$

On the other hand,

$$\begin{aligned}
&\int_{\mathbb{Z}_p}\int_{\mathbb{Z}_p} B_{k,n}(x_1,x_2|q)B_{k,m}(x_1,x_2|q)d\mu_{-1}(x_1)d\mu_{-1}(x_2) \\
&= \sum_{l=0}^{2k}\binom{n}{k}\binom{m}{k}\binom{2k}{l}(-1)^{2k-l} \\
&\times \int_{\mathbb{Z}_p}\int_{\mathbb{Z}_p}[1-x_1]_{q^{-1}}^{2k-l}[1-x_2]_{q^{-1}}^{n+m-2k}d\mu_{-1}(x_1)d\mu_{-1}(x_2) \\
&= \binom{n}{k}\binom{m}{k}\int_{\mathbb{Z}_p}[1-x_2]_{q^{-1}}^{n+m-2k}d\mu_{-1}(x_2) \\
&\times \left\{1+\sum_{l=0}^{2k-1}\binom{2k}{l}(-1)^{2k-l}\int_{\mathbb{Z}_p}[1-x_1]_{q^{-1}}^{2k-l}d\mu_{-1}(x_1)\right\}.
\end{aligned} \tag{39}$$

Hence, by (28), (38) and (39), we arrive at the following theorem.

Theorem 5. *For any $k \in \mathbb{N}$, we have*

$$E_{2k,q} = -2+\sum_{l=0}^{2k}\binom{2k}{l}(-1)^{2k-l}E_{2k-l,q^{-1}}. \tag{40}$$

Let $n_1, n_2, \ldots, n_s, k \in \mathbb{Z}_{\geq 0}$, with $s \in \mathbb{N}$. Then we clearly have

$$\begin{aligned}
&\int_{\mathbb{Z}_p}\int_{\mathbb{Z}_p}\prod_{i=1}^{s} B_{k,n_i}(x_1,x_2|q)d\mu_{-1}(x_1)d\mu_{-1}(x_2) \\
&= \prod_{i=1}^{s}\binom{n_i}{k}\int_{\mathbb{Z}_p}\int_{\mathbb{Z}_p}[x_1]_q^{sk}[1-x_2]_{q^{-1}}^{n_1+\cdots+n_s-sk}(x_1)d\mu_{-1}(x_2) \\
&= \prod_{i=1}^{s}\binom{n_i}{k}E_{sk,q}\int_{\mathbb{Z}_p}[1-x_2]_{q^{-1}}^{n_1+\cdots+n_s-sk}d\mu_{-1}(x_2).
\end{aligned} \tag{41}$$

On the other hand,

$$\begin{aligned}
&\int_{\mathbb{Z}_p}\int_{\mathbb{Z}_p}\prod_{i=1}^{s} B_{k,n_i}(x_1,x_2|q)d\mu_{-1}(x_1)d\mu_{-1}(x_2) \\
&= \sum_{l=0}^{sk}\prod_{i=1}^{s}\binom{n_i}{k}\binom{sk}{l}(-1)^{sk-l} \\
&\times \int_{\mathbb{Z}_p}\int_{\mathbb{Z}_p}[1-x_1]_{q^{-1}}^{sk-l}[1-x_2]_{q^{-1}}^{n_1+\cdots+n_s-sk}d\mu_{-1}(x_1)d\mu_{-1}(x_2).
\end{aligned} \tag{42}$$

By (41) and (42), we get

$$\begin{aligned}
E_{sk,q} &= \sum_{l=0}^{sk}\binom{sk}{l}(-1)^{sk-l}\int_{\mathbb{Z}_p}\int_{\mathbb{Z}_p}[1-x_1]_{q^{-1}}^{sk-l}d\mu_{-1}(x_1) \\
&= 1+\sum_{l=0}^{sk-1}\binom{sk}{l}(-1)^{sk-l}\int_{\mathbb{Z}_p}\int_{\mathbb{Z}_p}[1-x_1]_{q^{-1}}^{sk-l}d\mu_{-1}(x_1).
\end{aligned} \tag{43}$$

Hence (28) and (43) together yield the next theorem.

Theorem 6. *For any $s \in \mathbb{N}$, we have*

$$E_{sk,q} = -2+\sum_{l=0}^{sk}\binom{sk}{l}(-1)^{sk-l}E_{sk-l,q^{-1}}. \tag{44}$$

3. Conclusions

In the previous paper [8], the q-Bernstein polynomials were introduced as a generalization of the classical Bernstein polynomials. Here we studied some properties of a q-analogue of Euler numbers and polynomials arising from the p-adic fermionic integrals on $\mathbb{Z}_p$. Then we considered p-adic fermionic integrals on $\mathbb{Z}_p$ of the two variable q-Bernstein polynomials, recently introduced by Kim, and show that they can be expressed in terms of the q-analogues of Euler numbers. Along the same line, we can introduce a new q-Bernoulli numbers and polynomials, different from the classical Carlitz q-Bernoulli numbers $\beta_{n,q}$ and polynomials $\beta_{n,q}(x)$, by considering the Volkenborn integrals in lieu of the p-adic fermionic integrals on $\mathbb{Z}_p$. Then we may investigate Volkenborn integrals on $\mathbb{Z}_p$ of the q-Bernstein polynomials and unveil their connections with those new q-Bernoulli numbers which is our ongoing project.

Author Contributions: T.K. and D.S.K. conceived the framework and structured the whole paper; T.K. wrote the paper; L.C.J. and D.V.D. checked the results of the paper; D.S.K. and L.-C.J. completed the revision of the article.

Acknowledgments: The authors would like to express their sincere gratitude to the referees for their valuable comments which improved the original manuscript in its present form.

References

1. Açikgöz, M.; Erdal, D.; Araci, S. A new approach to q-Bernoulli numbers and q-Bernoulli polynomials related to q-Bernstein polynomials. *Adv. Differ. Equ.* **2010**, 2015, 272. [CrossRef]
2. Araci, S.; Acikgoz, M. A note on the Frobenius-Euler numbers and polynomials associated with Bernstein polynomials. *Adv. Stud. Contemp. Math. (Kyungshang)* **2012**, *22*, 399–406.
3. Bernstein, S. Démonstration du théorème de Weierstrass fondée sur le calcul des probabilities. *Commun. Soc. Math. Kharkov* **1912**, *2*, 1–2.
4. Bayad, A.; Kim, T. Identities involving values of Bernstein, q-Bernoulli, and q-Euler polynomials. *Russ. J. Math. Phys.* **2011**, *18*, 133–143. [CrossRef]
5. Carlitz, L. Degenerate Stirling, Bernoulli and Eulerian numbers. *Utilitas Math.* **1979**, *15*, 51–88.
6. Goodman, T.N.T.; Oruç, H.; Phillips, G.M. Convexity and generalized Bernstein polynomials. *Proc. Edinb. Math. Soc.* **1999**, *42*, 179–190. [CrossRef]
7. Kim, T.; Choi, J.; Kim, Y.-H. On the k-dimensional generalization of q-Bernstein polynomials. *Proc. Jangjeon Math. Soc.* **2011**, *14*, 199–207.
8. Kim, T. A note on q-Bernstein polynomials. *Russ. J. Math. Phys.* **2011**, *18*, 73–82. [CrossRef]
9. Kim, T.; Choi, J.; Kim, Y.-H. Some identities on the q-Bernstein polynomials, q-Stirling numbers and q-Bernoulli numbers. *Adv. Stud. Contemp. Math. (Kyungshang)* **2010**, *20*, 335–341.
10. Kim, T.; Lee, B.; Choi, J.; Kim, Y.-H.; Rim, S.H. On the q-Euler numbers and weighted q-Bernstein polynomials. *Adv. Stud. Contemp. Math. (Kyungshang)* **2011**, *21*, 13–18.
11. Kim, T. Identities on the weighted q-Euler numbers and q-Bernstein polynomials. *Adv. Stud. Contemp. Math. (Kyungshang)* **2012**, *22*, 7–12. [CrossRef]
12. Kim, T. A study on the q-Euler numbers and the fermionic q-integral of the product of several type q-Bernstein polynomials on $\mathbb{Z}_p$. *Adv. Stud. Contemp. Math. (Kyungshang)* **2013**, *23*, 5–11.
13. Kim, T.; Ryoo, C. S.; Yi, H. A note on q-Bernoulli numbers and q-Bernstein polynomials. *Ars Comb.* **2012**, *104*, 437–447.
14. Kurt, V. Some relation between the Bernstein polynomials and second kind Bernoulli polynomials. *Adv. Stud. Contemp. Math. (Kyungshang)* **2013**, *23*, 43–48.
15. Oruç, H.; Phillips, G.M. A generalization of Bernstein polynomials. *Proc. Edinb. Math. Soc.* **1999**, *42*, 403–413. [CrossRef]
16. Oruç, H.; Tuncer, N. On the convergence and iterates of q-Bernstein polynomials. *J. Approx. Theory* **2002**, *117*, 301–313. [CrossRef]
17. Ostrovska, S. q-Bernstein polynomials and their iterates *J. Approx. Theory* **2003** *123*, 232–255. [CrossRef]
18. Ostrovska, S. On the q-Bernstein polynomials. *Adv. Stud. Contemp. Math. (Kyungshang)* **2005**, *11*, 193–204.
19. Ostrovska, S. On the q-Bernstein polynomials of the logarithmic function in the case $q > 1$. *Math. Slovaca* **2016**, *66*, 73–78. [CrossRef]

20. Ostrovska, S. On the approximation of analytic functions by the q-Bernstein polynomials in the case $q > 1$. *Electron. Trans. Numer. Anal.* **2010**, *37*, 105–112.
21. Park, J.-W.; Pak, H.K.; Rim, S.-H.; Kim, T.; Lee, S.-H. A note on the q-Bernoulli numbers and q-Bernstein polynomials. *J. Comput. Anal. Appl.* **2013**, *15*, 722–729.
22. Phillips, G.M. Bernstein polynomials based on the q-integers. *Ann. Numer. Math.* **1997**, *4*, 511–518.
23. Phillips, G.M. A de Casteljau algorithm for generalized Bernstein polynomials. *BIT* **1997**, *37*, 232–236. [CrossRef]
24. Phillips, G.M. A generalization of the Bernstein polynomials based on the q-integers. *ANZIAM J.* **2000**, *42*, 79–86. [CrossRef]
25. Rim, S.-H.; Joung, J.; Jin, J.-H.; Lee, S.-J. A note on the weighted Carlitz's type q-Euler numbers and q-Bernstein polynomials. *Proc. Jangjeon Math. Soc.* **2012**, *15*, 195–201.
26. Ryoo, C.S. Some identities of the twisted q-Euler numbers and polynomials associated with q-Bernstein polynomials. *Proc. Jangjeon Math. Soc.* **2011**, *14*, 239–248.
27. Simsek, Y.; Gunay, M. On Bernstein type polynomials and their applications. *Adv. Differ. Equ.* **2015**, *2015*, 79. [CrossRef]
28. Siddiqui, M.A.; Agrawal, R.R.; Gupta, N. On a class of modified new Bernstein operators. *Adv. Stud. Contemp. Math. (Kyungshang)* **2014**, *24*, 97–107.
29. Srivastava, H.M.; Mursaleen, M.; Alotaibi, A.M.; Nasiruzzaman, M.; Al-Abied, A.A.H. Some approximation results involving the q-Szász-Mirakyan-Kantorovich type operators via Dunkl's generalization. *Math. Methods Appl. Sci.* **2017**, *40*, 5437–5452. [CrossRef]
30. Tunc, T.; Simsek, E. Some approximation properties of Szász-Mirakjan-Bernstein operators. *Eur. J. Pure Appl. Math.* **2014**, *7*, 419–428.

8

An Erdős-Ko-Rado Type Theorem via the Polynomial Method

Kyung-Won Hwang [1,†], Younjin Kim [2,*,†] and Naeem N. Sheikh [3,†]

1 Department of Mathematics, Dong-A University, Busan 49315, Korea; khwang@dau.ac.kr
2 Institute of Mathematical Sciences, Ewha Womans University, Seoul 03760, Korea
3 School of Sciences and Engineering, Al Akhawayn University, 53000 Ifrane, Morocco; n.sheikh@aui.ma
* Correspondence: younjinkim@ewha.ac.kr.
† These authors contributed equally to this work.

Abstract: A family $\mathcal{F}$ is an intersecting family if any two members have a nonempty intersection. Erdős, Ko, and Rado showed that $|\mathcal{F}| \leq \binom{n-1}{k-1}$ holds for a k-uniform intersecting family $\mathcal{F}$ of subsets of $[n]$. The Erdős-Ko-Rado theorem for non-uniform intersecting families of subsets of $[n]$ of size at most k can be easily proved by applying the above result to each uniform subfamily of a given family. It establishes that $|\mathcal{F}| \leq \binom{n-1}{k-1} + \binom{n-1}{k-2} + \cdots + \binom{n-1}{0}$ holds for non-uniform intersecting families of subsets of $[n]$ of size at most k. In this paper, we prove that the same upper bound of the Erdős-Ko-Rado Theorem for k-uniform intersecting families of subsets of $[n]$ holds also in the non-uniform family of subsets of $[n]$ of size at least k and at most $n-k$ with one more additional intersection condition. Our proof is based on the method of linearly independent polynomials.

Keywords: Erdős-Ko-Rado theorem; intersecting families; polynomial method

---

1. Introduction

Let $[n]$ be the set $\{1, 2, \cdots, n\}$. A family $\mathcal{F}$ of subsets of $[n]$ is *intersecting* if $F \cap F'$ is non-empty for all $F, F' \in \mathcal{F}$. A family $\mathcal{F}$ of subsets of $[n]$ is *t-intersecting* if $|F \cap F'| \geq t$ holds for any $F, F' \in \mathcal{F}$. A family $\mathcal{F}$ is *k-uniform* if it is a collection of k-subsets of $[n]$. In 1961, Erdős, Ko, and Rado [1] were interested in obtaining an upper bound on the maximum size that an intersecting k-uniform family can have and proved the following theorem which bounds the cardinality of an intersecting k-uniform family.

Theorem 1 (Erdős-Ko-Rado Theorem [1]). *If $n \geq 2k$ and $\mathcal{F}$ is an intersecting k-uniform family of subsets of $[n]$, then*

$$|\mathcal{F}| \leq \binom{n-1}{k-1}.$$

Erdős-Ko-Rado Theorem is an important result of extremal set theory and has been an inspiration for various generalizations by many authors for over 50 years. Erdős, Ko, and Rado [1] also proved that there exists an integer $n_0(k,t)$ such that if $n \geq n_0(k,t)$, then the maximum size of a t-intersecting k-uniform family of subsets of $[n]$ is $\binom{n-t}{k-t}$. The following generalization of the Erdős-Ko-Rado Theorem was proved by Frankl [2] for $t \geq 15$, and was completed by Wilson [3] for all t. It establishes that the generalized EKR theorem is true if $n \geq (k-t+1)(t+1)$.

Theorem 2 (Generalized Erdős-Ko-Rado Theorem [2,3]). *If $n \geq (k-t+1)(t+1)$ and $\mathcal{F}$ is a t-intersecting k-uniform family of subsets of $[n]$, then we have*

$$|\mathcal{F}| \leq \binom{n-t}{k-t}.$$

The Erdős-Ko-Rado Theorem can be restated as follows.

Theorem 3 (Erdős-Ko-Rado Theorem [1]). *If $\mathcal{F}$ is a family of subsets F_i of $[n]$ with $|F_i| = k$ and $|F_i| \leq n-k$ that satisfies the following two conditions, for $i \neq j$*

(a) $1 \leq |F_i \cap F_j| \leq k-1$
(b) $1 \leq |F_i \cap F_j^c| \leq k-1$

then we have

$$|\mathcal{F}| \leq \binom{n-1}{k-1}.$$

2. Results

The following EKR-type theorem for non-uniform intersecting families of subsets of $[n]$ of size at most k can be easily proved by applying Theorem 3 to each uniform subfamily of the given non-uniform family.

Theorem 4. *If $\mathcal{F}$ is a family of subsets F_i of $[n]$, with $|F_i| \leq k$ and $n \geq 2k$, that satisfies the following two conditions, for $i \neq j$*

(a) $1 \leq |F_i \cap F_j| \leq k-1$
(b) $1 \leq |F_i \cap F_j^c| \leq k-1$

then we have

$$|\mathcal{F}| \leq \binom{n-1}{k-1} + \binom{n-1}{k-2} + \cdots + \binom{n-1}{0}.$$

In 2014, Alon, Aydinian, and Huang [4] gave the following strengthening of the bounded rank Erdős-Ko-Rado theorem by obtaining the same upper bound under a weaker condition as follows.

Theorem 5 (Alon, Aydinian, and Huang [4]). *Let $\mathcal{F}$ be a family of subsets of $[n]$ of size at most k, $1 \leq k \leq n-1$. Suppose that for every two subsets $A, B \in \mathcal{F}$, if $A \cap B = \emptyset$, then $|A| + |B| \leq k$. Then we have*

$$|\mathcal{F}| \leq \binom{n-1}{k-1} + \binom{n-1}{k-2} + \cdots + \binom{n-1}{0}.$$

Since the bound $\binom{n-1}{k-1} + \binom{n-1}{k-2} + \cdots + \binom{n-1}{0}$ is much larger than $\binom{n-1}{k-1}$, this leads to the following interesting question: when is it possible to get the same bound as in the Erdős-Ko-Rado theorem for uniform intersecting families for the non-uniform intersecting families? We answer this question in the main result of this paper, where we prove that the same upper bound of the EKR Theorem for k-uniform intersecting families of subsets of $[n]$ also holds in the non-uniform family of subsets of $[n]$ of size at least k and at most $n-k$ with one more additional intersection condition, as follows.

Theorem 6. *If $\mathcal{F}$ is a family of subsets F_i of $[n]$ with $k \leq |F_i| \leq n-k$ that satisfies the following three conditions, for $i \neq j$*

(a) $1 \leq |F_i \cap F_j| \leq k-1$
(b) $1 \leq |F_i \cap F_j^c| \leq k-1$
(c) $1 \leq |F_i^c \cap F_j^c| \leq k-1$

then we have

$$|\mathcal{F}| \leq \binom{n-1}{k-1}.$$

Please note that if we remove the third condition in Theorem 6, we get the same bound of the Erdős-Ko-Rado theorem for k-uniform intersecting families under the same condition for subsets of $[n]$ that are of size at least k and at most $n-k$.

Erdős-Ko-Rado Theorem is a seminal result in extremal combinatorics and has been proved by various methods (see a survey in [5]). There have been many results that have generalized EKR in various ways over the decades. The aim of this paper is to give a generalization of the EKR Theorem to non-uniform families with some extra conditions. Our proof is based on the method of linearly independent multilinear polynomials.

Our paper is organized as follows. In Section 3, we will introduce our main tool, the method of linearly independent multilinear polynomials. In Section 4, we will give the proof of our main result, Theorem 6.

3. Polynomial Method

The method of linearly independent polynomials is one of the most powerful methods for counting the number of sets in various combinatorial settings. In this method, we correspond multilinear polynomials to the sets and then prove that these polynomials are linearly independent in some space. In 1975, Ray-Chaudhuri and Wilson [6] obtained the following result by using the method of linearly independent polynomials.

Theorem 7 (Ray-Chaudhuri and Wilson [6]). *Let $l_1, l_2, \cdots, l_s < n$ be nonnegative integers. If $\mathcal{F}$ is a k-uniform family of subsets of $[n]$ such that $|A \cap B| \in L = \{l_1, l_2, \cdots, l_s\}$ holds for every pair of distinct subsets $A, B \in \mathcal{F}$, then $|\mathcal{F}| \leq \binom{n}{s}$ holds.*

In 1981, Frankl and Wilson [7] obtained the following nonuniform version of the Ray-Chaudhuri-Wilson Theorem using the polynomial method. Their proof is given underneath.

Theorem 8 (Frankl and Wilson [7]). *Let $l_1, l_2, \cdots, l_s < n$ be nonnegative integers. If $\mathcal{F}$ is a family of subsets of $[n]$ such that $|A \cap B| \in L = \{l_1, l_2, \cdots, l_s\}$ holds for every pair of distinct subsets $A, B \in \mathcal{F}$, then $|\mathcal{F}| \leq \sum_{k=0}^{s} \binom{n}{k}$ holds.*

Proof. Let x be the n-tuple of variables $x_1, x_2, \cdots, x_n$, where x_i takes the values only 0 and 1. Then all the polynomials we will work with have the relation $x_i^2 = x_i$ in their domain. Let $F_1, F_2, \cdots, F_m$ be the distinct sets in $\mathcal{F}$, listed in non-decreasing order according to their sizes. We define the characteristic vector $v_i = (v_{i_1}, v_{i_2}, \cdots, v_{i_n})$ of F_i such that $v_{i_j} = 1$ if $j \in F_i$ and $v_{i_j} = 0$ if $j \notin F_i$. We consider the following multilinear polynomial

$$f_i(x) = \prod_{l \in L,\ l < |F_i|} (v_i \cdot x - l)$$

where $x = (x_1, x_2, \cdots, x_n)$.

Then we obtain that $f_i(v_i) \neq 0$ and $f_i(v_j) = 0$ for $j < i$. As the vectors v_i are $0-1$ vectors, we have an another multilinear polynomial $g_i(x)$ such that $f_i(x) = g_i(x)$ holds for all $x \in \{0,1\}^n$ by substituting x_k for the powers of x_k, where $k = 1, 2, \cdots, n$. Then it is easy to see that the polynomials $g_1, g_2, \cdots, g_m$ are linearly independent over $\mathbb{R}$. Since the dimension of n-variable multilinear polynomials of degree at most s is $\sum_{k=0}^{s} \binom{n}{k}$, we have

$$|\mathcal{F}| \leq \sum_{k=0}^{s} \binom{n}{k}$$

finishing the proof of Theorem 8. □

In the same paper, Frankl and Wilson [7] obtained the following modular version of Theorem 7.

Theorem 9 (Frankl and Wilson [7]). *If $\mathcal{F}$ is a family of subsets of $[n]$ such that $|A \cap B| \equiv l \in L$ (mod p) holds for every pair of distinct subsets $A, B \in \mathcal{F}$, then $|\mathcal{F}| \leq \binom{n}{|L|}$ holds.*

In 1983, Deza, Frankl and Singhi [8] obtained the following modular version of Theorem 8.

Theorem 10 (Deza, Frankl and Singhi [8]). *If $\mathcal{F}$ is a family of subsets of $[n]$ such that $|A \cap B| \equiv l \in L$ (mod p) holds for every pair of distinct subsets $A, B \in \mathcal{F}$ and $|A| \not\equiv l$ (mod p) for every $A \in \mathcal{F}$, then $|\mathcal{F}| \leq \sum_{i=0}^{|L|} \binom{n}{i}$ holds.*

In 1991, Alon, Babai, and Suzuki [9] gave another modular version of Theorem 8 by replacing the condition of nonuniformity with the condition that the members of $\mathcal{F}$ have r different sizes as follows. Their proof was also based on the polynomial method.

Theorem 11 (Alon-Babai-Suzuki [9]). *Let $K = \{k_1, k_2, \cdots, k_r\}$ and $L = \{l_1, l_2, \cdots, l_s\}$ be two disjoint subsets of $\{0, 1, \cdots, p-1\}$, where p is a prime, and let $\mathcal{F}$ be a family of subsets of $[n]$ whose sizes modulo p are in the set K, and $|A \cap B| \pmod{p} \in L$ holds for every distinct two subsets A, B in $\mathcal{F}$, then the largest size of such a family $\mathcal{F}$ is $\binom{n}{s} + \binom{n}{s-1} + \cdots + \binom{n}{s-r+1}$ under the conditions $r(s-r+1) \leq p-1$ and $n \geq s + \max_{1 \leq i \leq r} k_i$.*

In the same paper, Alon, Babai, and Suzuki [9] also conjectured that the statement of Theorem 11 remains true if the condition $r(s-r+1) \leq p-1$ is dropped. Recently Hwang and Kim [10] proved this conjecture of Alon, Babai and Suzuki (1991), using the method of linearly independent polynomials. This result is as follows.

Theorem 12 (Hwang and Kim [10]). *Let $K = \{k_1, k_2, \cdots, k_r\}$ and $L = \{l_1, l_2, \cdots, l_s\}$ be two disjoint subsets of $\{0, 1, \cdots, p-1\}$, where p is a prime, and let $\mathcal{F}$ be a family of subsets of $[n]$ whose sizes modulo p are in the set K, and $|A \cap B| \pmod{p} \in L$ for every distinct two subsets A, B in $\mathcal{F}$, then the largest size of such a family $\mathcal{F}$ is $\binom{n}{s} + \binom{n}{s-1} + \cdots + \binom{n}{s-r+1}$ under the only condition that $n \geq s + \max_{1 \leq i \leq r} k_i$.*

The method of linearly independent polynomials has also been used to prove many intersection theorems about set families by Blokhuis [11], Chen and Liu [12], Furedi, Hwang, and Weichsel [13], Liu and Yang [14], Qian and Ray-Chaudhuri [15], Ramanan [16], Snevily [17,18], Wang, Wei, and Ge [19], and others.

4. Proof of the Main Result

In this section, we prove Theorem 6. As we have mentioned before, our proof is based on the polynomial method. Let x be the n-tuple of variables $x_1, x_2, \cdots, x_n$, where x_i takes the values only 0 and 1. Then all the polynomials we will work with have the relation $x_i^2 = x_i$ in their domain.

Proof of Theorem 6. The result is immediate if $|\mathcal{F}| = 1$. Suppose $|\mathcal{F}| > 1$. Let $F_1, F_2, \cdots, F_f$ be the distinct sets in $\mathcal{F}$, listed in non-decreasing order of size. We define the characteristic vector $v_i = (v_{i_1}, v_{i_2}, \cdots, v_{i_n})$ of F_i such that $v_{i_j} = 1$ if $j \in F_i$ and $v_{i_j} = 0$ if $j \notin F_i$.

We consider the following family of multilinear polynomials

$$f_i(x) = \prod_{j=1}^{k-1} (v_i \cdot x - j)$$

where $x = (x_1, x_2, \cdots, x_n)$.

Since $|F_1| \leq |F_2|$, there exists some $p \in F_2$ such that $p \notin F_1$. Let $\mathcal{G} = \{G_1, G_2, \cdots, G_g\}$ be the family of subsets of $[n]$ with the size at most $k-2$, which is listed in non-decreasing order of size, and not containing p. Next, we consider the second family of multilinear polynomials

$$g_i(x) = (x_p - 1) \prod_{j \in G_i} x_j$$

where $1 \le i \le g$. Let $\mathcal{H} = \{H_1, H_2, \cdots, H_h\}$ be the family of subsets of $[n]$ with the size at most $k-1$, which is listed in non-decreasing order of size, and containing p. Then, we consider our third and last family of multilinear polynomials

$$h_i(x) = \prod_{j=0}^{|H_i|-1}(w_i \cdot x - j) - \sum_{l:p\notin F_l} \frac{\prod_{j=0}^{|H_i|-1}(w_i \cdot v_l^c - j)}{\prod_{j=1}^{k-1}(v_l^c \cdot v_l^c - j)} \prod_{j=1}^{k-1}(v_l^c \cdot x - j)$$
$$- \sum_{l:p\in F_l} \frac{\prod_{j=0}^{|H_i|-1}(w_i \cdot v_l - j)}{\prod_{j=1}^{k-1}(v_l \cdot v_l - j)} \prod_{j=1}^{k-1}(v_l \cdot x - j)$$

where w_i is the characteristic vector of H_i.

We claim that the functions $f_i(x)$, $g_i(x)$, and $h_i(x)$ taken together are linearly independent. Assume that

$$\sum_{i=1}^{f} \alpha_i f_i(x) + \sum_{i=1}^{g} \beta_i g_i(x) + \sum_{i=1}^{h} \gamma_i h_i(x) = 0 \qquad (1)$$

We substitute the characteristic vector v_s of F_s containing p into Equation (1). Because of the $(x_p - 1)$ factor, we have

$$g_i(v_s) = 0 \quad \text{for all} \quad 1 \le i \le g.$$

Let v_l^c be the characteristic vector of F_l^c. Next, let us consider $h_i(v_s)$:

$$h_i(v_s) = \prod_{j=0}^{|H_i|-1}(w_i \cdot v_s - j) - \sum_{l:p\notin F_l} \frac{\prod_{j=0}^{|H_i|-1}(w_i \cdot v_l^c - j)}{\prod_{j=1}^{k-1}(v_l^c \cdot v_l^c - j)} \prod_{j=1}^{k-1}(v_l^c \cdot v_s - j)$$
$$- \sum_{l:p\in F_l} \frac{\prod_{j=0}^{|H_i|-1}(w_i \cdot v_l - j)}{\prod_{j=1}^{k-1}(v_l \cdot v_l - j)} \prod_{j=1}^{k-1}(v_l \cdot v_s - j).$$

Since $1 \le |F_l \cap F_s| \le k-1$, we have $\prod_{j=1}^{k-1}(v_l \cdot v_s - j) = 0$ except when $s = l$. Since $|F_i| \ge k$ for all i, we have

$$- \sum_{l:p\in F_l} \frac{\prod_{j=0}^{|H_i|-1}(w_i \cdot v_l - j)}{\prod_{j=1}^{k-1}(v_l \cdot v_l - j)} \prod_{j=1}^{k-1}(v_l \cdot v_s - j) = - \frac{\prod_{j=0}^{|H_i|-1}(w_i \cdot v_s - j)}{\prod_{j=1}^{k-1}(v_s \cdot v_s - j)} \prod_{j=1}^{k-1}(v_s \cdot v_s - j)$$
$$= - \prod_{j=0}^{|H_i|-1}(w_i \cdot v_s - j).$$

Since $1 \le |F_l^c \cap F_s| \le k-1$ for $s \ne l$, we have $\prod_{j=1}^{k-1}(v_l^c \cdot v_s - j) = \prod_{j=1}^{k-1}(|F_l^c \cap F_s| - j) = 0$. Thus, we have

$$h_i(v_s) = \prod_{j=0}^{|H_i|-1}(w_i \cdot v_s - j) - \prod_{j=0}^{|H_i|-1}(w_i \cdot v_s - j) = 0 \quad \text{for all} \quad 1 \le i \le h.$$

Finally, we consider $f_i(v_s)$. Since $f_s(v_s) \ne 0$ and $1 \le |F_i \cap F_s| \le k-1$ for $i \ne s$, we get $\alpha_s = 0$ whenever $p \in F_s$.

Next, we substitute the characteristic vector v_s^c of F_s^c into Equation (1), where $p \notin F_s$. Because of the $(x_p - 1)$ factor, we have

$$g_i(v_s^c) = 0 \quad \text{for all} \quad 1 \le i \le g.$$

Next, let us consider $h_i(v_s^c)$. Since $1 \le |F_l^c \cap F_s^c| \le k-1$, we have $\prod_{j=1}^{k-1}(v_l^c \cdot v_s^c - j) = 0$ except when $s = l$. Since $n - |F_i| \ge k$, we have

$$- \sum_{l:p\notin F_l} \frac{\prod_{j=0}^{|H_i|-1}(w_i \cdot v_l^c - j)}{\prod_{j=1}^{k-1}(v_l^c \cdot v_l^c - j)} \prod_{j=1}^{k-1}(v_l^c \cdot v_s^c - j) = - \prod_{j=0}^{|H_i|-1}(w_i \cdot v_s^c - j).$$

Since $1 \leq |F_l \cap F_s^c| \leq k-1$ for $s \neq l$, we have $\prod_{j=1}^{k-1}(v_l \cdot v_s^c - j) = \prod_{j=1}^{k-1}(|F_l \cap F_s^c| - j) = 0$. Thus, we have

$$h_i(v_s^c) = \prod_{j=0}^{|H_i|-1} (w_i \cdot v_s^c - j) - \prod_{j=0}^{|H_i|-1} (w_i \cdot v_s^c - j) = 0 \quad \text{for all } 1 \leq i \leq h.$$

Finally we consider $f_i(v_s^c)$. Since $1 \leq |F_i \cap F_s^c| \leq k-1$, by the hypothesis $f_i(v_s^c)$ is also 0 except for $f_s(v_s^c)$. Since $f_s(v_s^c) \neq 0$, we get $\alpha_s = 0$ whenever $p \notin F_s$.

So Equation (1) is reduced to :

$$\sum_{i=1}^{g} \beta_i g_i(x) + \sum_{i=1}^{h} \gamma_i h_i(x) = 0 \tag{2}$$

Next, we substitute the characteristic vector w_s of H_s in order of increasing size into Equation (2). Now we note that $p \in H_s$. Because of the $(x_p - 1)$ factor, we have $g_i(w_s) = 0$ for all $1 \leq i \leq g$. Since the size of H_i is at most $k-1$ for all i, we have $1 \leq |F_l^c \cap H_s| \leq k-1$ for $p \in F_l^c$. Thus, the factor $\prod_{j=1}^{k-1}(v_l^c \cdot w_s - j)$ is 0. Similarly, the factor $\prod_{j=1}^{k-1}(v_l \cdot w_s - j)$ is 0 for $p \in F_l$. Thus, we have $h_i(w_s) = \prod_{j=0}^{|H_i|-1}(w_i \cdot w_s - j)$. Since $h_s(w_s) \neq 0$, and $h_i(w_s) = 0$ for $i > s$, we have $\sum_{i=1}^{h} \gamma_i h_i(w_s) = \sum_{i=1}^{s} \gamma_i h_i(w_s)$.

Recall that we substitute the vector w_s in order of increasing size. When we first plug w_1 into Equation (2), we have $\gamma_1 h_1(w_1) = 0$, and thus $\gamma_1 = 0$. Next, we plug w_2 into (2) after dropping $\gamma_1 h_1(w_1)$ term from (2). Then we have $\gamma_2 h_2(w_2) = 0$, and thus $\gamma_2 = 0$. Similarly, we have $\gamma_i = 0$ for all i.

Thus, Equation (1) becomes

$$\sum_i \beta_i g_i(x) = 0. \tag{3}$$

Next, we substitute the characteristic vector y_s of G_s in order of increasing size into Equation (3). Thus, we have

$$g_i(y_s) = (y_{s_p} - 1) \prod_{j \in G_i} y_{s_j} = - \prod_{j \in G_i} y_{s_j} \quad \text{for all } 1 \leq i \leq g.$$

Recall that we substitute the vector y_s in order of increasing size. Please note that $g_i(0)$ is the empty product, which is taken to be 1. When we first plug y_1 into Equation (3), we have $g_1(y_1) \neq 0$ and $g_i(y_1) = 0$ for all $i > 1$, and thus $\beta_1 = 0$. Next, we plug y_2 into (3) after dropping $\beta_1 g_1(x)$ term from (3). Then we have $g_2(y_2) \neq 0$ and $g_i(y_2) = 0$ for all $i > 2$, and thus $\beta_2 = 0$. Similarly, we have $\beta_i = 0$ for all i.

This concludes that all the polynomials $f_i(x)$, $g_i(x)$, and $h_i(x)$ are linearly independent. We found $|\mathcal{F}| + |\mathcal{G}| + |\mathcal{H}|$ linearly independent polynomials. All these polynomials are of degree less than or equal to $k-1$. The space of these multilinear polynomials has dimension $\sum_{i=0}^{k-1} \binom{n}{i}$. We have

$$|\mathcal{F}| + |\mathcal{G}| + |\mathcal{H}| \leq \sum_{i=0}^{k-1} \binom{n}{i}.$$

Since $|\mathcal{G}| = \sum_{i=0}^{k-2} \binom{n-1}{i}$ and $|\mathcal{H}| = \sum_{i=0}^{k-2} \binom{n-1}{i}$, we have $|\mathcal{F}| + 2\sum_{i=0}^{k-2} \binom{n-1}{i} \leq \sum_{i=0}^{k-1} \binom{n}{i}$. This gives us

$$|\mathcal{F}| \leq \binom{n-1}{k-1}$$

finishing the proof of Theorem 6. □

5. Conclusions

We have answered the following question: when is it possible to get the same bound of the Erdős-Ko-Rado theorem for uniform intersecting families in the non-uniform intersecting families?

Since the EKR-type bound for the non-uniform family of subsets of $[n]$, which is $\binom{n-1}{k-1} + \binom{n-1}{k-2} + \cdots + \binom{n-1}{0}$, is much larger than $\binom{n-1}{k-1}$, this question is interesting and deserves further study.

Please note that if we can delete the condition (c) in Theorem 6, we can get the same bound of the Erdős-Ko-Rado theorem for k-uniform intersecting families under the same condition for non-uniform intersecting families of size at least k and at most $n-k$. Another intriguing question motivated by our result is the problem of getting the same bound of Theorem 6 without the condition (c) or finding a better bound for the non-uniform intersecting families than the previous results by the others.

Author Contributions: All authors have contributed equally to this work. All authors have read and agreed to the published version of the manuscript.

Acknowledgments: All authors sincerely appreciate the reviewers for their valuable comments and suggestions to improve the paper.

References

1. Erdős, P.; Ko, C.; Rado, R. Intersection theorem for systems of finite sets. *Q. J. Math. Oxf. Ser.* **1961**, *12*, 313–320. [CrossRef]
2. Frankl, P. The Erdős-Ko-Rado theorem is true for $n = ckt$. In *Combinatorics, Proceedings of the Fifth Hungarian Colloquium, Keszthely*; North-Holland Publishing Company: Amsterdam, The Netherlands, 1978; Volume 1, pp. 365–375.
3. Wilson, R. The exact bound in the Erdős-Ko-Rado theorem. *Combinatorica* **1984**, *4*, 247–257. [CrossRef]
4. Alon, N.; Aydinian, H.; Huang, H. Maximizing the number of nonnegative subsets. *SIAM J. Discret. Math.* **2014**, *28*, 811–816. [CrossRef]
5. Deza, M.; Frankl, P. Erdős-Ko-Rado theorem–22 years later. *SIAM J. Algebr. Discret. Methods* **1983**, *4*, 419–431. [CrossRef]
6. Ray-Chaudhuri, D.; Wilson, R. On t-designs. *Osaka J. Math.* **1975**, *12*, 737–744.
7. Frankl, P.; Wilson, R. Intersection theorems with geometric consequences. *Combinatorica* **1981**, *1*, 357–368. [CrossRef]
8. Deza, M.; Frankl, P.; Singhi, N. On functions of strength t. *Combinatorica* **1983**, *3*, 331–339. [CrossRef]
9. Alon, N.; Babai, L.; Suzuki, H. Multilinear polynomials and Frankl-Ray-Chaudhuri-Wilson type intersection theorems. *J. Comb. Theory Ser. A* **1991**, *58*, 165–180. [CrossRef]
10. Hwang, K.; Kim, Y. A proof of Alon-Babai-Suzuki's Conjecture and Multilinear Polynomials. *Eur. J. Comb.* **2015**, *43*, 289–294. [CrossRef]
11. Blokhuis, A. Solution of an extremal problem for sets using resultants of polynomials. *Combinatorica* **1990**, *10*, 393–396. [CrossRef]
12. Chen, W.Y.C.; Liu, J. Set systems with L-intersections modulo a prime number. *J. Comb. Theory Ser. A* **2009**, *116*, 120–131. [CrossRef]
13. Füredi, Z.; Hwang, K.; Weichsel, P. A proof and generalization of the Erős-Ko-Rado theorem using the method of linearly independent polynomials. In *Topics in Discrete Mathematics*; Algorithms Combin. 26; Springer: Berlin, Germnay, 2006; pp. 215–224.
14. Liu, J.; Yang, W. Set systems with restricted k-wise L-intersections modulo a prime number. *Eur. J. Comb.* **2014**, *36*, 707–719. [CrossRef]
15. Qian, J.; Ray-Chaudhuri, D. On mod p Alon-Babai-Suzuki inequality. *J. Algebr. Comb.* **2000**, *12*, 85–93. [CrossRef]
16. Ramanan, G. Proof of a conjecture of Frankl and Füredi. *J. Comb. Theory Ser. A* **1997**, *79*, 53–67. [CrossRef]
17. Snevily, H. On generalizations of the de Bruijn-Erdős theorem. *J. Comb. Theory Ser. A* **1994**, *68*, 232–238. [CrossRef]
18. Snevily, H. A sharp bound for the number of sets that pairwise intersect at k positive values. *Combinatorica* **2003**, *23*, 527–532. [CrossRef]
19. Wang, X.; Wei, H.; Ge, G. A strengthened inequality of Alon-Babai-Suzuki's conjecture on set systems with restricted intersections modulo p. *Discret. Math.* **2018**, *341*, 109–118. [CrossRef]

9

On Central Complete and Incomplete Bell Polynomials I

Taekyun Kim [1,*], Dae San Kim [2] and Gwan-Woo Jang [1]

[1] Department of Mathematics, Kwangwoon University, Seoul 139-701, Korea; gwjang@kw.ac.kr
[2] Department of Mathematics, Sogang University, Seoul 121-742, Korea; dskim@sogang.ac.kr
[*] Correspondence: tkkim@kw.ac.kr.

Abstract: In this paper, we introduce central complete and incomplete Bell polynomials which can be viewed as generalizations of central Bell polynomials and central factorial numbers of the second kind, and also as 'central' analogues for complete and incomplete Bell polynomials. Further, some properties and identities for these polynomials are investigated. In particular, we provide explicit formulas for the central complete and incomplete Bell polynomials related to central factorial numbers of the second kind.

Keywords: central incomplete Bell polynomials; central complete Bell polynomials; central complete Bell numbers

---

1. Introduction

In this paper, we introduce central incomplete Bell polynomials $T_{n,k}(x_1, x_2, \cdots, x_{n-k+1})$ given by

$$\frac{1}{k!}\Big(\sum_{m=1}^{\infty}\frac{1}{2^m}(x_m-(-1)^m x_m)\frac{t^m}{m!}\Big)^k=\sum_{n=k}^{\infty}T_{n,k}(x_1,x_2,\cdots,x_{n-k+1})\frac{t^n}{n!}$$

and central complete Bell polynomials $B_n^{(c)}(x|x_1, x_2, \cdots, x_n)$ given by

$$exp\Big(x\sum_{i=1}^{\infty}\frac{1}{2^i}(x_i-(-1)^i x_i)\frac{t^i}{i!}\Big)=\sum_{n=0}^{\infty}B_n^{(c)}(x|x_1,x_2,\cdots,x_n)\frac{t^n}{n!}$$

and investigate some properties and identities for these polynomials. They can be viewed as generalizations of central Bell polynomials and central factorial numbers of the second kind, and also as 'central' analogues for complete and incomplete Bell polynomials.

Here, we recall that the central factorial numbers $T(n,k)$ of the second kind and the central Bell polynomials $B_n^{(c)}(x)$ are given in terms of generating functions by

$$\frac{1}{k!}(e^{\frac{t}{2}}-e^{-\frac{t}{2}})^k=\sum_{n=k}^{\infty}T(n,k)\frac{t^n}{n!},\quad e^{x(e^{\frac{t}{2}}-e^{-\frac{t}{2}})}=\sum_{n=0}^{\infty}B_n^{(c)}(x)\frac{t^n}{n!},$$

so that $T_{n,k}(1,1,\cdots,1)=T(n,k)$ and $B_n^{(c)}(x|1,1,\cdots,1)=B_n^{(c)}(x)$.

The incomplete and complete Bell polynomials have applications in such diverse areas as combinatorics, probability, algebra, modules over a *-algebra (see [1,2]), quasi local algebra and analysis. Here, we recall some applications of them and related works. The incomplete Bell polynomials $B_{n,k}(x_1, x_2, \cdots, x_{n-k+1})$ (see [3,4]) arise naturally when we want to find higher-order derivatives of composite functions. Indeed, such higher-order derivatives can be expressed in terms of incomplete Bell polynomials, which is known as Faà di Bruno formula given as in the following (see [3]):

$$\frac{d^n}{dt^n} g(f(t)) = \sum_{k=0}^{n} g^{(k)}(f(t)) B_{n,k}(f'(t), f''(t), \cdots, f^{(n-k+1)}(t)).$$

For the curious history on this formula, we let the reader refer to [5].

In addition, the number of monomials appearing in $B_{n,k} = B_{n,k}(x_1, x_2, \cdots, x_{n-k+1})$ is the number of partitioning a set with n elements into k blocks and the coefficient of each monomial is the number of partitioning a set with n elements as the corresponding k blocks. For example,

$$B_{10,7} = 3150 x_2^3 x_1^4 + 2520 x_3 x_2 x_1^5 + 210 x_4 x_1^6$$

shows that there are three ways of partitioning a set with 10 elements into seven blocks, and 3150 partitions with blocks of size 2, 2, 2, 1, 1, 1, 1, 2520 partitions with blocks of size 3, 2, 1, 1, 1, 1, 1, and 210 partitions with blocks of size 4, 1,1, 1, 1, 1, 1. This example is borrowed from [4], which gives a practical way of computing $B_{n,k}$ for any given n, k (see [4], (1.5)).

Furthermore, the incomplete Bell polynomials can be used in constructing sequences of binomial type (also called associated sequences). Indeed, for any given scalars $c_1, c_2, \cdots, c_n, \cdots$ the following form a sequence of binomial type

$$s_n(x) = \sum_{k=0}^{n} B_{n,k}(c_1, c_2, \cdots, c_{n-k+1}) x^k, \quad (n = 0, 1, 2, \cdots)$$

and, conversely, any sequence of binomial type arises in this way for some scalar sequence $c_1, c_2, \cdots, c_n \cdots$. For these, the reader may want to look at the paper [6].

There are certain connections between incomplete Bell polynomials and combinatorial Hopf algebras such as the Hopf algebra of word symmetric functions, the Hopf algebra of symmetric functions, the Faà di Bruno algebra, etc. The details can be found in [7].

The complete Bell polynomials $B_n(x_1, x_2, \cdots, x_n)$ (see [3,8–10]) have applications to probability theory. Indeed, the nth moment $\mu_n = E[X^n]$ of the random variable X is the nth complete Bell polynomial in the first n cumulants. Namely,

$$\mu_n = B_n(\kappa_1, \kappa_2, \cdots, \kappa_n).$$

For many applications to probability theory and combinatorics, the reader can refer to the Ph. D. thesis of Port [10].

Many special numbers, like Stirling numbers of both kinds, Lah numbers and idempotent numbers, appear in many combinatorial and number theoretic identities involving complete and incomplete Bell polynomials. For these, the reader refers to [3,8].

The central factorial numbers have received less attention than Stirling numbers. However, according to [11], they are at least as important as Stirling numbers, said to be "as important as Bernoulli numbers, or even more so". A systematic treatment of these important numbers was given in [11], including their properties and applications to difference calculus, spline theory, and to approximation theory, etc. For some other related references on central factorial numbers, we let the reader refer to [1,2,12–14]. Here, we note that central Bell polynomials and central factorial numbers of the second kind are respectively 'central' analogues for Bell polynomials and Stirling numbers of the second kind. They have been studied recently in [13,15].

The complete Bell polynomials and the incomplete Bell polynomials are respectively mutivariate versions for Bell polynomials and Stirling numbers of the second kind. This paper deals with central complete and incomplete Bell polynomials which are 'central' analogues for the complete and incomplete Bell polynomials. In addition, they can be viewed as generalizations of central Bell polynomials and central factorial numbers of the second kind (see [15]). The outline of the paper is as follows. After giving an introduction to the present paper in Section 1, we review some known properties and results about Bell polynomials, and incomplete and complete Bell polynomials in Section 2. We state the new and main results of this paper in Section 3, where we introduce central incomplete and complete Bell polynomials and investigate some properties and identities for them. In particular, Theorems 1 and 3 give basic formulas for computing central incomplete Bell polynomials and central complete Bell polynomials, respectively. We remark that the number of monomials appearing in $T_{n,k}(x_1, 2x_2, \cdots, 2^{n-k}x_{n-k+1})$ is the number of partitioning a set with n elements into k blocks with odd sizes and the coefficient of each monomial is the number of partitioning a set with n elements as the corresponding k blocks with odd sizes. This is illustrated by an example. Furthermore, we give expressions for the central incomplete and complete Bell polynomials with some various special arguments and also for the connection between the two Bell polynomials. We defer more detailed study of the central incomplete and complete Bell polynomials to a later paper.

2. Preliminaries

The Stirling numbers of the second kind are given in terms of generating function by (see [3,16])

$$\frac{1}{k!}(e^t-1)^k = \sum_{n=k}^{\infty} S_2(n,k)\frac{t^n}{n!}. \tag{1}$$

The Bell polynomials are also called Tochard polynomials or exponential polynomials and defined by (see [9,13,15,17])

$$e^{x(e^t-1)} = \sum_{n=0}^{\infty} B_n(x)\frac{t^n}{n!}. \tag{2}$$

From Equations (1) and (2), we immediately see that (see [3,18])

$$\begin{aligned} B_n(x) &= e^{-x}\sum_{k=0}^{\infty}\frac{k^n}{k!}x^k \\ &= \sum_{k=0}^{n} x^k S_2(n,k), \quad (n \geq 0). \end{aligned} \tag{3}$$

When $x = 1$, $B_n = B_n(1)$ are called Bell numbers.

The (exponential) incomplete Bell polynomials are also called (exponential) partial Bell polynomials and defined by the generating function (see [9,15])

$$\frac{1}{k!}\left(\sum_{m=1}^{\infty} x_m\frac{t^m}{m!}\right)^k = \sum_{n=k}^{\infty} B_{n,k}(x_1, \cdots, x_{n-k+1})\frac{t^n}{n!}, \quad (k \geq 0). \tag{4}$$

Thus, by Equation (4), we get

$$\begin{aligned} B_{n,k}(x_1, \cdots, x_{n-k+1}) = \sum \frac{n!}{i_1! i_2! \cdots i_{n-k+1}!}\left(\frac{x_1}{1!}\right)^{i_1}\left(\frac{x_2}{2!}\right)^{i_2} \times \cdots \\ \times \left(\frac{x_{n-k+1}}{(n-k+1)!}\right)^{i_{n-k+1}}, \end{aligned} \tag{5}$$

where the summation runs over all integers $i_1, \cdots, i_{n-k+1} \geq 0$ such that $i_1 + i_2 + \cdots + i_{n-k+1} = k$ and $i_1 + 2i_2 + \cdots + (n-k+1)i_{n-k+1} = n$.

From (1) and (4), we easily see that

$$B_{n,k}\underbrace{(1,1,\cdots,1)}_{n-k+1-times} = S_2(n,k), \quad (n,k \geq 0). \tag{6}$$

We easily deduce from (5) the next two identities:

$$B_{n,k}(\alpha x_1, \alpha x_2, \cdots, \alpha x_{n-k+1}) = \alpha^k B_{n,k}(x_1, x_2, \cdots, x_{n-k+1}) \tag{7}$$

and

$$B_{n,k}(\alpha x_1, \alpha^2 x_2, \cdots, \alpha^{n-k+1} x_{n-k+1}) = \alpha^n B_{n,k}(x_1, x_2, \cdots, x_{n-k+1}), \tag{8}$$

where $\alpha \in \mathbb{R}$ (see [15]).

From (4), it is not difficult to note that

$$\begin{aligned}
&\sum_{n=k}^{\infty} B_{n,k}(x,1,0,0,\cdots,0)\frac{t^n}{n!} = \frac{1}{k!}(xt + \frac{t^2}{2})^k \\
&= \frac{t^k}{k!}\sum_{n=0}^{k}\binom{k}{n}\left(\frac{t}{2}\right)^n x^{k-n} \\
&= \sum_{n=0}^{k}\frac{(n+k)!}{k!}\binom{k}{n}\frac{1}{2^n}x^{k-n}\frac{t^{n+k}}{(n+k)!},
\end{aligned} \tag{9}$$

and

$$\sum_{n=k}^{\infty} B_{n,k}(x,1,0,0,\cdots,0)\frac{t^n}{n!} = \sum_{n=0}^{\infty} B_{n+k,k}(x,1,0,\cdots,0)\frac{t^{n+k}}{(n+k)!}. \tag{10}$$

Combining (9) with (10), we have

$$B_{n+k,k}(x,1,0,\cdots,0) = \frac{(n+k)!}{k!}\binom{k}{n}\frac{1}{2^n}x^{k-n}, \quad (0 \leq n \leq k). \tag{11}$$

Replacing n by $n-k$ in (11) yields the following identity

$$B_{n,k}(x,1,0,\cdots,0) = \frac{n!}{k!}\binom{k}{n-k}x^{2k-n}\left(\frac{1}{2}\right)^{n-k}, \quad (k \leq n \leq 2k). \tag{12}$$

We recall here that the (exponential) complete Bell polynomials are defined by

$$\exp\Big(\sum_{i=1}^{\infty} x_i \frac{t^i}{i!}\Big) = \sum_{n=0}^{\infty} B_n(x_1, x_2, \cdots, x_n)\frac{t^n}{n!}. \tag{13}$$

Then, by (4) and (13), we get

$$B_n(x_1, x_2, \cdots, x_n) = \sum_{k=0}^{n} B_{n,k}(x_1, x_2, \cdots, x_{n-k+1}). \tag{14}$$

From (3), (6), (7) and (14), we have

$$B_n(x,x,\cdots,x) = \sum_{k=0}^{n} x^k B_{n,k}(1,1,\cdots,1) \\ = \sum_{k=0}^{n} x^k S_2(n,k) = B_n(x), \quad (n \geq 0). \tag{15}$$

We recall that the central factorial numbers of the second kind are given by (see [19,20])

$$\frac{1}{k!}(e^{\frac{t}{2}} - e^{-\frac{t}{2}})^k = \sum_{n=k}^{\infty} T(n,k)\frac{t^n}{n!}, \tag{16}$$

where $k \geq 0$.

From (16), it is not difficult to derive the following expression

$$T(n,k) = \frac{1}{k!}\sum_{j=0}^{k}\binom{k}{j}(-1)^{k-j}(j-\frac{k}{2})^n, \tag{17}$$

where $n,k \in \mathbb{Z}$ with $n \geq k \geq 0$, (see [16,20]).

In [20], the central Bell polynomials $B_n^{(c)}(x)$ are defined by

$$B_n^{(c)}(x) = \sum_{k=0}^{n} T(n,k)x^k, \quad (n \geq 0). \tag{18}$$

When $x = 1$, $B_n^{(c)} = B_n^{(c)}(1)$ are called the central Bell numbers.

It is not hard to derive the generating function for the central Bell polynomials from (18) as follows (see [15]):

$$e^{x\left(e^{\frac{t}{2}}-e^{-\frac{t}{2}}\right)} = \sum_{n=0}^{\infty} B_n^{(c)}(x)\frac{t^n}{n!}. \tag{19}$$

By making use of (19), the following Dobinski-like formula was obtained earlier in [15]:

$$B_n^{(c)}(x) = \sum_{l=0}^{\infty}\sum_{j=0}^{\infty}\binom{l+j}{j}(-1)^j\frac{1}{(l+j)!}\left(\frac{l}{2}-\frac{j}{2}\right)^n x^{l+j}, \tag{20}$$

where $n \geq 0$.

Motivated by (4) and (13), we will introduce central complete and incomplete Bell polynomials and investigate some properties and identities for these polynomials. Also, we present explicit formulas for the central complete and incomplete Bell polynomials related to central factorial numbers of the second kind.

3. On Central Complete and Incomplete Bell Polynomials

In view of (13), we may consider the *central incomplete Bell polynomials* which are given by

$$\frac{1}{k!}\left(\sum_{m=1}^{\infty}\frac{1}{2^m}(x_m-(-1)^m x_m)\frac{t^m}{m!}\right)^k = \sum_{n=k}^{\infty} T_{n,k}(x_1,x_2,\cdots,x_{n-k+1})\frac{t^n}{n!}, \tag{21}$$

where $k = 0,1,2,3,\cdots$.

For $n,k \geq 0$ with $n-k \equiv 0 \pmod 2$, by (4) and (5), we get

$$\begin{aligned} T_{n,k}(x_1,x_2,\cdots,x_{n-k+1}) = \sum \frac{n!}{i_1!i_2!\cdots i_{n-k+1}!}\Big(\frac{x_1}{1!}\Big)^{i_1}\Big(\frac{0}{2\cdot 2!}\Big)^{i_2} \\ \times\Big(\frac{x_3}{2^2\cdot 3!}\Big)^{i_3}\cdots\Big(\frac{x_{n-k+1}}{2^{n-k}(n-k+1)!}\Big)^{i_{n-k+1}}, \end{aligned} \tag{22}$$

where the summation is over all integers $i_1,i_2,\cdots,i_{n-k+1} \geq 0$ such that $i_1+\cdots+i_{n-k+1} = k$ and $i_1+2i_2+\cdots+(n-k+1)i_{n-k+1} = n$.

From (5) and (22), we note that

$$T_{n,k}(x_1,x_2,\cdots,x_{n-k+1}) = B_{n,k}(x_1,0,\frac{x_3}{2^2},0,\cdots,\frac{x_{n-k+1}}{2^{n-k}}), \tag{23}$$

where $n,k \geq 0$ with $n-k \equiv 0 \pmod 2$ and $n \geq k$.
Therefore, from (22) and (23), we obtain the following theorem.

Theorem 1. *For $n,k \geq 0$ with $n \geq k$ and $n-k \equiv 0 \pmod 2$, we have*

$$\begin{aligned} T_{n,k}(x_1,x_2,\cdots,x_{n-k+1}) &= B_{n,k}(x_1,0,\frac{x_3}{2^2},0,\cdots,\frac{x_{n-k+1}}{2^{n-k}}) \\ &= \sum \frac{n!}{i_1!i_3!\cdots i_{n-k+1}!}\Big(\frac{x_1}{1!}\Big)^{i_1}\Big(\frac{x_3}{2^2\cdot 3!}\Big)^{i_3}\times\cdots\times\Big(\frac{x_{n-k+1}}{2^{n-k}(n-k+1)!}\Big)^{i_{n-k+1}}, \end{aligned} \tag{24}$$

where the summation is over all integers $i_1,i_2,\cdots,i_{n-k+1} \geq 0$ such that $i_1+i_3+\cdots+i_{n-k+1} = k$ and $i_1+3i_3+\cdots+(n-k+1)i_{n-k+1} = n$.

Remark 1. *Theorem 1 shows in particular that we have*

$$T_{n,k}(x_1,2x_2,\cdots,2^{n-k}x_{n-k+1}) = B_{n,k}(x_1,0,x_3,0,\cdots,x_{n-k+1}).$$

From this, we note that the number of monomials appearing in $T_{n,k}(x_1,2x_2,\cdots,2^{n-k}x_{n-k+1})$ is the number of partitioning a set with n elements into k blocks with odd sizes and the coefficient of each monomial is the number of partitioning a set with n elements as the corresponding k blocks with odd sizes. For example, from the example in Section 3 of [4], we have

$$T_{13,7}(x_1,2x_2,2^2x_3,2^3x_4,2^4x_5,2^5x_6,2^6x_7) = 200,200x_3^3x_1^4+72,072x_5x_3x_1^5+1716x_7x_1^6.$$

Thus, there are three ways of partitioning a set with 13 elements into seven blocks with odd sizes, and 200,200 partitions with blocks of size 3, 3, 3, 1, 1, 1, 1, 72,072 partitions with blocks of size 5, 3, 1, 1, 1, 1, 1, and 1716 partitions with blocks of size 7, 1, 1, 1, 1, 1, 1.

For $n, k \geq 0$ with $n \geq k$ and $n - k \equiv 0 \pmod 2$, by (21), we get

$$\begin{aligned}\sum_{n=k}^{\infty} T_{n,k}(x, x^2, x^3, \cdots, x^{n-k+1})\frac{t^n}{n!} &= \frac{1}{k!}\Big(xt + \frac{x^3}{2^2}\frac{t^3}{3!} + \frac{x^5}{2^4}\frac{t^5}{5!} + \cdots\Big)^k \\ &= \frac{1}{k!}\Big(e^{\frac{x}{2}t} - e^{-\frac{x}{2}t}\Big)^k = \frac{1}{k!}e^{-\frac{kx}{2}t}\Big(e^{xt} - 1\Big)^k \\ &= \frac{1}{k!}\sum_{l=0}^{k}\binom{k}{l}(-1)^{k-l}e^{(l-\frac{k}{2})xt} \\ &= \frac{1}{k!}\sum_{l=0}^{k}\binom{k}{l}(-1)^{k-l}\sum_{n=0}^{\infty}(l - \frac{k}{2})^n x^n \frac{t^n}{n!} \\ &= \sum_{n=0}^{\infty}\Big(\frac{x^n}{k!}\sum_{l=0}^{k}\binom{k}{l}(-1)^{k-l}(l - \frac{k}{2})^n\Big)\frac{t^n}{n!}.\end{aligned} \tag{25}$$

Now, the next theorem follows by comparing the coefficients on both sides of (25).

Theorem 2. *For $n, k \geq 0$ with $n - k \equiv 0 \pmod 2$, we have*

$$\frac{x^n}{k!}\sum_{l=0}^{k}\binom{k}{l}(-1)^{k-l}(l - \frac{k}{2})^n = \begin{cases} T_{n,k}(x, x^2, \cdots, x^{n-k+1}), & \text{if } n \geq k, \\ 0, & \text{if } n < k. \end{cases} \tag{26}$$

In particular,

$$\frac{1}{k!}\sum_{l=0}^{k}\binom{k}{l}(-1)^{k-l}(l - \frac{k}{2})^n = \begin{cases} T_{n,k}(1, 1, \cdots, 1), & \text{if } n \geq k, \\ 0, & \text{if } n < k. \end{cases} \tag{27}$$

For $n, k \geq 0$ with $n - k \equiv 0 \pmod 2$ and $n \geq k$, by (17) and (27), we get

$$T_{n,k}(1, 1, \cdots, 1) = T(n, k). \tag{28}$$

Therefore, by (26)–(28) and Theorem 1, we obtain the following corollary

Corollary 1. *For $n, k \geq 0$ with $n - k \equiv 0 \pmod 2$, $n \geq k$, we have*

$$T_{n,k}(x, x^2, \cdots, x^{n-k+1}) = x^n T_{n,k}(1, 1, \cdots, 1)$$

and

$$\begin{aligned}T_{n,k}(1, 1, \cdots, 1) &= T(n, k) = B_{n,k}(1, 0, \frac{1}{2^2}, \cdots, \frac{1}{2^{n-k}}) \\ &= \sum \frac{n!}{i_1! i_3! \cdots i_{n-k+1}!}\Big(\frac{1}{1!}\Big)^{i_1}\Big(\frac{1}{2^2 3!}\Big)^{i_3}\cdots\Big(\frac{1}{2^{n-k}(n-k+1)!}\Big)^{i_{n-k+1}},\end{aligned}$$

where $i_1 + i_3 + \cdots + i_{n-k+1} = k$ and $i_1 + 3i_3 + \cdots + (n-k+1)i_{n-k+1} = n$.

For $n, k \geq 0$ with $n \geq k$ and $n - k \equiv 0 \pmod 2$, we observe that

$$\sum_{n=k}^{\infty} T_{n,k}(x, 1, 0, 0, \cdots, 0)\frac{t^n}{n!} = \frac{1}{k!}(xt)^k. \tag{29}$$

Thus, we have

$$T_{n,k}(x,1,0,\cdots,0)=x^k\binom{0}{n-k}.$$

The next two identities follow easily from (24):

$$T_{n,k}(x,x,\cdots,x)=x^kT_{n,k}(1,1,\cdots,1), \tag{30}$$

and

$$T_{n,k}(\alpha x_1,\alpha x_2,\cdots,\alpha x_{n-k+1})=\alpha^kT_{n,k}(x_1,x_2,\cdots,x_{n-k+1}),$$

where $n,k\geq 0$ with $n-k\equiv 0 \pmod 2$ and $n\geq k$.

Now, we observe that

$$\begin{aligned}
&\exp\Big(x\sum_{i=1}^{\infty}\big(\frac{1}{2}\big)^i(x_i-(-1)^ix_i)\frac{t^i}{i!}\Big)\\
&=\sum_{k=0}^{\infty}x^k\frac{1}{k!}\Big(\sum_{i=1}^{\infty}\big(\frac{1}{2}\big)^i(x_i-(-1)^ix_i)\frac{t^i}{i!}\Big)^k\\
&=1+\sum_{k=1}^{\infty}x^k\frac{1}{k!}\Big(\sum_{i=1}^{\infty}\big(\frac{1}{2}\big)^i(x_i-(-1)^ix_i)\frac{t^i}{i!}\Big)^k\\
&=1+\sum_{k=1}^{\infty}x^k\sum_{n=k}^{\infty}T_{n,k}(x_1,x_2,\cdots,x_{n-k+1})\frac{t^n}{n!}\\
&=1+\sum_{n=1}^{\infty}\Big(\sum_{k=1}^{n}x^kT_{n,k}(x_1,x_2,\cdots,x_{n-k+1})\Big)\frac{t^n}{n!}.
\end{aligned} \tag{31}$$

In view of (13), it is natural to define the *central complete Bell polynomials* by

$$\exp\Big(x\sum_{i=1}^{\infty}\big(\frac{1}{2}\big)^i(x_i-(-1)^ix_i)\frac{t^i}{i!}\Big)=\sum_{n=0}^{\infty}B_n^{(c)}(x|x_1,x_2,\cdots,x_n)\frac{t^n}{n!}. \tag{32}$$

Thus, by (31) and (32), we get

$$B_n^{(c)}(x|x_1,x_2,\cdots,x_n)=\sum_{k=0}^{n}x^kT_{n,k}(x_1,x_2,\cdots,x_{n-k+1}). \tag{33}$$

When $x=1$, $B_n^{(c)}(1|x_1,x_2,\cdots,x_n)=B_n^{(c)}(x_1,x_2,\cdots,x_n)$ are called the *central complete Bell numbers*.

For $n\geq 0$, we have

$$B_n^{(c)}(x_1,x_2,\cdots,x_n)=\sum_{k=0}^{n}T_{n,k}(x_1,x_2,\cdots,x_{n-k+1}) \tag{34}$$

and

$$B_0^{(c)}(x_1,x_2,\cdots,x_n)=1.$$

By (18) and (33), we get

$$B_n^{(c)}(1,1,\cdots,1)=\sum_{k=0}^{n}T_{n,k}(1,1,\cdots,1)=\sum_{k=0}^{n}T(n,k)=B_n^{(c)}, \tag{35}$$

and

$$B_n^{(c)}(x|1,1,\cdots,1)=\sum_{k=0}^{n}x^kT_{n,k}(1,1,\cdots,1)=\sum_{k=0}^{n}x^kT(n,k)=B_n^{(c)}(x). \tag{36}$$

From (31), we note that

$$\begin{aligned}
&\exp\Big(\sum_{i=1}^{\infty}\big(\frac{1}{2}\big)^i(x_i-(-1)^ix_i)\frac{t^i}{i!}\Big)\\
&=1+\sum_{n=1}^{\infty}\frac{1}{n!}\Big(\sum_{i=1}^{\infty}\big(\frac{1}{2}\big)^i(x_i-(-1)^ix_i)\frac{t^i}{i!}\Big)^n\\
&=1+\frac{1}{1!}\sum_{i=1}^{\infty}\big(\frac{1}{2}\big)^i(x_i-(-1)^ix_i)\frac{t^i}{i!}+\frac{1}{2!}\Big(\sum_{i=1}^{\infty}\big(\frac{1}{2}\big)^i(x_i-(-1)^i\\
&\times x_i)\frac{t^i}{i!}\Big)^2+\frac{1}{3!}\Big(\sum_{i=1}^{\infty}\big(\frac{1}{2}\big)^i(x_i-(-1)^ix_i)\frac{t^i}{i!}\Big)^3+\cdots\\
&=1+\frac{1}{1!}x_1t+\frac{1}{2!}x_1^2t^2+\frac{1}{3!}\Big(x_1^3+\frac{1}{2^2}x_3\Big)t^3+\cdots\\
&=\sum_{n=0}^{\infty}\Big(\sum_{m_1+2m_2+\cdots+nm_n=n}\frac{n!}{m_1!m_2!\cdots m_n!}\Big(\frac{x_1}{1!}\Big)^{m_1}\Big(\frac{0}{2!2}\Big)^{m_2}\\
&\times\Big(\frac{x_3}{3!2^2}\Big)^{m_3}\cdots\Big(\frac{x_n(1-(-1)^n)}{n!2^n}\Big)^{m_n}\Big)\frac{t^n}{n!}.
\end{aligned} \tag{37}$$

Now, for $n\in\mathbb{N}$ with $n\equiv 1 \pmod 2$, by (32), (34) and (37), we get

$$\begin{aligned}
B_n^{(c)}(x_1,x_2,\cdots,x_n)&=\sum_{k=0}^{n}T_{n,k}(x_1,x_2,\cdots,x_{n-k+1})\\
&=\sum_{m_1+3m_3+\cdots+nm_n=n}\frac{n!}{m_1!m_3!\cdots m_n!}\Big(\frac{x_1}{1!}\Big)^{m_1}\Big(\frac{x_3}{3!2^2}\Big)^{m_3}\cdots\Big(\frac{x_n}{n!2^{n-1}}\Big)^{m_n}.
\end{aligned} \tag{38}$$

Therefore, Equation (38) yields the following theorem.

Theorem 3. *For $n\in\mathbb{N}$ with $n\equiv 1$ (mod 2), we have*

$$\begin{aligned}
B_n^{(c)}(x_1,x_2,\cdots,x_n)&=\sum_{k=0}^{n}T_{n,k}(x_1,x_2,\cdots,x_{n-k+1})\\
&=\sum_{m_1+3m_3+\cdots+nm_n=n}\frac{n!}{m_1!m_3!\cdots m_n!}\Big(\frac{x_1}{1!}\Big)^{m_1}\Big(\frac{x_3}{3!2^2}\Big)^{m_3}\cdots\Big(\frac{x_n}{n!2^{n-1}}\Big)^{m_n}.
\end{aligned}$$

Example 1. *Here, we illustrate Theorem 3 with the following example:*

$$\begin{aligned}
B_5^{(c)}(x_1,2x_2,2^2x_3,2^3x_4,2^4x_5)&=\frac{5!}{0!0!1!}\Big(\frac{x_1}{1!}\Big)^0\Big(\frac{x_3}{3!}\Big)^0\Big(\frac{x_5}{5!}\Big)^1+\frac{5!}{2!1!0!}\Big(\frac{x_1}{1!}\Big)^2\Big(\frac{x_3}{3!}\Big)^1\Big(\frac{x_5}{5!}\Big)^0\\
&+\frac{5!}{5!0!0!}\Big(\frac{x_1}{1!}\Big)^5\Big(\frac{x_3}{3!}\Big)^0\Big(\frac{x_5}{5!}\Big)^0=x_1^5+10x_1^2x_3+x_5,
\end{aligned}$$

$$T_{5,1}(x_1,2x_2,2^2x_3,2^3x_4,2^4x_5)=\frac{5!}{0!0!1!}\Big(\frac{x_1}{1!}\Big)^0\Big(\frac{x_3}{3!}\Big)^0\Big(\frac{x_5}{5!}\Big)^1=x_5,$$
$$T_{5,3}(x_1,2x_2,2^2x_3)=\frac{5!}{2!1!}\Big(\frac{x_1}{1!}\Big)^2\Big(\frac{x_3}{3!}\Big)^1=10x_1^2x_3,\ T_{5,5}(x_1)=\frac{5!}{5!}\Big(\frac{x_1}{1!}\Big)^5=x_1^5,$$
$$T_{5,0}(x_1,2x_2,2^2x_3,2^3x_4,2^4x_5,2^5x_6)=0,\ T_{5,2}(x_1,2x_2,2^2x_3,2^3x_4)=0,\ T_{5,4}(x_1,2x_2)=0.$$

On the one hand, we have

$$\begin{aligned}\exp\Big(x\sum_{i=1}^{\infty}\big(\tfrac{1}{2}\big)^i(1-(-1)^i)\frac{t^i}{i!}\Big)&=1+\sum_{k=1}^{\infty}\frac{x^k}{k!}\Big(\sum_{n=k}^{\infty}\big(\tfrac{1}{2}\big)^i(1-(-1)^i)\frac{t^i}{i!}\Big)^k\\&=1+\sum_{k=1}^{\infty}x^k\sum_{n=k}^{\infty}T_{n,k}(1,1,\cdots,1)\frac{t^n}{n!}\\&=1+\sum_{n=1}^{\infty}\Big(\sum_{k=1}^{n}x^kT_{n,k}(1,1,\cdots,1)\Big)\frac{t^n}{n!}.\end{aligned}\tag{39}$$

On the other hand, from (19), we have

$$\begin{aligned}\exp\Big(x\sum_{i=1}^{\infty}\big(\tfrac{1}{2}\big)^i(1-(-1)^i)\frac{t^i}{i!}\Big)&=\exp\Big(x\big(t+\frac{1}{2^2}t^3+\frac{1}{2^4}t^5+\cdots\big)\Big)\\&=\exp\Big(x\big(e^{\frac{t}{2}}-e^{-\frac{t}{2}}\big)\Big)=\sum_{n=0}^{\infty}B_n^{(c)}(x)\frac{t^n}{n!}.\end{aligned}\tag{40}$$

Therefore, by (39) and (40), we obtain the following theorem.

Theorem 4. *For $n,k\geq 0$ with $n\geq k$, we have*

$$\sum_{k=0}^{n}x^kT_{n,k}(1,1,\cdots,1)=B_n^{(c)}(x).$$

We note from Theorem 4 the next identities:

$$\sum_{k=0}^{n}x^kT_{n,k}(1,1,\cdots,1)=\sum_{k=0}^{n}T_{n,k}(x,x,\cdots,x)=B_n^{(c)}(x,x,\cdots,x).\tag{41}$$

Thus, Theorem 4 and (41) together give us the following corollary.

Corollary 2. *For $n\geq 0$, we have*

$$B_n^{(c)}(x,x,\cdots,x)=B_n^{(c)}(x).$$

The Stirling numbers of the first kind are given in terms of the generating function by (see [3,21])

$$\frac{1}{k!}\big(\log(1+t)\big)^k=\sum_{n=k}^{\infty}S_1(n,k)\frac{t^n}{n!},\quad(k\geq 0).\tag{42}$$

In order to get the following result and using (42), we first observe that

$$\begin{aligned}\frac{1}{k!}\Big(\log\Big(1+\frac{x}{1-\frac{x}{2}}\Big)\Big)^k &= \sum_{l=k}^{\infty} S_1(l,k)\frac{1}{l!}\Big(\frac{x}{1-\frac{x}{2}}\Big)^l \\ &= \sum_{l=k}^{\infty} S_1(l,k)\frac{x^l}{l!}(1-\frac{x}{2})^{-l} \\ &= \sum_{l=k}^{\infty}\frac{1}{l!}S_1(l,k)\sum_{n=l}^{\infty}\binom{n-1}{l-1}(\frac{1}{2})^{n-l}x^n \\ &= \sum_{n=k}^{\infty}\Big(\sum_{l=k}^{n}\frac{1}{l!}S_1(l,k)\binom{n-1}{l-1}(\frac{1}{2})^{n-l}\Big)x^n.\end{aligned} \tag{43}$$

The following equation can be derived from (21) and (43):

$$\begin{aligned}&\sum_{n=k}^{\infty} T_{n,k}(0!,1!,2!,\cdots,(n-k)!)\frac{t^n}{n!} \\ &= \frac{1}{k!}\Big(t+(\frac{1}{2})^2\frac{t^3}{3}+(\frac{1}{2})^4\frac{t^5}{5}+(\frac{1}{2})^6\frac{t^7}{7}+\cdots\Big)^k \\ &= \frac{1}{k!}\Big(\log(1+\frac{t}{2})-\log(1-\frac{t}{2})\Big)^k = \frac{1}{k!}\Big(\log\Big(\frac{1+\frac{t}{2}}{1-\frac{t}{2}}\Big)\Big)^k \\ &= \frac{1}{k!}\Big(\log(1+\frac{t}{1-\frac{t}{2}})\Big)^k = \sum_{n=k}^{\infty}\Big(\sum_{l=k}^{n}\frac{S_1(l,k)}{l!}\binom{n-1}{l-1}(\frac{1}{2})^{n-l}\Big)t^n.\end{aligned} \tag{44}$$

Now, we obtain the following theorem by comparing the coefficients on both sides of (44).

Theorem 5. *For $n,k \geq 0$ with $n \geq k$, we have*

$$T_{n,k}(0!,1!,2!,\cdots,(n-k)!) = n!\sum_{l=k}^{n}\frac{S_1(l,k)}{l!}\binom{n-1}{l-1}(\frac{1}{2})^{n-l}.$$

4. Conclusions

In this paper, we introduced central complete and incomplete Bell polynomials which can be viewed as generalizations of central Bell polynomials and central factorial numbers of the second kind, and also as 'central' analogues for complete and incomplete Bell polynomials. As examples and recalling some relevant works, we reminded the reader that the incomplete and complete Bell polynomials appearing in a Faà di Bruno formula, which encode integer partition information, can be used in constructing sequences of binomial type, have connections with combinatorial Hopf algebras, have applications in probability theory and arise in many combinatorial and number theoretic identities. One additional thing we want to mention here is that the Faà di Bruno formula has been proved to be very useful in finding explicit expressions for many special numbers arising from many different families of linear and nonlinear differential equations having generating functions of some special numbers and polynomials as solutions (see [22]).

The main results of the present paper are stated in Section 3, in which we introduced central incomplete and complete Bell polynomials and investigated some properties and identities. In particular, in Theorems 1 and 3, we gave basic formulas for computing central incomplete Bell polynomials and central complete Bell polynomials, respectively. We remarked that the number of monomials appearing in $T_{n,k}(x_1,2x_2,\cdots,2^{n-k}x_{n-k+1})$ is the number of partitioning n into k odd parts and the coefficient of each monomial is the number of partitioning n as the corresponding k odd parts. This was illustrated by an example. Furthermore, we gave expressions for the central incomplete and complete Bell polynomials with some various special arguments and also for the connection between the two Bell polynomials. In the near future, we hope to find some further properties, identities and various applications for central complete and incomplete Bell polynomials.

Author Contributions: T.K. and D.S.K. conceived the framework and structured the whole paper; T.K. wrote the paper; D.S.K. and G.-W.J. checked the results of the paper; T.K. and D.S.K. completed the revision of the article.

Acknowledgments: This paper is dedicated to the 70th birthday of Gradimir V. Milovanovic. In addition, we would like to express our sincere condolences on the death of Simsek's mother. The authors thank the referees for their helpful suggestions and comments which improved the original manuscript greatly.

References

1. Bagarello, F.; Trapani, C.; Triolo, S. Representable states on quasilocal quasi*-algebras. *J. Math. Phys.* **2011**, *52*, 013510. [CrossRef]
2. Trapani, C.; Triolo, S. Representations of modules over a *-algebra and related seminorms. *Stud. Math.* **2008**, *184*, 133–148. [CrossRef]
3. Comtet, L. *Advanced Combinatorics: The Art of Finite and Infinite Expansions*; D. Reidel Publishing Co.: Dordrecht, The Netherlands, 1974.
4. Cvijović, D. New identities for the partial Bell polynomials. *Appl. Math. Lett.* **2011**, *24*, 1544–1547. [CrossRef]
5. Johnson, W.P. The curious history of Faà di Bruno's formula. *Am. Math. Mon.* **2002**, *109*, 217–234.
6. Mihoubi, M. Bell polynomials and binomial type sequences. *Discret. Math.* **2008**, *308*, 2450–2459. [CrossRef]
7. Aboud, A.; Bultel, J.-P.; Chouria, A.; Luque, J.-G.; Mallet, O. Word Bell polynomials. *Sém. Lothar. Combin.* **2017**, *75*, B75h.
8. Connon, D.F.; Various applications of the (exponential) complete Bell polynomials. *arXiv* **2010**, arXiv:1001.2835.
9. Kolbig, K.S. The complete Bell polynomials for certain arguments in terms of Stirling numbers of the first kind. *J. Comput. Appl. Math.* **1994**, *51*, 113–116. [CrossRef]
10. Port, D. Polynomial Maps with Applications to Combinatorics and Probability Theory. Ph.D. Thesis, Massachusetts Institute of Technology, Cambridge, MA, USA, February 1994.
11. Butzer, P.L.; Schmidt, M.; Stark, E.L.; Vogt, L. Central factorial numbers, their main properties and some applications. *Numer. Funct. Anal. Optim.* **1989**, *10*, 419–488. [CrossRef]
12. Charalambides, C.A. Central factorial numbers and related expansions. *Fibonacci Quart.* **1981**, *19*, 451–456.
13. Kim, T. A note on central factorial numbers. *Proc. Jangjeon. Math. Soc.* **2018**, *21*, 575–588.
14. Zhang, W. Some identities involving the Euler and the central factorial numbers. *Fibonacci Quart.* **1998**, *36*, 154–157.
15. Kim, T.; Kim, D.S. A note on central Bell numbers and polynomials. *Russ. J. Math. Phys.* **2019**, *26*, in press.
16. Kim, D.S.; Kwon, J.; Dolgy, D.V.; Kim, T. On central Fubini polynomials associated with central factorial numbers of the second kind. *Proc. Jangjeon Math. Soc.* **2018**, *21*, 589–598.
17. Bouroubi, S.; Abbas, M. New identities for Bell's polynomials: New approaches. *Rostock. Math. Kolloq.* **2006**, *61*, 49–55.
18. Carlitz, L. Some remarks on the Bell numbers. *Fibonacci Quart.* **1980**, *18*, 66–73.
19. Carlitz, L.; Riordan, J. The divided central differences of zero. *Can. J. Math.* **1963**, *15*, 94–100. [CrossRef]
20. Kim, T.; Kim, D.S. On λ-Bell polynomials associated with umbral calculus. *Russ. J. Math. Phys.* **2017**, *24*, 69–78. [CrossRef]
21. Kim, D.S.; Kim, T. Some identities of Bell polynomials. *Sci. China Math.* **2015**, *58*, 2095–2104. [CrossRef]
22. Kim, T.; Kim, D.S. Differential equations associated with degenerate Changhee numbers of the second kind. *Rev. R. Acad. Cienc. Exactas Fís. Nat. Ser. A Math.* **2018**, doi:10.1007/s13398-018-0576-y. [CrossRef]

10

Representing Sums of Finite Products of Chebyshev Polynomials of Third and Fourth Kinds by Chebyshev Polynomials

Taekyun Kim [1], Dae San Kim [2], Dmitry V. Dolgy [3] and Cheon Seoung Ryoo [4,*]

1 Department of Mathematics, Kwangwoon University, Seoul 139-701, Korea; tkkim@kw.ac.kr
2 Department of Mathematics, Sogang University, Seoul 121-742, Korea; dskim@sogang.ac.kr
3 Institute of Natural Sciences, Far Eastern Federal University, 690950 Vladivostok, Russia; d_dol@mail.ru
4 Department of Mathematics, Hannam University, Daejeon 306-791, Korea
* Correspondence: ryoocs@hnu.kr

Abstract: Here, we consider the sums of finite products of Chebyshev polynomials of the third and fourth kinds. Then, we represent each of those sums of finite products as linear combinations of the four kinds of Chebyshev polynomials, which involve the hypergeometric function ${}_3F_2$.

Keywords: Chebyshev polynomials; sums of finite products; hypergeometric function

MSC: 11B68; 33C45

---

1. Introduction and Preliminaries

We first recall here that, for any nonnegative integer n, the falling factorial polynomials $(x)_n$ and the rising factorial polynomials $< x >_n$ are respectively given by:

$$(x)_n = x(x-1)\cdots(x-n+1), \quad (n \geq 1), \quad (x)_0 = 1, \tag{1}$$

$$< x >_n = x(x+1)\cdots(x+n-1), \quad (n \geq 1), \quad < x >_0 = 1. \tag{2}$$

The two factorial polynomials are related by:

$$(x)_n = (-1)^n < -x >_n, \quad < x >_n = (-1)^n(-x)_n. \tag{3}$$

We will make use of the following.

$$\frac{(2n-2s)!}{(n-s)!} = \frac{2^{2n-2s}(-1)^s < \frac{1}{2} >_n}{< \frac{1}{2} - n >_s}, \tag{4}$$

for any integers n, s with $n \geq s \geq 0$.

$$B(x,y) = \int_0^1 t^{x-1}(1-t)^{y-1}dt = \frac{\Gamma(x)\Gamma(y)}{\Gamma(x+y)}, \quad (\mathrm{Re}(x), \mathrm{Re}(y) > 0), \tag{5}$$

$$\Gamma\left(n+\frac{1}{2}\right) = \frac{(2n)!\Gamma(\frac{1}{2})}{2^{2n}n!}, \quad (n \geq 0). \tag{6}$$

Here, $B(x,y)$ and $\Gamma(x)$ are respectively the Beta and Gamma functions.

The hypergeometric function ${}_pF_q\left(\begin{smallmatrix} a_1, & \cdots, & a_p \\ b_1, & \cdots, & b_q \end{smallmatrix}; x\right)$ is defined by (see [1]):

$$
{}_pF_q\left(\begin{matrix} a_1, & \cdots, & a_p \\ b_1, & \cdots, & b_p \end{matrix}; x\right) = \sum_{n=0}^{\infty} \frac{< a_1 >_n \cdots < a_p >_n}{< b_1 >_n \cdots < b_q >_n} \frac{x^n}{n!} \tag{7}
$$
$$
(p \leq q+1, \quad |x| < 1).
$$

In this paper, we will need only some basic knowledge about Chebyshev polynomials, which we recall here in below. The interested reader may want to refer to [1–3] for full accounts of this fascinating area of orthogonal polynomials.

The Chebyshev polynomials of the first, second, third and fourth kinds are respectively defined by the following generating functions.

$$
\frac{1-xt}{1-2xt+t^2} = \sum_{n=0}^{\infty} T_n(x)t^n, \tag{8}
$$

$$
\frac{1}{1-2xt+t^2} = \sum_{n=0}^{\infty} U_n(x)t^n, \tag{9}
$$

$$
F(t,x) = \frac{1-t}{1-2xt+t^2} = \sum_{n=0}^{\infty} V_n(x)t^n, \tag{10}
$$

$$
G(t,x) = \frac{1+t}{1-2xt+t^2} = \sum_{n=0}^{\infty} W_n(x)t^n. \tag{11}
$$

One way of deriving their generating functions is from their trigonometric formulas. For example, those formulas for $V_n(x)$ and $W_n(x)$ are given by:

$$
\begin{aligned}
V_n(\cos\theta) &= \frac{\cos(n+\frac{1}{2})\theta}{\cos\frac{\theta}{2}}, \\
W_n(\cos\theta) &= \frac{\sin(n+\frac{1}{2})\theta}{\sin\frac{\theta}{2}}.
\end{aligned}
$$

They are explicitly expressed as in the following.

$$
\begin{aligned}
T_n(x) &= {}_2F_1\left(-n, n; \frac{1}{2}; \frac{1-x}{2}\right) \\
&= \frac{n}{2}\sum_{l=0}^{[\frac{n}{2}]}(-1)^l \frac{1}{n-l}\binom{n-l}{l}(2x)^{n-2l}, \quad (n \geq 1),
\end{aligned} \tag{12}
$$

$$
\begin{aligned}
U_n(x) &= (n+1)\,{}_2F_1\left(-n, n+2; \frac{3}{2}; \frac{1-x}{2}\right) \\
&= \sum_{l=0}^{[\frac{n}{2}]}(-1)^l \binom{n-l}{l}(2x)^{n-2l}, \quad (n \geq 0),
\end{aligned} \tag{13}
$$

$$
\begin{aligned}
V_n(x) &= {}_2F_1\left(-n, n+1; \frac{1}{2}; \frac{1-x}{2}\right) \\
&= \sum_{l=0}^{n}\binom{2n-l}{l}2^{n-l}(x-1)^{n-l}, \quad (n \geq 0),
\end{aligned} \tag{14}
$$

$$W_n(x) = (2n+1)\,{}_2F_1\left(-n, n+1; \frac{3}{2}; \frac{1-x}{2}\right)$$
$$= (2n+1)\sum_{l=0}^{n} \frac{2^{n-l}}{2n-2l+1}\binom{2n-l}{l}(x-1)^{n-l}, \quad (n \geq 0). \tag{15}$$

The Chebyshev polynomials of the first, second, third and fourth kinds are also given by Rodrigues' formulas.

$$T_n(x) = \frac{(-1)^n 2^n n!}{(2n)!}(1-x^2)^{\frac{1}{2}}\frac{d^n}{dx^n}(1-x^2)^{n-\frac{1}{2}}, \tag{16}$$

$$U_n(x) = \frac{(-1)^n 2^n (n+1)!}{(2n+1)!}(1-x^2)^{-\frac{1}{2}}\frac{d^n}{dx^n}(1-x^2)^{n+\frac{1}{2}}, \tag{17}$$

$$(1-x)^{-\frac{1}{2}}(1+x)^{\frac{1}{2}}V_n(x) = \frac{(-1)^n 2^n n!}{(2n)!}\frac{d^n}{dx^n}(1-x)^{n-\frac{1}{2}}(1+x)^{n+\frac{1}{2}}, \tag{18}$$

$$(1-x)^{\frac{1}{2}}(1+x)^{-\frac{1}{2}}W_n(x) = \frac{(-1)^n 2^n n!}{(2n)!}\frac{d^n}{dx^n}(1-x)^{n+\frac{1}{2}}(1+x)^{n-\frac{1}{2}}. \tag{19}$$

They have the following orthogonalities with respect to various weight functions.

$$\int_{-1}^{1}(1-x^2)^{-\frac{1}{2}}T_n(x)T_m(x)dx = \frac{\pi}{\epsilon_n}\delta_{n,m}, \tag{20}$$

$$\int_{-1}^{1}(1-x^2)^{\frac{1}{2}}U_n(x)U_m(x)dx = \frac{\pi}{2}\delta_{n,m}, \tag{21}$$

$$\int_{-1}^{1}\left(\frac{1+x}{1-x}\right)^{\frac{1}{2}}V_n(x)V_m(x)dx = \pi\delta_{n,m}, \tag{22}$$

$$\int_{-1}^{1}\left(\frac{1-x}{1+x}\right)^{\frac{1}{2}}W_n(x)W_m(x)dx = \pi\delta_{n,m}, \tag{23}$$

where:

$$\epsilon_n = \begin{cases} 1, & \text{if } n = 0, \\ 2, & \text{if } n \geq 1, \end{cases} \quad \delta_n = \begin{cases} 0, & \text{if } n \neq m, \\ 1, & \text{if } n = m. \end{cases} \tag{24}$$

To proceed further, we let:

$$\alpha_{n,r}(x) = \sum_{l=0}^{n}\sum_{i_1+i_2+\cdots+i_{r+1}=l}\binom{r-1+n-l}{r-1}V_{i_1}(x)V_{i_2}(x)\cdots V_{i_{r+1}}(x),$$
$$(n \geq 0, r \geq 1), \tag{25}$$

$$\beta_{n,r}(x) = \sum_{l=0}^{n}\sum_{i_1+i_2+\cdots+i_{r+1}=l}(-1)^{n-l}\binom{r-1+n-l}{r-1}W_{i_1}(x)W_{i_2}(x)\cdots W_{i_{r+1}}(x),$$
$$(n \geq 0, r \geq 1). \tag{26}$$

We note here that both $\alpha_{n,r}(x)$ and $\beta_{n,r}(x)$ are polynomials of degree n.

In the following, we assume that the polynomials with subscript n, like $p_n(x), q_n(x)$ and $r_n(x)$, have degree n.

The linearization problem in general consists of determining the coefficients $c_{nm}(k)$ in the expansion of the product of two polynomials $q_n(x)$ and $r_m(x)$ in terms of an arbitrary polynomial sequence $\{p_k(x)\}_{k\geq 0}$:

$$q_n(x)r_m(x) = \sum_{k=0}^{n+m} c_{nm}(k)p_k(x).$$

A special problem of this is the case when $p_n(x) = q_n(x) = r_n(x)$, which is called either the standard linearization or the Clebsch–Gordan-type problem.

Another particular case is when $r_m(x) = 1$, which is the so-called connection problem. If further $q_n(x) = x^n$, it is called the inversion problem for the sequence $\{p_k(x)\}_{k\geq 0}$.

In this paper, we will consider the sums of finite products of Chebyshev polynomials of the third and fourth kinds in (25) and (26). Then, we are going to express each of them as linear combinations of the four kinds of Chebyshev polynomials $T_n(x)$, $U_n(x)$, $V_n(x)$ and $W_n(x)$. Thus, our problem may be regarded as a generalization of the linearization problem. We obtain them by explicit computations and using Propositions 1 and Lemma 1. The general formulas in Proposition 1 can be derived by using orthogonalities and Rodrigues' formulas for Chebyshev polynomials and integration by parts.

Finally, we note that many problems in physics and engineering can be solved with the help of special functions; for instance, we let the reader refer to the excellent papers [4–6] in this direction.

The next two theorems are our main results in which the terminating hypergeometric functions ${}_3F_2\left(\begin{smallmatrix} -n,\ a,\ b \\ d,\ e \end{smallmatrix};1\right)$ appear.

Theorem 1. *Let n, r be integers with $n \geq 0$, $r \geq 1$. Then we have following.*

$$\sum_{l=0}^{n} \sum_{i_1+i_2+\cdots+i_{r+1}=l} \binom{r-1+n-l}{r-1} V_{i_1}(x)V_{i_2}(x)\cdots V_{i_{r+1}}(x)$$
$$= \frac{(-1)^n(2n+2r)!}{r!2^{2r}(n+r-\frac{1}{2})_r} \times \sum_{k=0}^{n} \frac{(-1)^k \epsilon_k}{(n-k)!(n+k)!} {}_3F_2\left(\begin{matrix} k-n,\ -k-n,\ \frac{1}{2}-n-r \\ \frac{1}{2}-n,\ -2n-2r \end{matrix};1\right) T_k(x) \tag{27}$$

$$= \frac{(-1)^n(2n+2r)!}{r!2^{2r-2}(n+r-\frac{1}{2})_{r-1}} \times \sum_{k=0}^{n} \frac{(-1)^k (k+1)}{(n-k)!(n+k+2)!} {}_3F_2\left(\begin{matrix} k-n,\ -k-n-2,\ \frac{1}{2}-n-r \\ -\frac{1}{2}-n,\ -2n-2r \end{matrix};1\right) U_k(x) \tag{28}$$

$$= \frac{(-1)^n(2n+2r)!}{r!2^{2r}(n+r-\frac{1}{2})_r} \times \sum_{k=0}^{n} \frac{(-1)^k (2k+1)}{(n-k)!(n+k+1)!} {}_3F_2\left(\begin{matrix} k-n,\ -k-n-1,\ \frac{1}{2}-n-r \\ \frac{1}{2}-n,\ -2n-2r \end{matrix};1\right) V_k(x) \tag{29}$$

$$= \frac{(-1)^n(2n+2r)!}{r!2^{2r-1}(n+r-\frac{1}{2})_{r-1}} \times \sum_{k=0}^{n} \frac{(-1)^k}{(n-k)!(n+k+1)!} {}_3F_2\left(\begin{matrix} k-n,\ -k-n-1,\ \frac{1}{2}-n-r \\ -\frac{1}{2}-n,\ -2n-2r \end{matrix};1\right) W_k(x). \tag{30}$$

Theorem 2. *Let n, r be integers with $n \geq 0$, $r \geq 1$. Then we have following.*

$$\sum_{l=0}^{n} \sum_{i_1+i_2+\cdots+i_{r+1}=l} (-1)^{n-l} \binom{r-1+n-l}{r-1} W_{i_1}(x)W_{i_2}(x)\cdots W_{i_{r+1}}(x)$$
$$= \frac{(-1)^n(2n+2r)!}{r!2^{2r}(n+r+\frac{1}{2})_r} \times \sum_{k=0}^{n} \frac{(-1)^k \epsilon_k}{(n-k)!(n+k)!} {}_3F_2\left(\begin{matrix} k-n,\ -k-n,\ -\frac{1}{2}-n-r \\ \frac{1}{2}-n,\ -2n-2r \end{matrix};1\right) T_k(x) \tag{31}$$

$$=\frac{(-1)^n(2n+1)(2n+2r)!}{r!2^{2r-1}(n+r+\frac{1}{2})_r}$$
$$\times\sum_{k=0}^{n}\frac{(-1)^k(k+1)}{(n-k)!(n+k+2)!}{}_3F_2\left(\begin{matrix}k-n, & -k-n-2, & -\frac{1}{2}-n-r\\ & -\frac{1}{2}-n, & -2n-2r\end{matrix};1\right)U_k(x) \tag{32}$$
$$=\frac{(-1)^n(2n+2r)!}{r!2^{2r}(n+r+\frac{1}{2})_r}$$
$$\times\sum_{k=0}^{n}\frac{(-1)^k(2k+1)}{(n-k)!(n+k+1)!}{}_3F_2\left(\begin{matrix}k-n, & -k-n-1, & -\frac{1}{2}-n-r\\ & \frac{1}{2}-n, & -2n-2r\end{matrix};1\right)V_k(x) \tag{33}$$
$$=\frac{(-1)^n(2n+1)(2n+2r)!}{r!2^{2r}(n+r+\frac{1}{2})_r}$$
$$\times\sum_{k=0}^{n}\frac{(-1)^k}{(n-k)!(n+k+1)!}{}_3F_2\left(\begin{matrix}k-n, & -k-n-1, & -\frac{1}{2}-n-r\\ & -\frac{1}{2}-n, & -2n-2r\end{matrix};1\right)W_k(x). \tag{34}$$

As we know, the Bernoulli polynomials are not orthogonal polynomials, but Appell polynomials. In [7], the sums of finite products of Chebyshev polynomials in (25) and (26) were expressed as linear combinations of Bernoulli polynomials. Furthermore, the same has been done for the sums of finite products of Bernoulli, Euler and Genocchi polynomials in [8–10]. All of these were found by deriving Fourier series expansions for the functions closely connected with those various sums of finite products. For some other applications of Chebyshev polynomials, we let the reader refer to [11–13].

2. Proof of Theorem 1

Here, we will prove Theorem 1. For this purpose, we first state Proposition 1 and Lemma 1 that will be used in Sections 2 and 3.

The results in Proposition 1 can be derived by using the orthogonalities in (20)–(23) and the Rodrigues formulas in (16)–(19). The statements (a) and (b) in Proposition 1 are respectively from the Equations (23) and (35) of [14], while (c) and (d) are respectively from the Equations (22) and (37) of [15].

Proposition 1. *Let $q(x)\in\mathbb{R}[x]$ be a polynomial of degree n. Then, we have the following.*

$$(a)\quad q(x)=\sum_{k=0}^{n}c_{k,1}T_k(x),$$
$$\text{where}\quad c_{k,1}=\frac{(-1)^k2^kk!\epsilon_k}{(2k)!\pi}\int_{-1}^{1}q(x)\frac{d^k}{dx^k}(1-x^2)^{k-\frac{1}{2}}dx,$$

$$(b)\quad q(x)=\sum_{k=0}^{n}c_{k,2}U_k(x),$$
$$\text{where}\quad c_{k,2}=\frac{(-1)^k2^{k+1}(k+1)!}{(2k+1)!\pi}\int_{-1}^{1}q(x)\frac{d^k}{dx^k}(1-x^2)^{k+\frac{1}{2}}dx,$$

$$(c)\quad q(x)=\sum_{k=0}^{n}c_{k,3}V_k(x),$$
$$\text{where}\quad c_{k,3}=\frac{(-1)^k2^kk!}{(2k)!\pi}\int_{-1}^{1}q(x)\frac{d^k}{dx^k}(1-x)^{k-\frac{1}{2}}(1+x)^{k+\frac{1}{2}}dx,$$

$$(d)\quad q(x)=\sum_{k=0}^{n}c_{k,4}W_k(x),$$
$$\text{where}\quad c_{k,4}=\frac{(-1)^k 2^k k!}{(2k)!\pi}\int_{-1}^{1}q(x)\frac{d^k}{dx^k}(1-x)^{k+\frac{1}{2}}(1+x)^{k-\frac{1}{2}}dx.$$

Lemma 1. *Let l, m be nonnegative integers. Then, we have the following.*

$$\begin{aligned}&\int_{-1}^{1}(1-x)^{m-\frac{1}{2}}(1+x)^{l-\frac{1}{2}}dx\\&=\frac{2^{l+m}}{(l+m)!}\Gamma(l+\frac{1}{2})\Gamma(m+\frac{1}{2})\\&=\frac{(2l)!\ (2m)!\ \pi}{2^{l+m}\ (l+m)!\ l!\ m!}.\end{aligned}\tag{35}$$

Proof. By changing the variables $1+x=2y$, the integral in (35) becomes:

$$\begin{aligned}2^{l+m}\int_{0}^{1}y^{l+\frac{1}{2}-1}(1-y)^{m+\frac{1}{2}-1}dy&=2^{l+m}\frac{\Gamma(l+\frac{1}{2})\Gamma(m+\frac{1}{2})}{\Gamma(l+m+1)}\\&=\frac{2^{l+m}\ (2l)!\ \Gamma(\frac{1}{2})\ (2m)!\ \Gamma(\frac{1}{2})}{(l+m)!\ 2^{2l}\ l!\ 2^{2m}\ m!},\end{aligned}$$

where we used (5) and (6). □

As was shown in [7], the following lemma can be obtained by differentiating Equation (10). It expresses the sums of finite products in (25) very neatly, which plays an important role in the following discussion.

Lemma 2. *Let n, r be integers with $n \geq 0$, $r \geq 1$. Then, we have the identity.*

$$\sum_{l=0}^{n}\sum_{i_1+i_2+\cdots+i_{r+1}=l}\binom{r-1+n-l}{r-1}V_{i_1}(x)\cdots V_{i_{r+1}}(x)=\frac{1}{2^r\ r!}V_{n+r}^{(r)}(x),\tag{36}$$

where the inner sum runs over all nonnegative integers $i_1, i_2, \cdots, i_{r+1}$, with $i_1+i_2+\cdots+i_{r+1}=l$.

From (14), the r-th derivative of $V_n(x)$ is given by:

$$V_n^{(r)}(x)=\sum_{l=0}^{n-r}\binom{2n-l}{l}2^{n-l}(n-l)_r(x-1)^{n-l-r}.\tag{37}$$

In particular, we have:

$$V_{n+r}^{(r+k)}(x)=\sum_{l=0}^{n-k}\binom{2n+2r-l}{l}2^{n+r-l}(n+r-l)_{r+k}(x-1)^{n-k-l}.\tag{38}$$

$$V_{n+r}^{(r+k)}(x)=\sum_{l=0}^{n-k}\binom{2n+2r-l}{l}2^{n+r-l}(n+r-l)_{r+k}(x-1)^{n-k-l}.\tag{38}$$

Here, we will show only (28) of Theorem 1, since (27), (29) and (30) can be proved similarly to (28). With $\alpha_{n,r}(x)$ as in (25), we let:

$$\alpha_{n,r}(x)=\sum_{k=0}^{n}c_{k,2}U_k(x).\tag{39}$$

Then, from (b) of Proposition 1, (36), (38) and integration by parts k times, we have:

$$\begin{aligned} c_{k,2} &= \frac{(-1)^k 2^{k+1}(k+1)!}{(2k+1)!\pi} \int_{-1}^{1} \alpha_{n,r}(x) \frac{d^k}{dx^k}(1-x^2)^{k+\frac{1}{2}} dx \\ &= \frac{(-1)^k 2^{k+1}(k+1)!}{(2k+1)!\pi 2^r r!} \int_{-1}^{1} V_{n+r}^{(r)}(x) \frac{d^k}{dx^k}(1-x^2)^{k+\frac{1}{2}} dx \\ &= \frac{2^{k+1}(k+1)!}{(2k+1)!\pi 2^r r!} \int_{-1}^{1} V_{n+r}^{(r+k)}(x)(1-x^2)^{k+\frac{1}{2}} dx \\ &= \frac{2^{k+1}(k+1)!}{(2k+1)!\pi 2^r r!} \sum_{l=0}^{n-k} (-1)^{n-k-l} \binom{2n+2r-l}{l} 2^{n+r-l} \\ &\quad \times (n+r-l)_{r+k} \int_{-1}^{1} (1-x)^{n-l+1-\frac{1}{2}}(1+x)^{k+1-\frac{1}{2}} dx. \end{aligned} \tag{40}$$

From (40), (35), we get:

$$\begin{aligned} c_{k,2} &= \frac{2^{k+1}(k+1)!}{(2k+1)!\pi 2^r r!} \\ &\times \sum_{l=0}^{n-k} \frac{(-1)^{n-k-l}(2n+2r-l)!2^{n+r-l}(n+r-l)!(2k+2)!(2n-2l+2)!\pi}{l!(2n+2r-2l)!(n-k-l)!2^{n-l+k+2}(n-l+k+2)!(n-l+1)!(k+1)!} \\ &= \frac{(-1)^{n-k}(k+1)}{r!} \\ &\times \sum_{l=0}^{n-k} \frac{(-1)^l(2n+2r-l)!(n+r-l)!(2n+2-2l)!}{l!(n-k-l)!(n+k-l+2)!(2n+2r-2l)!(n+1-l)!}. \end{aligned} \tag{41}$$

Using (3) and (4), (41) is equal to:

$$\begin{aligned} c_{k,2} &= \frac{(-1)^{n-k}(k+1)(2n+2r)!}{r!(n-k)!(n+k+2)!} \\ &\times \sum_{l=0}^{n-k} \frac{(-1)^l(n-k)_l(n+k+2)_l < \frac{1}{2}-n-r >_l 2^{2n-2l+2}(-1)^l < \frac{1}{2} >_{n+1}}{l!(2n+2r)_l 2^{2n+2r-2l}(-1)^l < \frac{1}{2} >_{n+r} < \frac{1}{2}-n-1 >_l}. \\ &= \frac{(-1)^n(2n+2r)!}{r!2^{2r-2}(n+r-\frac{1}{2})_{r-1}} \\ &\times \quad \frac{(-1)^k(k+1)}{(n-k)!(n+k+2)!} \sum_{l=0}^{n-k} \frac{< k-n >_l < -k-n-2 >_l < \frac{1}{2}-n-r >_l}{< -\frac{1}{2}-n >_l < -2n-2r >_l \, l!} \\ &= \frac{(-1)^n(2n+2r)!}{r!2^{2r-2}(n+r-\frac{1}{2})_{r-1}} \\ &\times \quad \frac{(-1)^k(k+1)}{(n-k)!(n+k+2)!} {}_3F_2\left(\begin{matrix} k-n, \; -k-n-2, \; \frac{1}{2}-n-r \\ -\frac{1}{2}-n, \; -2n-2r \end{matrix} ; 1 \right). \end{aligned} \tag{42}$$

Now, the Equation (28) in Theorem 1 follows from (39) and (42).

3. Proof of Theorem 2

In this section, we will show (31) of Theorem 2, as (32)–(34) can be treated analogously to (31). The following lemma can be obtained by differentiating (11) and is stated as Lemma 3 in [7].

Lemma 3. *Let n, r be integers with $n \geq 0$, $r \geq 1$. Then, we have the following identity.*

$$\begin{aligned}&\sum_{l=0}^{n} \sum_{i_1+i_2+\cdots+i_{r+1}=l} (-1)^{n-l} \binom{r-1+n-l}{r-1} W_{i_1}(x) W_{i_2}(x) \cdots W_{i_{r+1}}(x) \\ &= \frac{1}{2^r r!} W_{n+r}^{(r)}(x),\end{aligned} \tag{43}$$

where the inner sum runs over all nonnegative integers $i_1, i_2, \cdots, i_{r+1}$, with $i_1 + i_2 + \cdots + i_{r+1} = l$.

From (15), the r-th derivative of $W_n(x)$ is given by:

$$W_n^{(r)}(x) = (2n+1) \sum_{l=0}^{n-r} \frac{2^{n-l}}{2n+1-2l} \binom{2n-l}{l} (n-l)_r (x-1)^{n-l-r}. \tag{44}$$

In particular,

$$\begin{aligned}&W_{n+r}^{(r+k)}(x) \\ &= (2n+1) \sum_{l=0}^{n-k} \frac{2^{n+r-l}}{2n+2r+1-2l} \binom{2n+2r-l}{l} (n+r-l)_{r+k} (x-1)^{n-k-l}.\end{aligned} \tag{45}$$

Here, we will show only (31) of Theorem 2, since (32)–(34) can be proven analogously to (31). With $\beta_{n,r}(x)$ as in (26), we put:

$$\beta_{n,r}(x) = \sum_{k=0}^{n} c_{k,1} T_k(x). \tag{46}$$

Then, from (a) of Proposition 1, (43), (45) and integration by parts k times, we have:

$$\begin{aligned}c_{k,1} &= \frac{(-1)^k 2^k k! \epsilon_k}{(2k)!\pi} \int_{-1}^{1} \beta_{n,r}(x) \frac{d^k}{dx^k} (1-x^2)^{k-\frac{1}{2}} dx \\ &= \frac{(-1)^k 2^k k! \epsilon_k}{(2k)!\pi 2^r r!} \int_{-1}^{1} W_{n+r}^{(r)}(x) \frac{d^k}{dx^k} (1-x^2)^{k-\frac{1}{2}} dx \\ &= \frac{2^k k! \epsilon_k}{(2k)!\pi 2^r r!} \int_{-1}^{1} W_{n+r}^{(r+k)}(x) (1-x^2)^{k-\frac{1}{2}} dx \\ &= \frac{(2n+1) 2^k k! \epsilon_k}{(2k)!\pi 2^r r!} \sum_{l=0}^{n-k} \frac{(-1)^{n-k-l} 2^{n+r-l}}{2n+2r+1-2l} \binom{2n+2r-l}{l} \\ &\quad \times (n+r-l)_{r+k} \int_{-1}^{1} (1-x)^{n-l-\frac{1}{2}} (1+x)^{k-\frac{1}{2}} dx.\end{aligned} \tag{47}$$

From (47), (35) and after some simplifications, we get:

$$\begin{aligned}c_{k,2} &= \frac{(2n+1)\epsilon_k (-1)^{n-k}}{r!} \\ &\quad \times \sum_{l=0}^{n-k} \frac{(-1)^l (2n+2r-l)!(n+r-l)!}{l!(n-k-l)!(n+k-l)!(2n+2r-2l+1)!(n-l)!} \\ &= \frac{2(2n+1)\epsilon_k (-1)^{n-k}}{r!} \\ &\quad \times \sum_{l=0}^{n-k} \frac{(-1)^l (2n+2r-l)!(n+r-l+1)!(2n-2l)!}{l!(n-k-l)!(n+k-l)!(2n+2r-2l+2)!(n-l)!}.\end{aligned} \tag{48}$$

Using (3) and (4), (48) is equal to:

$$\begin{aligned} c_{k,1} &= \frac{2(2n+1)(2n+2r)!\epsilon_k(-1)^{n-k}}{r!(n-k)!(n+k)!} \\ &\times \sum_{l=0}^{n-k} \frac{(-1)^l(n-k)_l(n+k)_l < \frac{1}{2}-n-r-1 >_l 2^{2n-2l}(-1)^l < \frac{1}{2} >_n}{l!(2n+2r)_l 2^{2n+2r+2-2l}(-1)^l < \frac{1}{2} >_{n+r+1} < \frac{1}{2}-n >_l} \\ &= \frac{(2n+1)(-1)^n(2n+2r)!}{r!2^{2r+1}(n+r+\frac{1}{2})_{r+1}} \\ &\times \frac{(-1)^k\epsilon_k}{(n-k)!(n+k)!} \sum_{l=0}^{k} \frac{< k-n >_l < -k-n >_l < -\frac{1}{2}-n-r >_l}{< \frac{1}{2}-n >_l < -2n-2r >_l} \\ &= \frac{(-1)^n(2n+2r)!}{r!2^{2r}(n+r+\frac{1}{2})_r} \\ &\times \frac{(-1)^k\epsilon_k}{(n-k)!(n+k)!} {}_3F_2\left(\begin{matrix} k-n, & -k-n, & -\frac{1}{2}-n-r \\ & \frac{1}{2}-n, & -2n-2r \end{matrix} ;1 \right). \end{aligned} \tag{49}$$

Now, Equation (31) in Theorem 2 follows from (46) and (49).

Remark 1. *As we noted earlier, Lemmas 2 and 3 play crucial roles and express sums of finite products in (25) and (26) very neatly as higher-order derivatives of $V_n(x)$ and $W_n(x)$. These could be derived by noting that Chebyshev polynomials are special cases of Jacobi polynomials and using the general formula for the derivative of Jacobi polynomials. Indeed, their Jacobi polynomial expressions and the derivatives of the Jacobi polynomials are as follows:*

$$\begin{aligned} V_n(x) &= P_n^{(-1/2,1/2)}(x)/P_n^{(-1/2,1/2)}(1), \\ W_n(x) &= P_n^{(1/2,-1/2)}(x)/P_n^{(1/2,-1/2)}(1), \\ \frac{d}{dx}P_n^{(a,b)}(x) &= \frac{1}{2}(n+a+b+1)P_{n-1}^{(a+1,b+1)}(x). \end{aligned}$$

4. Conclusions

The linearization problem in general consists of determining the coefficients $c_{nm}(k)$ in the expansion of the product of two polynomials $q_n(x)$ and $r_m(x)$ in terms of an arbitrary polynomial sequence $\{p_k(x)\}_{k\geq 0}$:

$$q_n(x)r_m(x) = \sum_{k=0}^{n+m} c_{nm}(k)p_k(x).$$

Along this line and as a generalization of this, we considered sums of finite products of Chebyshev polynomials of the third and fourth kinds and represented each of those sums of finite products as linear combinations of the four kinds of Chebyshev polynomials, which involve the hypergeometric function ${}_3F_2$. It is certainly possible to represent such sums of finite products by other orthogonal polynomials, which is our ongoing project.

Author Contributions: T.K. and D.S.K. conceived the framework and structured the whole paper; T.K. wrote the paper; C.S.R. and D.D.V. checked the results of the paper; D.S.K. and C.S.R. completed the revision of the article.

Acknowledgments: The authors would like to thank the referees for their valuable comments, which improved the original manuscript in its present form.

References

1. Andrews, G.E.; Askey, R.; Roy, R. *Special Functions*; Encyclopedia of Mathematics and Its Applications 71; Cambridge University Press: Cambridge, UK, 1999.
2. Beals, R.; Wong, R. *Special Functions and Orthogonal Polynomials*; Cambridge Studies in Advanced Mathematics 153; Cambridge University Press: Cambridge, UK, 2016.
3. Wang, Z.X.; Guo, D.R. *Special Functions*; Guo, D.R., Xia, X.J., Trans.; World Scientific Publishing Co., Inc.: Teaneck, NJ, USA, 1989.
4. Marin, M. A temporally evolutionary equation in elasticity of micropolar bodies with voids. *Politehn. Univ. Buchar. Sci. Bull. Ser. A Appl. Math. Phys.* **1998**, *60*, 3–12.
5. Marin, M.; Stan, G. Weak solutions in elasticity of dipolar bodies with stretch. *Carpath. J. Math.* **2013**, *29*, 33–40.
6. Marin, D.; Baleanu, D. On vibrations in thermoelasticity without energy dissipation for micropolar bodies. *Bound. Value Probl.* **2016**, *2016*, 111. [CrossRef]
7. Kim, T.; Kim, D.S.; Dolgy, D.V.; Kwon, J. Representing Sums of finite products of Chebyshev polynomials of the third and fourth kinds by Chebyshev Polynomials. *Preprints* **2018**, 2018060079. [CrossRef]
8. Agarwal, R.P.; Kim, D.S.; Kim, T.; Kwon, J. Sums of finite products of Bernoulli functions. *Adv. Differ. Equ.* **2017**, *2017*, 15. [CrossRef]
9. Kim, T.; Kim, D.S.; Jang, G.-W.; Kwon, J. Sums of finite products of Euler functions. In *Advances in Real and Complex Analysis with Applications, Trends in Mathematics*; Springer: New York, NY, USA, 2017; pp. 243–260.
10. Kim, T.; Kim, D.S.; Jang, L.C.; Jang, G.-W. Sums of finite products of Genocchi functions. *Adv. Differ. Equ.* **2017**, *2017*, 17. [CrossRef]
11. Doha, E.H.; Abd-Elhameed, W.M.; Alsuyuti, M.M. On using third and fourth kinds Chebyshev polynomials for solving the integrated forms of high odd-order linear boundary value problems. *J. Egyptian Math. Soc.* **2015**, *23*, 397–405. [CrossRef]
12. Kruchinin, D.V.; Kruchinin, V.V. Application of a composition of generating functions for obtaining explicit formulas of polynomials. *J. Math. Anal. Appl.* **2013**, *404*, 161–171. [CrossRef]
13. Mason, J.C. Chebyshev polynomials of the second, third and fourth kinds in approximation, indefinite integration, and integral transforms. *J. Comput. Appl. Math.* **1993**, *49*, 169–178. [CrossRef]
14. Kim, D.S.; Kim, T.; Lee, S.-H. Some identities for Berounlli polynomials involving Chebyshev polynomials. *J. Comput. Anal. Appl.* **2014**, *16*, 172–180.
15. Kim, D.S.; Dolgy, D.V.; Kim, T.; Rim, S.-H. Identities involving Bernoulli and Euler polynomials arising from Chebyshev polynomials. *Proc. Jangjeon Math. Soc.* **2012**, *15*, 361–370.

11

The Solution Equivalence to General Models for the RIM Quantifier Problem

Dug Hun Hong

Department of Mathematics, Myongji University, Yongin 449-728, Kyunggido, Korea; dhhong@mju.ac.kr

Abstract: Hong investigated the relationship between the minimax disparity minimum variance regular increasing monotone (RIM) quantifier problems. He also proved the equivalence of their solutions to minimum variance and minimax disparity RIM quantifier problems. Hong investigated the relationship between the minimax ratio and maximum entropy RIM quantifier problems and proved the equivalence of their solutions to the maximum entropy and minimax ratio RIM quantifier problems. Liu proposed a general RIM quantifier determination model and proved it analytically by using the optimal control technique. He also gave the equivalence of solutions to the minimax problem for the RIM quantifier. Recently, Hong proposed a modified model for the general minimax RIM quantifier problem and provided correct formulation of the result of Liu. Thus, we examine the general minimum model for the RIM quantifier problem when the generating functions are Lebesgue integrable under the more general assumption of the RIM quantifier operator. We also provide a solution equivalent relationship between the general maximum model and the general minimax model for RIM quantifier problems, which is the corrected and generalized version of the equivalence of solutions to the general maximum model and the general minimax model for RIM quantifier problems of Liu's result.

Keywords: OWA operator; RIM quantifier; maximum entropy; minimax ratio; generating function; minimal variability; minimax disparity; solution equivalence

---

1. Introduction

One of the important topics in the theory of ordered weighted averaging (OWA) operators is the determination of the associated weights. Several authors have suggested a number of methods for obtaining associated weights in various areas such as decision-making, approximate reasoning, expert systems, data mining, fuzzy systems and control [1–22]. Yager [12] proposed RIM quantifiers as a method for finding OWA weight vectors through fuzzy linguistic quantifiers. Liu [15] and Liu and Da [16] gave solutions to the maximum-entropy RIM quantifier model when the generating functions are differentiable. Liu and Lou [9] studied the equivalence of solutions to the minimax ratio and maximum-entropy RIM quantifier models, and the equivalence of solutions to the minimax disparity and minimum-variance RIM quantifier problems. Hong [17,18] gave the proof of the minimax ratio RIM quantifier problem and the minimax disparity RIM quantifier model when the generating functions are absolutely continuous. He also gave solutions to the maximum-entropy RIM quantifier model and the minimum-variance RIM quantifier model when the generating functions are Lebesgue integrable.

Based on these results, Hong [17,18] provided a relationship between the minimax disparity and minimum-variance RIM quantifier problems. He also provided a correct relationship between the minimax ratio and maximum-entropy RIM quantifier models. Liu [19] suggested a general RIM quantifier determination model and proved it analytically using the optimal control methods. He also studied the solution equivalence to the minimax problem for the RIM quantifier.

This paper investigates the general minimax model for the RIM quantifier problem for the case in which the generating functions are absolutely continuous and a generalized solution to the general minimum model for the RIM quantifier problem for the case in which the generating functions are Lebesgue integrable. Moreover, this paper provides a solution equivalent relationship between the general maximum model and the general minimax model for RIM quantifier problems and generalizes the results of Hong [17,18]. In this paper, we improve and extend Liu's theorems to be suitable for absolutely continuous generating functions. We have corrected and improved Theorem 13 [19] by using the absolutely continuous condition of generating functions and the absolute continuity condition of F' for the general minimax model for the RIM quantifier problem. Theorem 9 [19] has been improved using the Lebesgue integrability condition of generating functions and the continuity condition of F' for the general maximum model for the RIM quantifier problem.

Based on these results, we give a correct relationship between the general minimum model and the general minimax model for RIM quantifier problems.

2. Preliminaries

Yager [11] proposed a new aggregation technique based on OWA operators. An OWA operator of dimension n is a mapping $F : R^n \to R$ that has an associated weight vector $W = (w_1, \cdots, w_n)^T$ with the properties $w_1 + \cdots + w_n = 1,\ 0 \le w_i \le 1,\ i = 1, \cdots, n$, such that

$$F(a_1, \cdots, a_n) = \sum_{i=1}^{n} w_i b_i,$$

where b_j is the jth largest element of the collection of the aggregated objects $\{a_1, \cdots, a_n\}$. In [11], Yager introduced a measure of "orness" associated with the weight vector W of an OWA operator:

$$orness(W) = \sum_{i=1}^{n} \frac{n-i}{n-1} w_i.$$

This measure characterizes the degree to which the aggregation is like an *OR* operation.

Here, *min*, *max*, and *average* correspond to W^*, W_* and W_A respectively, where $W^* = (1, 0, \cdots, 0), W_* = (0, 0, \cdots, 1)$ and $W_A = (1/n, 1/n, \cdots, 1/n)$. Clearly, $orness(W^*) = 1, orness(W_*) = 0$ and $orness(W_A) = 1/2$.

Yager [12] introduced RIM quantifiers as a method for obtaining OWA weight vectors through fuzzy linguistic quantifiers.

Definition 1 ([12]). *A fuzzy subset Q on the real line is called a RIM quantifier if $Q(0) = 0, Q(1) = 1$ and $Q(x) \ge Q(y)$ for $x > y$.*

The quantifier $for\ \ all$ is represented by the fuzzy set

$$Q_*(r) = \begin{cases} 1, & \text{if } x = 1, \\ 0, & \text{if } x \neq 1. \end{cases}$$

The quantifier *there exists* is defined as

$$Q^*(r) = \begin{cases} 0, & \text{if } x = 0, \\ 1, & \text{if } x \neq 0. \end{cases}$$

Both of these are examples of the RIM quantifier. A generating function representation of RIM quantifiers has been proposed for analyzing the relationship between OWA operators and RIM quantifiers.

Definition 2. *For $f(t)$ on [0, 1] and the RIM quantifier $Q(x)$, $f(t)$ is called the generating function of $Q(x)$, if it satisfies*

$$Q(x) = \int_0^x f(t)dt,$$

where $f(t) \geq 0$ and $\int_0^1 f(t)dt = 1$.

If the RIM quantifier $Q(x)$ is smooth, then $f(x)$ should be continuous; if $Q(x)$ is a piecewise linear function, then $f(x)$ is a jump piecewise function of some constants; and if $Q(x)$ is an absolutely continuous function, then $f(x)$ is a Lesbegue integrable function and unique in the sense of being "almost everywhere" [23].

Yager extended the *orness* measure of OWA operators, and defined the *orness* of RIM quantifiers [10] as:

$$orness(Q) = \int_0^1 Q(x)dx = \int_0^1 (1-t)f(t)dt.$$

We see that Q_* leads to the weight vector W_*, Q^* leads to the weight vector W^*, and the ordinary *average* RIM quantifier $Q_A(x) = x$ leads to the weight vector W_A. We also have $orness(Q^*) = 1$, $rness(Q_*) = 0$, and $orness(Q_A) = 1/2$.

As the RIM quantifier can be seen as a continuous form of OWA, an operator with a generating function, the OWA optimization problem can be extended to the case of the RIM quantifier.

3. The General Model for the Minimax RIM Quantifier Problem

In this section, we consider the general model for the minimax RIM quantifier problem and generalize some results of Hong [17,18]. Hong [7] provided a modified model for the minimax RIM quantifier problem and the correct formulation of a result of Liu [19]. We summarize briefly.

∗ **The minimax disparity RIM quantifier problem [15,17].**

The minimax disparity RIM quantifier problem with a given orness level $0 < \alpha < 1$ consists of finding a solution $f : [0,1] \to [0,1]$ to the following optimization problem:

$$\begin{aligned}
\text{Minimize} \quad & \max_{t\in(0,1)} \mid f'(t) \mid, \\
\text{subject to} \quad & \int_0^1 (1-r)f(r)dr = \alpha, \;\; 0<\alpha<1, \\
& \int_0^1 f(r)dr = 1, \\
& f(r) \geq 0.
\end{aligned}$$

∗ **The minimax ratio RIM quantifier problem [9,18].**

The minimax ratio RIM quantifier problem with a given orness level $0 < \alpha < 1$ consists of finding a solution $f : [0,1] \to [0,1]$ to the following optimization problem:

$$\begin{aligned}
\text{Minimize} \quad & \max_{t\in(0,1)} \left| \frac{f'(t)}{f(t)} \right|, \\
\text{subject to} \quad & \int_0^1 (1-r)f(r)dr = \alpha, \;\; 0<\alpha<1, \\
& \int_0^1 f(r)dr = 1, \\
& f(r) > 0.
\end{aligned}$$

In regard to the above optimization problem, Liu [19] considered a general model for the minimax RIM quantifier problem:

$$\begin{aligned} \text{Minimize} \quad & M_f = \max_{r \in (0,1)} \left| F''(f(r))f'(r) \right|, \\ \text{subject to} \quad & \int_0^1 rf(r)dr = \alpha, \;\; 0 < \alpha < 1, \\ & \int_0^1 f(r)dr = 1, \\ & f(r) > 0, \end{aligned} \tag{1}$$

where the generating functions are continuous and F is a strictly convex function on $[0, \infty)$, which is differentiable to at least the 2nd order.

The above two cases are special cases of this model with $F(x) = x^2$ and $F(x) = x \ln x$. Hong [7] gave a corrected and modified general model for the minimax RIM quantifier problem as follows:

∗ **The general model for the minimax RIM quantifier problem.**

$$\begin{aligned} \text{Minimize} \quad & M_f = ess\ sup_{r \in (0,1)} \left| F''(f(x))f'(x) \right|, \\ \text{subject to} \quad & \int_0^1 rf(r)dr = \alpha, \;\; 0 < \alpha < 1, \\ & \int_0^1 f(r)dr = 1, \\ & f(r) > 0. \end{aligned} \tag{2}$$

Theorem 1. *Supposing that the generating functions are absolutely continuous, F is a strictly convex function on $[0, \infty)$, and F' is absolutely continuous, then there is a unique optimal solution for problem (2), and that the optimal solution has the form*

$$f^*(r) = \max \left\{ (F')^{-1} (a^* r + b^*), 0 \right\},$$

where a^* and b^* are determined by the constraints:

$$\begin{cases} \int_0^1 rf^*(r)dr = \alpha, \\ \int_0^1 f^*(r)dr = 1, \\ f^*(r) \geq 0. \end{cases}$$

The next example shows that the condition of F' being *absolutely continuous* on $[0, \infty)$ in Theorem 1 is essential.

Example 1. *Letting $F_1(x) = \int_0^x (C(r) + r)dr$ where $C(x)$ is a Cantor function, then $F'(x) = C(x) + x$ and $F_1''(x) = 1$ a.e. but $F_1'(x) \neq \int_0^x F_1''(r)dr$, that is, F_1' is not absolutely continuous on $[0, \infty)$. Let $F_2(x) = (1/2)x^2$, then $F_2''(x) = 1$. Since*

$$ess\ sup_{r \in (0,1)} \left| F_1''(f(x))f'(x) \right| = ess\ sup_{r \in (0,1)} \left| f'(x) \right| = ess\ sup_{r \in (0,1)} \left| F_2''(f(x))f'(x) \right|,$$

the optimal solution of problem (2) with respect to F_1 and F_1 are the same. However, since $F_1'(x) \neq F_2'(x)$, the optimal solution of problem (2) with respect to F_1 and F_1 cannot be the same by Theorem 2, which is a contradiction. This example shows the Theorem 2 is incorrect if F' is not absolutely continuous on $[0, \infty)$.

4. The General Model for the Minimum RIM Quantifier Problem

In this section, we consider the general model for the minimum RIM quantifier problem. We improve the results of Liu [19] and generalize Theorem 4 of Hong [17] and Theorem 5 of Hong [18]. Liu [19] obtained solutions to the general minimum RIM quantifier problem for the case in which the generating functions are continuous and F is differentiable to at least the 2nd order by considering a variational optimization problem using the Lagrangian multiplier method ([24], Chapter 2). In this section, we consider a generalized result for this problem.

$*$ **The minimum variance RIM quantifier problem [17,18].**

The minimum variance RIM quantifier problem under a given orness level is

$$\begin{aligned} \text{Minimize} \quad & D_f = \int_0^1 f^2(r)dr, \\ \text{subject to} \quad & \int_0^1 rf(r)dr = \alpha, \ \ 0 < \alpha < 1, \\ & \int_0^1 f(r)dr = 1, \\ & f(r) > 0. \end{aligned}$$

$*$ **The maximum entropy RIM quantifier problem [9,18].**

The maximum entropy RIM quantifier problem with a given orness level $0 < \alpha < 1$ consists of finding a solution $f : [0,1] \to [0,1]$ to the following optimization problem:

$$\begin{aligned} \text{Maximize} \quad & -\int_0^1 f(r)\ln f(r)dr, \\ \text{subject to} \quad & \int_0^1 rf(r)dr = \alpha, \ \ 0 < \alpha < 1, \\ & \int_0^1 f(r)dr = 1, \\ & f(r) > 0. \end{aligned}$$

Recently, Liu [19] considered the general model for the minimum variance and maximum entropy RIM quantifier problems, under a given orness level formulated as follows:

$*$ **The general model for the minimum RIM quantifier problem.**

$$\begin{aligned} \text{Minimize} \quad & V_f = \int_0^1 F(f(r))dr, \\ \text{subject to} \quad & \int_0^1 rf(r)dr = \alpha, \ \ 0 < \alpha < 1, \\ & \int_0^1 f(r)dr = 1, \\ & f(r) > 0, \end{aligned} \tag{3}$$

where F is a strictly convex function on $[0, \infty)$, and differentiable to at least the 2nd order.

The above two cases are special cases of the model where $F(x) = x^2$ and $F(x) = xlnx$.

Liu (Theorem 9, [19]) proved the following problem for the case in which generating functions are continuous and F is differentiable to at least the 2nd order:

Theorem 2 (Theorem 9, [19]). *There is a unique optimal solution for (3), and the optimal solution has the form*

$$f^*(r) = \begin{cases} (F')^{-1}(a^*r+b^*), & \textit{if } (F')^{-1}(a^*r+b^*) \geq 0, \\ 0, & \textit{elsewhere,} \end{cases}$$

where a^*, b^* are determined by the constraints:

$$\begin{cases} \int_0^1 rf^*(r)dr = \alpha, \\ \int_0^1 f^*(r)dr = 1, \\ f^*(r) \geq 0. \end{cases}$$

Here, we consider a generalized result for Theorem 3 when $f(x)$ is Lebesgue integrable and F' is continuous.

Theorem 3. *Suppose that the generating functions are Lebesgue integrable, F is a strictly convex function on $[0, \infty)$, and F' is continuous. Then, there is a unique optimal solution for problem (3), and that optimal solution has the form*

$$f^*(r) = \begin{cases} (F')^{-1}(a^*r+b^*) \ a.e., & \textit{if } (F')^{-1}(a^*r+b^*) > 0, \\ 0 \ a.e., & \textit{elsewhere,} \end{cases}$$

where a^* and b^* are determined by the constraints:

$$\begin{cases} \int_0^1 rf^*(r)dr = \alpha, \\ \int_0^1 f^*(r)dr = 1, \\ f^*(r) \geq 0. \end{cases}$$

Proof. As shown in Theorem 2, we consider the case where $\alpha \in (0, 1/2]$ and assume that $\{r < 1 : f^*(r) > 0\} = [0, t)$ for some $t \in (0, 1)$ and $\{r < 1 : f^*(r) = 0\} = [t, 1)$. We also note that for $r \in [0, t]$,

$$F'(f^*(r)) = a^*r + b^*$$

and for $r \in (t, 1)$,

$$a^*r + b^* < F'(0)$$

if $F'(0)$ exists. Let the nonnegative function f satisfy $1 = \int_0^1 f(r)dr$ and $\int_0^1 rf(r)dr = \alpha$. We set $f(r) = f^*(r) + g(r)$, $r \in [0, 1]$. Then, noting that $f(r) = g(r)$, $r \in [t, 1]$, we have

$$\int_0^t g(r)dr + \int_t^1 f(r)dr = \int_0^1 g(r)dr = 0, \tag{4}$$

since $1 = \int_0^1 f(r)dr = \int_0^1 f^*(r)dr + \int_0^1 g(r)dr = 1 + \int_0^1 g(r)dr$. We also have

$$\int_0^t rg(r)dr + \int_t^1 rf(r)dr = \int_0^1 rg(r)dr = 0, \tag{5}$$

since $\alpha = \int_0^1 rf(r)dr = \int_0^1 rf^*(r)dr + \int_0^1 rg(r)dr = \alpha + \int_0^1 rg(r)dr$. We now show that

$$\int_0^1 F(f(r))\, dr \geq \int_0^1 F(f^*(r))\, dr.$$

Since $F(x) - F(x_0) \geq F'(x_0)(x - x_0)$ (the equality holds if and only if $x = x_0$), we have that

$$\begin{aligned}
& \int_0^1 F(f(r))dr - \int_0^1 F(f^*(r))dr \\
= & \int_0^1 F((f^*(r) + g(r)))dr - \int_0^1 F(f^*(r))dr \\
\geq & \int_0^1 F'(f^*(r))g(r)dr \\
= & \int_0^t (a^*r + b^*)g(r)dr + \int_t^1 F'(0)g(r)dr \\
= & a^* \int_0^t rg(r)dr + b^* \int_0^t g(r)dr + \int_t^1 F'(0)g(r)dr \\
= & a^*(-\int_t^1 rf(r)dr) + b^*(-\int_t^1 f(r)dr) + \int_t^1 F'(0)g(r)dr \\
= & \int_t^1 (F'(0) - a^*r - b^*)f(r)dr \\
\geq & 0,
\end{aligned}$$

where the fourth equality comes from (4) and (5) and the second inequality comes from the fact that $a^*r + b^* \leq F'(0)$ *a.e.* for $r \in [t, 1]$. The equalities hold if and only if $f^* = f$ *a.e.* This completes the proof. □

Combining Theorems 2 and 4, we now have a solution equivalent relationship between the general minimum RIM quantifier problem and the general minimax RIM quantifier problem. This result generalizes Theorem 6 of Hong [17] and Theorem 5 of Hong [18] and provides a corrected version of Theorem 13 [19].

Theorem 4. *Suppose that the generating functions are absolutely continuous and F' is increasing and absolutely continuous. Then, the general minimum RIM quantifier problem has the same solution as the general minimax RIM quantifier problem.*

5. Numerical Example

We consider a RIM quantifier operator F which is not differentiable to at least the second order, but F' is absolutely continuous, and find an optimal solution of two RIM quantifier problems.

Let a RIM quantifier operator F be

$$F(x) = \begin{cases} \frac{x^2}{2}, & \text{if } 0 \leq x < \frac{1}{2}, \\ x^2 - \frac{1}{2}x + \frac{1}{8}, & \text{if } \frac{1}{2} \leq x \leq 1. \end{cases}$$

Then,

$$F'(x) = \begin{cases} x, & \text{if } 0 \leq x < \frac{1}{2}, \\ 2x - \frac{1}{2}, & \text{if } \frac{1}{2} \leq x \leq 1. \end{cases}$$

Hence, $F(x)$ is strictly convex and $F'(x)$ is absolutely continuous, but $F(x)$ is not the second order differentiable. Let

$$f^*(r) = \begin{cases} (F')^{-1}(a^*r + b^*), & \text{if } (F')^{-1}(a^*r + b^*) > 0, \\ 0, & \text{elsewhere}, \end{cases}$$

where a^* and b^* are determined by the constraints:

$$\begin{cases} \int_0^1 rf^*(r)dr = \alpha, \\ \int_0^1 f^*(r)dr = 1, \\ f^*(r) \geq 0. \end{cases} \tag{6}$$

We consider the case for $0 < \alpha \leq 1/2$. Then, $a^* \leq 0$ and $b^* > 0$.

Case (1) (See Figure 1) There exists $m, d \in [0,1]$ such that $m < d$ and

$$f^*(r) = \begin{cases} \frac{1}{2}(a^*r + b^*) + \frac{1}{4}, & \text{if } 0 \leq r \leq m, \\ a^*r + b^*, & m < r \leq d, \\ 0, & d < r \leq 1. \end{cases}$$

Since $a^*m + b^* = \frac{1}{2}$ and $a^*d + b^* = 0$, $b^* = -a^*m + \frac{1}{2}$ and $d = m - \frac{1}{2a^*}$. Hence,

$$f^*(r) = \begin{cases} \frac{1}{2}a^*(r - m) + \frac{1}{2}, & \text{if } 0 \leq r \leq m, \\ a^*(r - m) + \frac{1}{2}, & m < r \leq m - \frac{1}{2a^*}, \\ 0, & m - \frac{1}{2a^*} < r \leq 1. \end{cases}$$

From (6),

$$\begin{aligned} a^* &= \frac{2m - 4 - \sqrt{2m^2 - 16m + 16}}{2m^2}, \\ \alpha &= -\frac{4m^3a^{*3} - 12m^2a^{*2} + 6ma^* - 1}{48a^{*2}} \end{aligned}$$

hold. In addition, since $a^* < 0$ and $f^*(1) < 0$,

$$0 < m < 4 - \sqrt{10}, \quad 0 < \alpha < \frac{17 - 4\sqrt{10}}{12}.$$

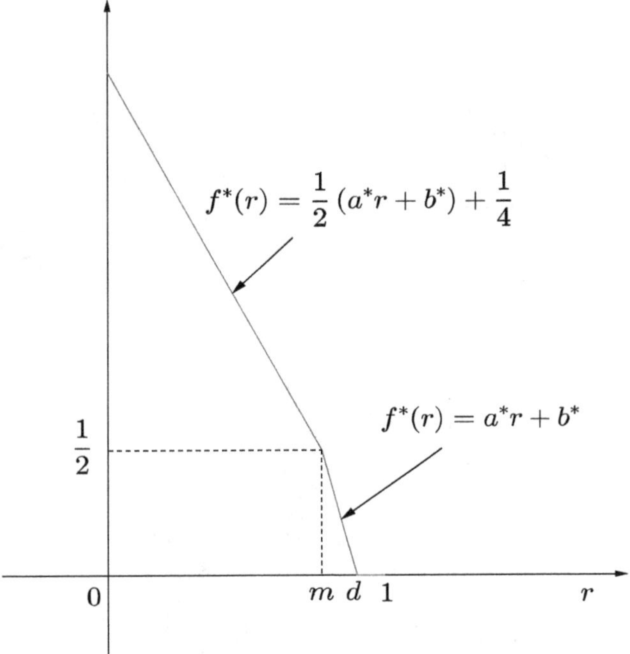

Figure 1. The graph of f^* $(0 < \alpha < \frac{17-4\sqrt{10}}{12})$.

Case (2) (See Figure 2) There exists $m \in [0,1]$ such that

$$f^*(r) = \begin{cases} \frac{1}{2}(a^*r + b^*) + \frac{1}{4}, & \text{if } 0 \le r \le m, \\ a^*r + b^*, & m < r \le 1. \end{cases}$$

Since $a^*m + b^* = \frac{1}{2}$,

$$f^*(r) = \begin{cases} \frac{1}{2}a^*(r - m) + \frac{1}{2}, & \text{if } 0 \le r \le m, \\ a^*(r - m) + \frac{1}{2}, & m < r \le 1. \end{cases}$$

From (6),

$$\begin{aligned} a^* &= \frac{2}{m^2 - 4m + 2}, \\ \alpha &= \frac{2m^3 + 3m^2 - 24m + 14}{12(m^2 - 4m + 2)} \end{aligned}$$

hold. In addition, since $a^* < 0$ and $f^*(1) \ge 0$,

$$4 - \sqrt{10} \le m \le 1, \quad \frac{17 - 4\sqrt{10}}{12} \le \alpha \le \frac{5}{12}.$$

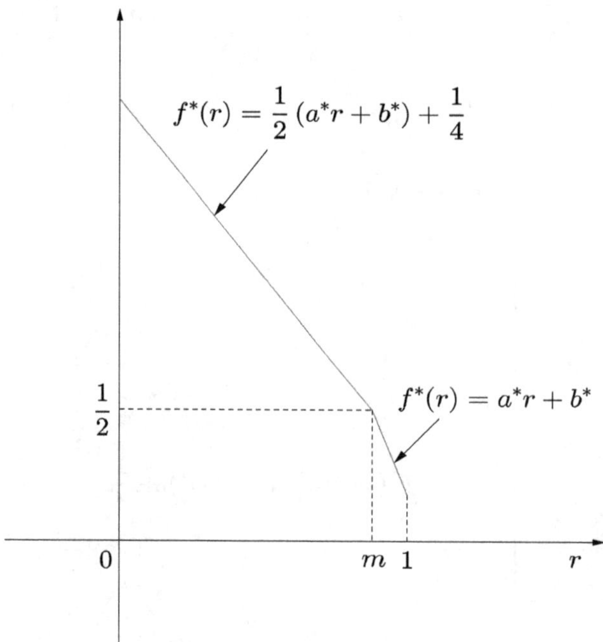

Figure 2. The graph f^* ($\frac{17-4\sqrt{10}}{12} \le \alpha \le \frac{5}{12}$).

Case (3) (See Figure 3) For all $0 \le r \le 1$,

$$f^*(r) = \frac{1}{2}(a^*r + b^*) + \frac{1}{4}.$$

From (6),

$$\begin{aligned} a^* &= -12 + 24\alpha, \\ b^* &= \frac{15}{2} - 12\alpha \end{aligned}$$

hold. In addition, since $a^* \leq 0$ and $f^*(1) > \frac{1}{2}$, $\frac{5}{12} < \alpha \leq \frac{1}{2}$.

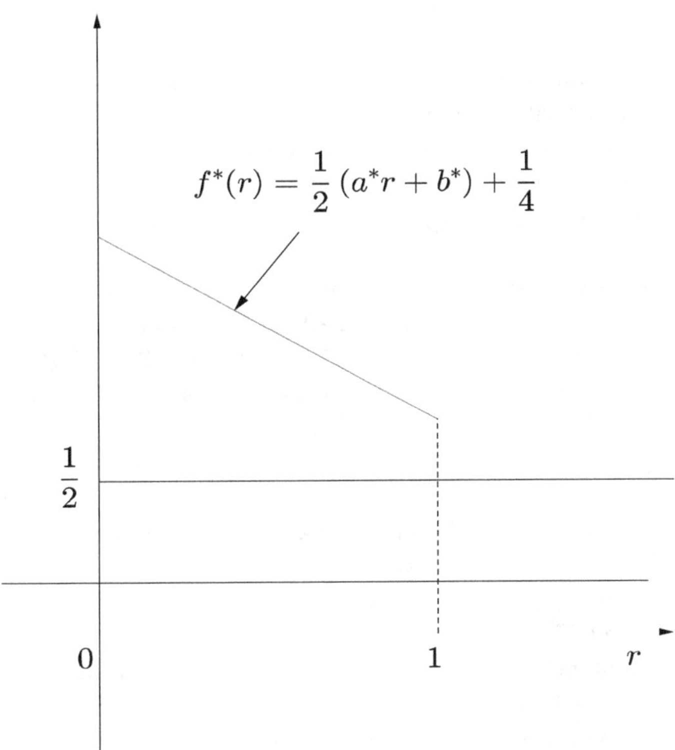

Figure 3. The graph f^* ($\frac{5}{12} < \alpha \leq \frac{1}{2}$).

6. Conclusions

In this paper, we examined the general minimax model for the RIM quantifier problem for the case in which the generating functions are absolutely continuous and a generalized solution to the general minimum model for the RIM quantifier problem for the case in which the generating functions are Lebesgue integrable. In addition, we provided a solution equivalent relationship between the general maximum model and the general minimax model for RIM quantifier problems and generalizes results of Hong based on these results. We also corrected Liu's theorems from a mathematical perspective as their theorems are not suitable for absolutely continuous generating functions.

References

1. Amin, G.R.; Emrouznejad, A. An extended minimax disparity to determine the OWA operator weights. *Comput. Ind. Eng.* **2006**, *50*, 312–316. [CrossRef]
2. Amin, G.R. Notes on priperties of the OWA weights determination model. *Comput. Ind. Eng.* **2007**, *52*, 533–538. [CrossRef]
3. Emrouznejad, A.; Amin, G.R. Improving minimax disparity model to determine the OWA operator weights. *Inf. Sci.* **2010**, *180*, 1477–1485. [CrossRef]
4. Filev, D.; Yager, R.R. On the issue of obtaining OWA operator weights. *Fuzzy Sets Syst.* **1988**, *94*, 157–169. [CrossRef]
5. Fullér, R.; Majlender, P. An analytic approach for obtaining maximal entropy OWA operators weights. *Fuzzy Sets Syst.* **2001**, *124*, 53–57. [CrossRef]
6. Hagan, M.O. Aggregating template or rule antecedents in real-time expert systems with fuzzy set logic. In Proceedings of the 22nd Annual IEEE Asilomar Conference on Signals, Systems, Computers, Pacific Grove, CA, USA, 31 October–2 November 1988; pp. 681–689.

7. Hong, D.H. A note on solution equivalence to general models for RIM quantifier problems. *Fuzzy Sets Syst.* **2018**, *332*, 25–28. [CrossRef]
8. Hong, D.H. On proving the extended minimax disparity OWA problem. *Fuzzy Sets Syst.* **2011**, *168*, 35–46. [CrossRef]
9. Liu, X.; Lou, H. On the equivalence of some approaches to the OWA operator and RIM quantifier determination. *Fuzzy Sets Syst.* **2007**, *159*, 1673–1688. [CrossRef]
10. Wang, Y.M.; Parkan, C. A minimax disparity approach obtaining OWA operator weights. *Inf. Sci.* **2005**, *175*, 20–29. [CrossRef]
11. Yager, R.R. Ordered weighted averaging aggregation operators in multi-criteria decision making. *IEEE Trans. Syst. Man Cybern.* **1988**, *18*, 183–190. [CrossRef]
12. Yager, R.R. OWA aggregation over a continuous interval argument with application to decision making. *IEEE Trans. Syst. Man Cybern. Part B* **2004**, *34*, 1952–1963. [CrossRef]
13. Yager, R.R. Families of OWA operators. *Fuzzy Sets Syst.* **1993**, *59*, 125–148. [CrossRef]
14. Yager, R.R.; Filev, D. Induced ordered weighted averaging operators. *IEEE Trans. Syst. Man Cybern. Part B* **1999**, *29*, 141–150. [CrossRef] [PubMed]
15. Liu, X. On the maximum entropy parameterized interval approximation of fuzzy numbers. *Fuzzy Sets Syst.* **2006**, *157*, 869–878. [CrossRef]
16. Liu, X.; Da, Q. On the properties of regular increasing monotone (RIM) quantifiers with maximum entropy. *Int. J. Gen. Syst.* **2008**, *37*, 167–179. [CrossRef]
17. Hong, D.H. The relationship between the minimum variance and minimax disparity RIM quantifier problems. *Fuzzy Sets Syst.* **2011**, *181*, 50–57. [CrossRef]
18. Hong, D.H. The relationship between the maximum entropy and minimax ratio RIM quantifier problems. *Fuzzy Sets Syst.* **2012**, *202*, 110–117. [CrossRef]
19. Liu, X. A general model of parameterized OWA aggregation with given orness level. *Int. J. Approx. Reason.* **2008**, *48*, 598–627. [CrossRef]
20. Fullér, R.; Majlender, P. On obtaining minimal variability OWA operator weights. *Fuzzy Sets Syst.* **2003**, *136*, 203–215. [CrossRef]
21. Sang, X.; Liu, X. An analytic approach to obtain the least square deviation OWA operater weights. *Fuzzy Sets Syst.* **2014**, *240*, 103–116. [CrossRef]
22. Wang, Y.M.; Luo, Y.; Liu, X. Two new models for determining OWA operater weights. *Comput. Ind. Eng.* **2007**, *52*, 203–209. [CrossRef]
23. Wheeden, R.L.; Zygmund, A. *Measure and Integral: An Introduction to Real Analysis*; Marcel Dekker, Inc.: New York, NY, USA, 1977.
24. Rustagi, J.S. *Variational Methods in Statistics*; Academic Press: New York, NY, USA, 1976.

12

Fluctuation Theorem of Information Exchange between Subsystems that Co-Evolve in Time

Lee Jinwoo

Department of Mathematics, Kwangwoon University, Seoul 01897, Korea; jinwoolee@kw.ac.kr

Abstract: Sagawa and Ueda established a fluctuation theorem of information exchange by revealing the role of correlations in stochastic thermodynamics and unified the non-equilibrium thermodynamics of measurement and feedback control. They considered a process where a non-equilibrium system exchanges information with other degrees of freedom such as an observer or a feedback controller. They proved the fluctuation theorem of information exchange under the assumption that the state of the other degrees of freedom that exchange information with the system does not change over time while the states of the system evolve in time. Here we relax this constraint and prove that the same form of the fluctuation theorem holds even if both subsystems co-evolve during information exchange processes. This result may extend the applicability of the fluctuation theorem of information exchange to a broader class of non-equilibrium processes, such as a dynamic coupling in biological systems, where subsystems that exchange information interact with each other.

Keywords: fluctuation theorem; thermodynamics of information; stochastic thermodynamics; mutual information; non-equilibrium free energy; entropy production

---

1. Introduction

Biological systems possess information processing mechanisms for their survival and heredity [1–3]. They, for example, sense external ligand concentrations [4,5], transmit information through signaling networks [6–8], and coordinate gene expressions [9] by secreting and sensing signaling molecules [10]. Cells even implement time integration by copying states of environment into molecular states inside the cells to reduce their sensing errors [11,12]. Therefore it is crucial to reveal the role of information in thermodynamics to properly understand complex biological information processes.

Historically, information has entered into the realm of thermodynamics by the name of Maxwell's demon. The demon observes the speed of molecules in a box that is divided into two portions by a partition in which there is a small hole, and lets the fast particles pass from the lower-half of the box to the upper-half, and only the slow particles pass from the upper-half to the lower-half by opening/closing the hole without expenditure of work (see Figure 1a). This results in raising the temperature of the upper-half of the box and lower that of the lower-half, indicating that the second law of thermodynamics, which implies heat flows spontaneously from hotter to colder places, might hypothetically be violated [13]. This paradox shows that information can affect thermodynamics of a physical system, or information is a physical element [14].

Szilard has devised a much simpler model that carries the essential role of information in Maxwell's thought experiment. The Szilard engine consists of a single particle in a box which is surrounded by a heat reservoir of constant temperature. A cycle of the engine begins with inserting a partition in the middle of the box. Depending on whether the particle is in the left-half or in the right-half of the box, one controls a lever such that a weight can be lifted during the wall moves quasi-statically in the direction that the particle pushes (see Figure 1b). If the partition reaches an end of the box, the partition is removed and a new cycle begins again with inserting a partition at the center. Since the energy required for lifting the weight comes from the heat reservoir, this engine corresponds to a perpetual-motion machine of the second kind, where the single heat reservoir is spontaneously cooled and the corresponding thermal energy is converted into mechanical work cyclically, which is prohibited by the second-law of thermodynamics [15].

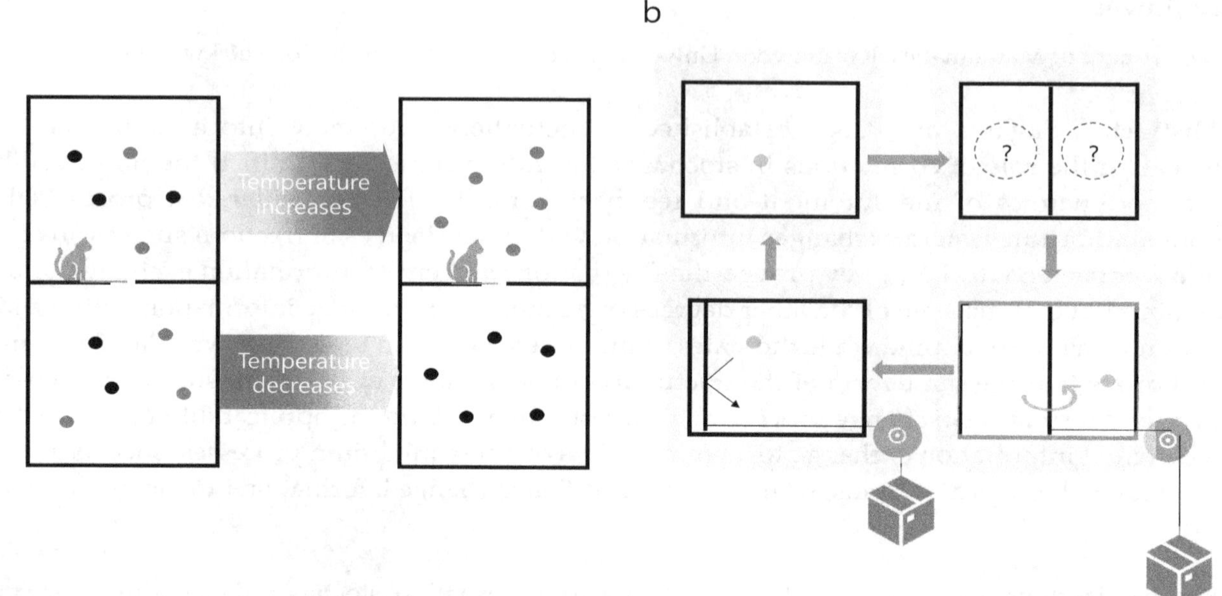

Figure 1. Paradox in thermodynamics of information (**a**) Maxwell's demon (orange cat) uses information on the speed of the particles in the box: He opens/closes the small hole (orange line) without expenditure of energy such that fast particles (red filled circles) are gathered in the upper-half of the box and slow particles (blue filled circles) are gathered in the lower-half of the box. Since temperature is the average velocity of the particles, the demon's action results in spontaneous flow of heat from colder places to hotter places, which violates the second-law of thermodynamics. (**b**) A cycle of Szilard's engine is represented. A lever (green curved arrow) is controlled such that a weight can be lifted during the wall moves quasi-statically in the direction that the particle pushes. This engine harnesses heat from the heat reservoir (yellow region around each boxes) and convert it into mechanical work, cyclically, and thus corresponds to a perpetual-motion engine of the second kind, which is prohibited by the second-law of thermodynamics.

Szilard interprets the coupling between the location of the particle and the direction of the lever as a sort of memory faculty and points out that the coupling is the main cause that enables an amount of work to be extracted from the heat reservoir. He infers, therefore, that establishing the coupling must be accompanied by a production of entropy (dissipation of heat into the environment) which compensates for the lost heat in the reservoir. In [16], Sagawa and Ueda have proved this idea in the form of a fluctuation theorem of information exchange, generalizing the second-law of thermodynamics by taking information into account:

$$\left\langle e^{-\sigma+\Delta I}\right\rangle = 1, \tag{1}$$

where σ is the entropy production of a system X, and ΔI is the change of mutual information between the system X and another system Y, such as a demon, during a process λ_t for $0 \leq t \leq \tau$. Here the bracket indicates the ensemble average over all microscopic trajectories of X and over all states of Y. By Jensen's inequality [17], Equation (1) implies

$$\langle\sigma\rangle \geq \langle\Delta I\rangle\,. \tag{2}$$

This tells indeed that establishing a correlation between the two subsystems, $\langle \Delta I \rangle > 0$, accompanies an entropy production, $\langle \sigma \rangle > 0$, and expenditure of this correlation, $\langle \Delta I \rangle < 0$, serves as a source of entropy decrease, $\langle \sigma \rangle < 0$. In proving this theorem, they have assumed that the state of system Y does not evolve in time. This assumption causes no problem for simple models of measurement and feedback control. However, in biological systems, it is not unusual that both subsystems that exchange information with each other co-evolve in time. For example, transmembrane receptor proteins transmit signals through thermodynamic coupling between extracellular ligands and conformation of intracellular parts of the receptors during a dynamic allosteric transition [18,19]. In this paper, we relax the constraint that Sagawa and Ueda have assumed, and generalize the fluctuation theorem of information exchange to be applicable to more involved situations, where the two subsystems can influence each other so that the states of both systems co-evolve in time.

2. Results

2.1. Theoretical Framework

We consider a finite classical stochastic system composed of subsystems X and Y that are in contact with a heat reservoir of inverse temperature $\beta \equiv 1/(k_B T)$ where k_B is the Boltzmann constant and T is the temperature of the reservoir. We allow both systems X and Y to be driven far from equilibrium by changing external parameter λ_t during time $0 \le t \le \tau$ [20–22]. We assume that time evolutions of subsystems X and Y are described by a classical stochastic dynamics from $t = 0$ to $t = \tau$ along trajectories $\{x_t\}$ and $\{y_t\}$, respectively, where x_t (y_t) denotes a specific microstate of X (Y) at time t for $0 \le t \le \tau$ on each trajectory. Since both trajectories fluctuate, we repeat the process λ_t with appropriate initial joint probability distribution $p_0(x, y)$ over all microstates (x, y) of systems X and Y. Then the joint probability distribution $p_t(x, y)$ would evolve for $0 \le t \le \tau$. Let $p_t(x) := \int p_t(x, y)\, dy$ and $p_t(y) := \int p_t(x, y)\, dx$ be the corresponding marginal probability distributions. We assume

$$p_0(x, y) \neq 0 \text{ for all } (x, y) \tag{3}$$

so that we have $p_t(x, y) \neq 0$, $p_t(x) \neq 0$, and $p_t(y) \neq 0$ for all x and y during $0 \le t \le \tau$.

Now, the entropy production σ during process λ_t for $0 \le t \le \tau$ is given by

$$\sigma := \Delta s + \beta Q_b, \tag{4}$$

where Δs is the sum of changes in stochastic entropy along $\{x_t\}$ and $\{y_t\}$, and Q_b is heat dissipated into the reservoir (entropy production in the reservoir) [23,24]. In detail, we have

$$\begin{aligned} \Delta s &:= \Delta s_x + \Delta s_y, \\ \Delta s_x &:= -\ln p_\tau(x_\tau) + \ln p_0(x_0), \\ \Delta s_y &:= -\ln p_\tau(y_\tau) + \ln p_0(y_0). \end{aligned} \tag{5}$$

We note that the stochastic entropy $s[p_t(\circ)] := -\ln p_t(\circ)$ of microstate $\circ$ at time t can be interpreted as uncertainty of occurrence of $\circ$ at time t: The greater the probability that state $\circ$ occurs, the smaller the uncertainty of occurrence of state $\circ$.

Now we consider situations where system X exchanges information with system Y during process λ_t. By this, we mean that trajectory $\{x_t\}$ of system X evolves depending on the trajectory $\{y_t\}$ of system Y. Then, information I_t at time t between x_t and y_t is characterized by the reduction of uncertainty of x_t due to given y_t [16]:

$$\begin{aligned} I_t(x_t, y_t) &:= s[p_t(x_t)] - s[p_t(x_t|y_t)] \\ &= \ln \frac{p_t(x_t, y_t)}{p_t(x_t)p_t(y_t)}, \end{aligned} \tag{6}$$

where $p_t(x_t|y_t)$ is the conditional probability distribution of x_t given y_t. We note that this is called the (time-dependent form of) thermodynamic coupling function [19]. The larger the value of $I_t(x_t, y_t)$ is, the more information is being shared between x_t and y_t for their occurrence. We note that $I_t(x_t, y_t)$ vanishes if x_t and y_t are independent at time t, and the average of $I_t(x_t, y_t)$ with respect to $p_t(x_t, y_t)$ over all microstates is the mutual information between the two subsystems, which is greater than or equal to zero [17].

2.2. *Proof of Fluctuation Theorem of Information Exchange*

Now we are ready to prove the fluctuation theorem of information exchange in this general setup. We define reverse process $\lambda'_t := \lambda_{\tau-t}$ for $0 \le t \le \tau$, where the external parameter is time-reversed [25,26]. Here we set the initial probability distribution $p'_0(x, y)$ for the reverse process as the final (time $t = \tau$) probability distribution for the forward process $p_\tau(x, y)$ so that we have

$$\begin{aligned} p'_0(x) &= \int p'_0(x,y)\,dy = \int p_\tau(x,y)\,dy = p_\tau(x), \\ p'_0(y) &= \int p'_0(x,y)\,dx = \int p_\tau(x,y)\,dx = p_\tau(y). \end{aligned} \tag{7}$$

Then, by Equation (3), we have $p'_t(x, y) \neq 0$, $p'_t(x) \neq 0$, and $p'_t(y) \neq 0$ for all x and y during $0 \le t \le \tau$. We also consider the time-reversed conjugate for each $\{x_t\}$ and $\{y_t\}$ for $0 \le t \le \tau$ as follows:

$$\begin{aligned} \{x'_t\} &:= \{x^*_{\tau-t}\}, \\ \{y'_t\} &:= \{y^*_{\tau-t}\}, \end{aligned} \tag{8}$$

where $*$ denotes momentum reversal. The microscopic reversibility condition connects the time-reversal symmetry of the microscopic dynamics to non-equilibrium thermodynamics, and reads in this framework as follows [23,27–29]:

$$\frac{p(\{x_t\},\{y_t\}|x_0,y_0)}{p'(\{x'_t\},\{y'_t\}|x'_0,y'_0)} = e^{\beta Q_b}, \tag{9}$$

where $p(\{x_t\}, \{y_t\}|x_0, y_0)$ is the conditional joint probability distribution of paths $\{x_t\}$ and $\{y_t\}$ conditioned at initial microstates x_0 and y_0, and $p'(\{x'_t\}, \{y'_t\}|x'_0, y'_0)$ is that for the reverse process. Now we have the following:

$$\begin{aligned} \frac{p'(\{x'_t\},\{y'_t\})}{p(\{x_t\},\{y_t\})} &= \frac{p'(\{x'_t\},\{y'_t\}|x'_0,y'_0)}{p(\{x_t\},\{y_t\}|x_0,y_0)} \cdot \frac{p'_0(x'_0,y'_0)}{p_0(x_0,y_0)} && (10) \\ &= \frac{p'(\{x'_t\},\{y'_t\}|x'_0,y'_0)}{p(\{x_t\},\{y_t\}|x_0,y_0)} \cdot \frac{p'_0(x'_0,y'_0)}{p'_0(x'_0)p'_0(y'_0)} \cdot \frac{p_0(x_0)p_0(y_0)}{p_0(x_0,y_0)} \cdot \frac{p'_0(x'_0)}{p_0(x_0)} \cdot \frac{p'_0(y'_0)}{p_0(y_0)} && (11) \\ &= \exp\{-\beta Q_b + I_\tau(x_\tau,y_\tau) - I_0(x_0,y_0) - \Delta s_x - \Delta s_y\} && (12) \\ &= \exp\{-\sigma + \Delta I\}. && (13) \end{aligned}$$

To obtain Equation (11) from Equation (10), we multiply Equation (10) by $\frac{p'_0(x'_0)p'_0(y'_0)}{p'_0(x'_0)p'_0(y'_0)}$ and $\frac{p_0(x_0)p_0(y_0)}{p_0(x_0)p_0(y_0)}$, which are 1. We obtain Equation (12) by applying Equations (5)–(7) and (9) consecutively to Equation (11). Finally, we set $\Delta I := I_\tau(x_\tau, y_\tau) - I_0(x_0, y_0)$, and use Equation (4) to obtain Equation (13) from Equation (12).

We note that Equation (13) generalizes the detailed fluctuation theorem in the presence of information exchange that is proved in [16]. Now we obtain the generalized version of Equation (1) by using Equation (13) as follows:

$$\begin{aligned}\left\langle e^{-\sigma+\Delta I}\right\rangle &= \int e^{-\sigma+\Delta I} p(\{x_t\},\{y_t\})\, d\{x_t\}d\{y_t\} \\ &= \int p'(\{x'_t\},\{y'_t\})\, d\{x'_t\}d\{y'_t\} = 1.\end{aligned} \tag{14}$$

Here we use the fact that there is a one-to-one correspondence between the forward and the reverse paths due to the time-reversal symmetry of the underlying microscopic dynamics such that $d\{x_t\} = d\{x'_t\}$ and $d\{y_t\} = d\{y'_t\}$ [30].

2.3. Corollary

Before discussing a corollary, we remark one thing: we have used similar notation to that used by Sagawa and Ueda in [16], but there is an important difference. Most importantly, their entropy production σ_{su} reads as follows:

$$\sigma_{\text{su}} := \Delta s_{\text{su}} + \beta Q_b,$$

where $\Delta s_{\text{su}} := \Delta s_x$. In [16], system X is in contact with the heat reservoir, but system Y is not. Nor does system Y evolve over time. Thus they have considered entropy production in system X and the bath. In this paper, both systems X and Y are in contact with the reservoir, and system Y also evolves in time. Thus both subsystems X and Y as well as the heat bath contribute to the entropy production as expressed in Equations (4) and (5). Keeping in mind this difference, we apply Jensen's inequality to Equation (14) to obtain

$$\langle\sigma\rangle \geq \langle\Delta I\rangle\,. \tag{15}$$

It tells us that firstly, establishing correlation between X and Y accompanies entropy production, and secondly, established correlation serves as a source of entropy decrease.

Now as a corollary, we refine the generalized fluctuation theorem in Equation (14) by including energetic terms. To this end, we define local free energy $\mathcal{F}_x$ of system X at x_t and $\mathcal{F}_y$ of system Y at y_t as follows:

$$\begin{aligned}\mathcal{F}_x(x_t,t) &:= E_x(x_t,t) - Ts[p_t(x_t)] \\ \mathcal{F}_y(y_t,t) &:= E_y(y_t,t) - Ts[p_t(y_t)],\end{aligned} \tag{16}$$

where E_x and E_y are internal energy of systems X and Y, respectively, and $s[p_t(\circ)] := -\ln p_t(\circ)$ is stochastic entropy [23,24]. Here T is the temperature of the heat bath and argument t indicates dependency of each terms on external parameter λ_t. During the process λ_t, work done on the systems is expressed by the first law of thermodynamics as follows:

$$W := \Delta E + Q_b, \tag{17}$$

where ΔE is the change in internal energy of the systems. If we assume that systems X and Y are weakly coupled, in that interaction energy between X and Y is negligible compared to internal energy of X and Y, we may have

$$\Delta E := \Delta E_x + \Delta E_y, \tag{18}$$

where $\Delta E_x := E_x(x_\tau,\tau) - E_x(x_0,0)$ and $\Delta E_y := E_y(y_\tau,\tau) - E_y(y_0,0)$ [31]. We rewrite Equation (12) by adding and subtracting the change of internal energy ΔE_x of X and ΔE_y of Y as follows:

$$\frac{p'(\{x'_t\},\{y'_t\})}{p(\{x_t\},\{y_t\})} = \exp\{-\beta(Q_b + \Delta E_x + \Delta E_y) + \Delta I + \beta\Delta E_x - \Delta s_x + \beta\Delta E_y - \Delta s_y\} \tag{19}$$

$$= \exp\{-\beta(W - \Delta\mathcal{F}_x - \Delta\mathcal{F}_y) + \Delta I\}, \tag{20}$$

where we have applied Equations (16)–(18) consecutively to Equation (19) to obtain Equation (20). Here $\Delta\mathcal{F}_x := \mathcal{F}_x(x_\tau, \tau) - \mathcal{F}_x(x_0, 0)$ and $\Delta\mathcal{F}_y := \mathcal{F}_y(y_\tau, \tau) - \mathcal{F}_y(y_0, 0)$. Now we obtain fluctuation theorem of information exchange with energetic terms as follows:

$$\begin{aligned} \left\langle e^{-\beta(W-\Delta\mathcal{F}_x-\Delta\mathcal{F}_y)+\Delta I} \right\rangle &= \int e^{-\beta(W-\Delta\mathcal{F}_x-\Delta\mathcal{F}_y)+\Delta I} p(\{x_t\},\{y_t\})\, d\{x_t\} d\{y_t\} \\ &= \int p'(\{x'_t\},\{y'_t\})\, d\{x'_t\} d\{y'_t\} = 1, \end{aligned} \tag{21}$$

which generalizes known relations in the literature [31–36]. We note that Equation (21) holds under the weak-coupling assumption between systems X and Y during the process λ_t. By Jensen's inequality, Equation (21) implies

$$\langle W \rangle \geq \left\langle \Delta\mathcal{F}_x + \Delta\mathcal{F}_y + \frac{\Delta I}{\beta} \right\rangle. \tag{22}$$

We remark that $\langle \Delta\mathcal{F}_x \rangle + \langle \Delta\mathcal{F}_y \rangle$ in Equation (22) is the difference in non-equilibrium free energy, which is different from the change in equilibrium free energy that appears in similar relations in the literature [32–36].

3. Examples

3.1. Measurement

Let X be a device (or a demon) which measures the state of other system and Y be a measured system, both of which are in contact with a heat bath of inverse temperature β (see Figure 2a). We consider a dynamic measurement process, which is described as follows: X and Y are prepared separately in equilibrium such that X and Y are not correlated initially, i.e., $I_0(x_0, y_0) = 0$ for all x_0 and y_0. At time $t = 0$, device X is put in contact with system Y so that the coupling of X and Y occurs due to their (weak) interactions until time $t = \tau$, at which a single measurement process finishes. We note that system Y is allowed to evolve in time during the process. Since each process fluctuates, we repeat the measurement many times to obtain probability distribution $p_t(x, y)$ for $0 \leq t \leq \tau$.

A distinguished feature of the framework in this paper is that mutual information $I_t(x_t, y_t)$ in Equation (6) enables us to obtain the time-varying amount of established information during the dynamic coupling process, unlike other approaches where they either provide the amount of information at a fixed time [31,36,37] or one of the system is fixed during the coupling process [16]. For example, let us assume that the probability distribution $p_t(x_t, y_t)$ at an intermediate time t is as shown in Table 1.

Table 1. The joint probability distribution of x and y at an intermediate time t: Here we assume for simplicity that both systems X and Y have two states, 0 (left) and 1 (right).

X\Y	0 (Left)	1 (Right)
0 (Left)	1/3	1/6
1 (Right)	1/6	1/3

Then we have the following:

$$\begin{aligned} I_t(x_t = 0, y_t = 0) &= \ln \frac{1/3}{(1/2)\cdot(1/2)} = \ln(4/3), \\ I_t(x_t = 0, y_t = 1) &= \ln \frac{1/6}{(1/2)\cdot(1/2)} = \ln(2/3), \\ I_t(x_t = 1, y_t = 0) &= \ln \frac{1/6}{(1/2)\cdot(1/2)} = \ln(2/3), \\ I_t(x_t = 1, y_t = 1) &= \ln \frac{1/3}{(1/2)\cdot(1/2)} = \ln(4/3), \end{aligned} \tag{23}$$

so that $\langle \Delta I \rangle = (1/3)\ln(4/3) + (1/6)\ln(2/3) + (1/6)\ln(2/3) + (1/3)\ln(4/3) \approx \ln(1.06)$. Thus by Equation (15) we obtain the lower bound of the average entropy production for the coupling that has been established until time t from the uncorrelated initial state, as follows: $\langle \sigma \rangle \geq \langle \Delta I \rangle \approx \ln 1.06$. If there is no measurement error at final time τ such that $p_\tau(x_\tau = 0, y_\tau = 1) = p_\tau(x_\tau = 1, y_\tau = 0) = 0$ and $p_\tau(x_\tau = 0, y_\tau = 0) = p_\tau(x_\tau = 1, y_\tau = 1) = 1/2$, then we may have $\langle \sigma \rangle \geq \langle \Delta I \rangle = \ln 2$, which is greater than $\ln 1.06$.

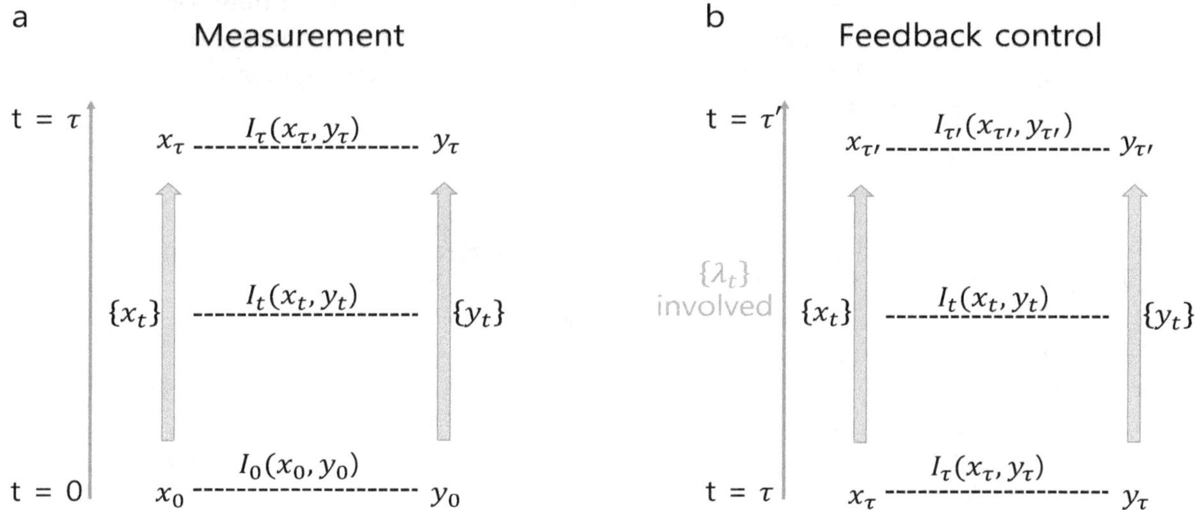

Figure 2. Measurement and feedback control: system X is, for example, a measuring device and system Y is a measured system. X and Y co-evolve, as they interact weakly, along trajectories $\{x_t\}$ and $\{y_t\}$, respectively. (**a**) Coupling is being established during the measurement process so that $I_t(x_t, y_t)$ for $0 \leq t \leq \tau$ may be increased (not necessarily monotonically). (**b**) Established correlation is being used as a source of work through external parameter λ_t so that $I_t(x_t, y_t)$ for $\tau \leq t \leq \tau'$ may be decreased (not necessarily monotonically).

3.2. Feedback Control

Unlike the case in [16], we need not to exchange subsystems X and Y to consider feedback control after the measurement. Thus we proceed continuously to feedback control immediately after each measurement process at time τ (see Figure 2b). We assume that correlation $I_\tau(x_\tau, y_\tau)$ at time τ is given by the values in Equation (23) and final correlation at later time τ' is zero, i.e., $I_{\tau'}(x_{\tau'}, y_{\tau'}) = 0$. By feedback control, we mean that external parameter λ_t for $\tau \leq t \leq \tau'$ is manipulated in a pre-determined manner [16], while systems X and Y co-evolve in time, such that the established correlation is used as a source of work while $I_t(x_t, y_t)$ for $\tau \leq t \leq \tau'$ is decreased, not necessarily monotonically. Equation (21) provides an exact relation on the energetics of this process. We rewrite its corollary, Equation (22), with respect to extractable work $W_{\text{ext}} := -W$ as follows:

$$\langle W_{\text{ext}} \rangle \leq -\left\langle \Delta \mathcal{F}_x + \Delta \mathcal{F}_y + \frac{\Delta I}{\beta} \right\rangle. \tag{24}$$

Then the extractable work on top of the conventional bound, $-\langle \Delta \mathcal{F}_x + \Delta \mathcal{F}_y \rangle$, is additionally given by $-\Delta I/\beta = \ln(1.06)$, which comes from the consumption of the established correlation.

4. Conclusions

We have proved the fluctuation theorem of information exchange, Equation (14), which holds even during the co-evolution of two systems that exchange information with each other. Equation (14) tells us that establishing correlation between two systems necessarily accompanies entropy production which is contributed by both systems and the heat reservoir, as expressed in Equations (4) and (5). We have also proved, as a corollary of Equation (14), the fluctuation theorem of information exchange with energetic terms, Equation (21), under the assumption of weak coupling between the two subsystems. Equation (21) reveals the exact relationship between non-equilibrium free energy of both sub-systems and mutual information that is established/consumed through their interactions. This more generalized framework than that in [16], enables us to apply thermodynamics of information to biological systems, where molecules generate/consume correlations through their information processing mechanisms [4–6]. Since the new framework is applicable to fully non-equilibrium situations, thermodynamic coupling during a dynamic allosteric transition, for example, may be analyzed based on this theoretical framework beyond current equilibrium thermodynamic approach [18,19].

References

1. Hartwell, L.H.; Hopfield, J.J.; Leibler, S.; Murray, A.W. From molecular to modular cell biology. *Nature* **1999**, *402*, C47. [CrossRef] [PubMed]
2. Crofts, A.R. Life, information, entropy, and time: Vehicles for semantic inheritance. *Complexity* **2007**, *13*, 14–50. [CrossRef] [PubMed]
3. Cheong, R.; Rhee, A.; Wang, C.J.; Nemenman, I.; Levchenko, A. Information transduction capacity of noisy biochemical signaling networks. *Science* **2011**, *334*, 354–358. [CrossRef]
4. McGrath, T.; Jones, N.S.; ten Wolde, P.R.; Ouldridge, T.E. Biochemical Machines for the Interconversion of Mutual Information and Work. *Phys. Rev. Lett.* **2017**, *118*, 028101. [CrossRef] [PubMed]
5. Ouldridge, T.E.; Govern, C.C.; ten Wolde, P.R. Thermodynamics of Computational Copying in Biochemical Systems. *Phys. Rev. X* **2017**, *7*, 021004. [CrossRef]
6. Becker, N.B.; Mugler, A.; ten Wolde, P.R. Optimal Prediction by Cellular Signaling Networks. *Phys. Rev. Lett.* **2015**, *115*, 258103. [CrossRef] [PubMed]
7. Cheng, F.; Liu, C.; Shen, B.; Zhao, Z. Investigating cellular network heterogeneity and modularity in cancer: A network entropy and unbalanced motif approach. *BMC Syst. Biol.* **2016**, *10*, 65. [CrossRef]
8. Whitsett, J.A.; Guo, M.; Xu, Y.; Bao, E.L.; Wagner, M. SLICE: Determining cell differentiation and lineage based on single cell entropy. *Nucleic Acids Res.* **2016**, *45*, e54.
9. Statistical Dynamics of Spatial-Order Formation by Communicating Cells. *iScience* **2018**, *2*, 27–40. [CrossRef]
10. Maire, T.; Youk, H. Molecular-Level Tuning of Cellular Autonomy Controls the Collective Behaviors of Cell Populations. *Cell Syst.* **2015**, *1*, 349–360. [CrossRef]
11. Mehta, P.; Schwab, D.J. Energetic costs of cellular computation. *Proc. Natl. Acad. Sci. USA* **2012**, *109*, 17978–17982. [CrossRef] [PubMed]
12. Govern, C.C.; ten Wolde, P.R. Energy dissipation and noise correlations in biochemical sensing. *Phys. Rev. Lett.* **2014**, *113*, 258102. [CrossRef] [PubMed]
13. Leff, H.S.; Rex, A.F. *Maxwell's Demon: Entropy, Information, Computing*; Princeton University Press: Princeton, NJ, USA, 2014.
14. Landauer, R. Information is physical. *Phys. Today* **1991**, *44*, 23–29. [CrossRef]
15. Szilard, L. On the decrease of entropy in a thermodynamic system by the intervention of intelligent beings. *Behav. Sci.* **1964**, *9*, 301–310. [CrossRef]
16. Sagawa, T.; Ueda, M. Fluctuation theorem with information exchange: Role of correlations in stochastic thermodynamics. *Phys. Rev. Lett.* **2012**, *109*, 180602. [CrossRef] [PubMed]
17. Cover, T.M.; Thomas, J.A. *Elements of Information Theory*; John Wiley & Sons: Hoboken, NJ, USA, 2012.
18. Tsai, C.J.; Nussinov, R. A unified view of how allostery works? *PLoS Comput. Biol.* **2014**, *10*, e1003394. [CrossRef]
19. Cuendet, M.A.; Weinstein, H.; LeVine, M.V. The allostery landscape: Quantifying thermodynamic couplings in biomolecular systems. *J. Chem. Theory Comput.* **2016**, *12*, 5758–5767. [CrossRef]
20. Jarzynski, C. Equalities and inequalities: Irreversibility and the second law of thermodynamics at the nanoscale. *Annu. Rev. Condens. Matter Phys.* **2011**, *2*, 329–351. [CrossRef]
21. Seifert, U. Stochastic thermodynamics, fluctuation theorems and molecular machines. *Rep. Prog. Phys.* **2012**, *75*, 126001. [CrossRef]

22. Spinney, R.; Ford, I. Fluctuation Relations: A Pedagogical Overview. In *Nonequilibrium Statistical Physics of Small Systems*; Wiley-VCH Verlag GmbH & Co. KGaA: Weinheim, Germany, 2013; pp. 3–56.
23. Crooks, G.E. Entropy production fluctuation theorem and the nonequilibrium work relation for free energy differences. *Phys. Rev. E* **1999**, *60*, 2721–2726. [CrossRef]
24. Seifert, U. Entropy production along a stochastic trajectory and an integral fluctuation theorem. *Phys. Rev. Lett.* **2005**, *95*, 040602. [CrossRef]
25. Ponmurugan, M. Generalized detailed fluctuation theorem under nonequilibrium feedback control. *Phys. Rev. E* **2010**, *82*, 031129. [CrossRef]
26. Horowitz, J.M.; Vaikuntanathan, S. Nonequilibrium detailed fluctuation theorem for repeated discrete feedback. *Phys. Rev. E* **2010**, *82*, 061120.
27. Kurchan, J. Fluctuation theorem for stochastic dynamics. *J. Phys. A Math. Gen.* **1998**, *31*, 3719. [CrossRef]
28. Maes, C. The fluctuation theorem as a Gibbs property. *J. Stat. Phys.* **1999**, *95*, 367–392. [CrossRef]
29. Jarzynski, C. Hamiltonian derivation of a detailed fluctuation theorem. *J. Stat. Phys.* **2000**, *98*, 77–102. [CrossRef]
30. Goldstein, H.; Poole, C., Jr.; Safko, J.L. *Classical Mechanics*, 3rd ed.; Pearson: London, UK, 2001.
31. Parrondo, J.M.; Horowitz, J.M.; Sagawa, T. Thermodynamics of information. *Nat. Phys.* **2015**, *11*, 131–139. [CrossRef]
32. Kawai, R.; Parrondo, J.M.R.; den Broeck, C.V. Dissipation: The phase-space perspective. *Phys. Rev. Lett.* **2007**, *98*, 080602. [CrossRef]
33. Generalization of the second law for a transition between nonequilibrium states. *Phys. Lett. A* **2010**, *375*, 88–92. [CrossRef]
34. Generalization of the second law for a nonequilibrium initial state. *Phys. Lett. A* **2010**, *374*, 1001–1004. [CrossRef]
35. Esposito, M.; Van den Broeck, C. Second law and Landauer principle far from equilibrium. *Europhys. Lett.* **2011**, *95*, 40004. [CrossRef]
36. Sagawa, T.; Ueda, M. Generalized Jarzynski equality under nonequilibrium feedback control. *Phys. Rev. Lett.* **2010**, *104*, 090602. [CrossRef] [PubMed]
37. Horowitz, J.M.; Parrondo, J.M. Thermodynamic reversibility in feedback processes. *Europhys. Lett.* **2011**, *95*, 10005. [CrossRef]

13

On the Catalan Numbers and Some of their Identities

Wenpeng Zhang [1,2] and Li Chen [2,*]

1 School of Mathematics and Statistics, Kashgar University, Xinjiang 844006, China; wpzhang888@163.com
2 School of Mathematics, Northwest University, Xi'an 710127, Shaanxi, China
* Correspondence: cl1228@stumail.nwu.edu.cn

Abstract: The main purpose of this paper is using the elementary and combinatorial methods to study the properties of the Catalan numbers, and give two new identities for them. In order to do this, we first introduce two new recursive sequences, then with the help of these sequences, we obtained the identities for the convolution involving the Catalan numbers.

Keywords: catalan numbers; elementary and combinatorial methods; recursive sequence; convolution sums

JEL Classification: 11B83; 11B75

---

1. Introduction

For any non-negative integer n, the famous Catalan numbers C_n are defined as $C_n = \frac{1}{n+1} \cdot \binom{2n}{n}$. For example, the first several Catalan numbers are $C_0 = 1$, $C_1 = 1$, $C_2 = 2$, $C_3 = 5$, $C_4 = 14$, $C_5 = 42$, $C_6 = 132$, $C_7 = 429$, $C_8 = 1430$, $\cdots$. The Catalan numbers C_n satisfy the recursive formula

$$C_n = \sum_{i=1}^{n} C_{i-1} \cdot C_i.$$

The generating function of the Catalan numbers C_n is

$$\frac{2}{1+\sqrt{1-4x}} = \sum_{n=0}^{\infty} \frac{\binom{2n}{n}}{n+1} \cdot x^n = \sum_{n=0}^{\infty} C_n \cdot x^n. \tag{1}$$

These numbers occupy a pivotal position in combinatorial mathematics, as many counting problems are closely related to Catalan numbers, and some famous examples can be found in R. P. Stanley [1]. Many papers related to the Catalan numbers and other special sequences can also be found in references [1–20], especially the works of T. Kim et al. give a series of new identities for the Catalan numbers, see [9–14], these are important results in the related field.

The main purpose of this paper is to consider the calculating problem of the following convolution sums involving the Catalan numbers:

$$\sum_{a_1+a_2+\cdots+a_h=n} C_{a_1} \cdot C_{a_2} \cdot C_{a_3} \cdots C_{a_h}, \tag{2}$$

where the summation is taken over all h-dimension non-negative integer coordinates $(a_1, a_2, \cdots, a_h)$ such that the equation $a_1 + a_2 + \cdots + a_h = n$.

About the convolution sums (2), it seems that none had studied it yet, at least we have not seen any related results before. We think this problem is meaningful. The reason is based on the following two aspects: First, it can reveal the profound properties of the Catalan numbers themselves. Second, for the other sequences, such as Fibonacci numbers, Fubini numbers, and Euler numbers, etc. (see [21–23]), there are corresponding results, so the Catalan numbers should have a corresponding identity. In this paper, we use the elementary and combinatorial methods to answer this question. That is, we shall prove the following:

Theorem 1. *For any positive integer h, we have the identity*

$$\sum_{a_1+a_2+\cdots+a_{2h+1}=n} C_{a_1}\cdot C_{a_2}\cdot C_{a_3}\cdots C_{a_{2h+1}}$$
$$= \frac{1}{(2h)!}\sum_{i=0}^{h} C(h,i)\sum_{j=0}^{\min(n,i)} \frac{(n-j+h+i)!\cdot C_{n-j+h+i}}{(n-j)!}\cdot\binom{i}{j}\cdot(-4)^j,$$

where $C(h,i)$ are defined as $C(1,0)=-2$, $C(h,h)=1$, $C(h+1,h)=C(h,h-1)-(8h+2)\cdot C(h,h)$, $C(h+1,0)=8\cdot C(h,1)-2\cdot C(h,0)$, and for all integers $1\le i\le h-1$, we have the recursive formula

$$C(h+1,i)=C(h,i-1)-(8i+2)\cdot C(h,i)+(4i+4)(4i+2)\cdot C(h,i+1).$$

Theorem 2. *For any positive integer h and non-negative n, we have*

$$\sum_{a_1+a_2+\cdots+a_{2h}=n} C_{a_1}\cdot C_{a_2}\cdot C_{a_3}\cdots C_{a_{2h}}$$
$$= \frac{1}{(2h-1)!}\sum_{i=0}^{h-1}\sum_{j=0}^{n} D(h,i+1)\cdot\binom{i+\frac{1}{2}}{j}\cdot(-4)^j\cdot\frac{(n-j+h+i)!\cdot C_{n-j+h+i}}{(n-j)!},$$

where $\binom{n+\frac{1}{2}}{i}=\left(n+\frac{1}{2}\right)\cdot\left(n-1+\frac{1}{2}\right)\cdots\left(n-i+1+\frac{1}{2}\right)/i!$, $D(k,i)$ are defined as $D(k,0)=0$, $D(k,k)=1$, $D(k+1,k)=D(k,k-1)-(8k-2)$, $D(k+1,1)=24D(k,2)-6D(k,1)$, and for all integers $1\le i\le k-1$,

$$D(k+1,i)=D(k,i-1)-(8i-2)\cdot D(k,i)+4i(4i+2)\cdot D(k,i+1).$$

To better illustrate the sequence $\{C(k,i)\}$ and $D(h,i)$, we compute them using mathematical software and list some values in the following Tables 1 and 2.

Table 1. Values of $C(k,i)$.

$C(k,i)$	$i=0$	$i=1$	$i=2$	$i=3$	$i=4$	$i=5$	$i=6$
$k=1$	−2	1					
$k=2$	12	−12	1				
$k=3$	−120	180	−30	1			
$k=4$	1680	−3360	840	−56	1		
$k=5$	−30,240	75,600	−25,200	2520	−90	1	
$k=6$	665,280	−1,995,840	831,600	−110,880	5940	−132	1

Table 2. Values of $D(k,i)$.

$D(k,i)$	$i=0$	$i=1$	$i=2$	$i=3$	$i=4$	$i=5$	$i=6$
$k=1$	0	1					
$k=2$	0	−6	1				
$k=3$	0	60	−20	1			
$k=4$	0	−840	420	−42	1		
$k=5$	0	15,120	−10,080	1512	−72	1	
$k=6$	0	−332,640	277,200	−55,440	3960	−110	1

Observing these two tables, we can easily find that if $2k-1=p$ is a prime, then for all integers $0 \leq i < k$, we have the congruences $C(k,i) \equiv 0 \bmod (2k-1)(2k)$ and $D(k,i) \equiv 0 \bmod (2k-1)(2k-2)$. So we propose the following two conjectures:

Conjecture 1. *Let p be a prime. Then for any integer $0 \leq i < \frac{p+1}{2}$, we have the congruence*

$$C\left(\frac{p+1}{2}, i\right) \equiv 0 \bmod p(p+1).$$

Conjecture 2. *Let p be a prime. Then for any integer $0 \leq i < \frac{p+1}{2}$, we have the congruence*

$$D\left(\frac{p+1}{2}, i\right) \equiv 0 \bmod p(p-1).$$

For some special integers n and h, from Theorem 1 and Theorem 2 we can also deduce several interesting corollaries. In fact if we take $n=0$ and $h=1$ in the theorems respectively, then we have the following four corollaries:

Corollary 1. *For any positive integer h, we have the identity*

$$\sum_{i=0}^{h} C(h,i) \cdot (h+i)! \cdot C_{h+i} = (2h)!.$$

Corollary 2. *For any positive integer h, we have the identity*

$$\sum_{i=1}^{h} D(h,i) \cdot (h+i-1)! \cdot C_{h+i-1} = (2h-1)!.$$

Corollary 3. *For any integer $n \geq 0$, we have the identity*

$$\sum_{a+b+d=n} C_a \cdot C_b \cdot C_d = (n+1) \cdot \left[\frac{1}{2} \cdot (n+2) \cdot C_{n+2} - (2n+1) \cdot C_{n+1}\right].$$

Corollary 4. *For any integer $n \geq 0$, we have the identity*

$$\sum_{u+v+w+x+y=n} C_u \cdot C_v \cdot C_w \cdot C_x \cdot C_y = \frac{(n+1)(n+2)(4n^2+8n+3)}{6} \cdot C_{n+2}$$
$$- \frac{(n+3)(n+2)(n+1)(2n+3)}{6} \cdot C_{n+3} + \frac{(n+4)(n+3)(n+2)(n+1)}{24} \cdot C_{n+4}.$$

2. Several Simple Lemmas

To prove our theorems, we need following four simple lemmas. First we have:

Lemma 1. *Let function $f(x) = \frac{2}{1+\sqrt{1-4x}}$. Then for any positive integer h, we have the identity*

$$(2h)! \cdot f^{2h+1}(x) = \sum_{i=0}^{h} C(h,i) \cdot (1-4x)^i \cdot f^{(h+i)}(x),$$

where $f^{(i)}(x)$ denotes the i-order derivative of $f(x)$ for x, and $\{C(h,i)\}$ are defined as the same as in Theorem 1.

Proof. In fact, this identity and its generalization had appeared in D. S. Kim and T. Kim's important work [9] (see Theorem 3.1), but only in different forms. For the completeness of our results, here we give a different proof by mathematical induction. First from the properties of the derivative we have

$$f'(x) = \frac{4}{\left(1+\sqrt{1-4x}\right)^2}\cdot\frac{1}{\sqrt{1-4x}} = \frac{f^2(x)}{\sqrt{1-4x}}$$

or identity

$$f^2(x) = (1-4x)^{\frac{1}{2}}\cdot f'(x). \tag{3}$$

From (3) and note that $C(1,0) = -2$ and $C(1,1) = 1$ we have

$$2f(x)\cdot f'(x) = -2\,(1-4x)^{-\frac{1}{2}}\cdot f'(x) + (1-4x)^{\frac{1}{2}}\cdot f''(x)$$

and

$$2!f^3(x) = -2f'(x) + (1-4x)\cdot f''(x) = \sum_{i=0}^{1} C(1,i)\cdot(1-4x)^i\cdot f^{(1+i)}(x).$$

That is, Lemma 1 is true for $h = 1$.

Assume that Lemma 1 is true for $h = k \geq 1$. That is,

$$(2k)!\cdot f^{2k+1}(x) = \sum_{i=0}^{k} C(k,i)\cdot(1-4x)^i\cdot f^{(k+i)}(x). \tag{4}$$

Then from (3), (4), the definition of $C(k,i)$, and the properties of the derivative we can deduce that

$$\begin{aligned}(2k+1)!\cdot f^{2k}(x)\cdot f'(x) &= \sum_{i=0}^{k} C(k,i)\cdot(1-4x)^i\cdot f^{(k+i+1)}(x)\\ &\quad - \sum_{i=1}^{k} 4i\cdot C(k,i)\cdot(1-4x)^{i-1}\cdot f^{(k+i)}(x)\end{aligned}$$

or

$$\begin{aligned}rrl(2k+1)!\cdot f^{2k+2}(x) = &\sum_{i=0}^{k} C(k,i)\cdot(1-4x)^{i+\frac{1}{2}}\cdot f^{(k+i+1)}(x)\\ &- \sum_{i=1}^{k} 4i\cdot C(k,i)\cdot(1-4x)^{i-\frac{1}{2}}\cdot f^{(k+i)}(x).\end{aligned} \tag{5}$$

Applying (5) and the properties of the derivative we also have

$$\begin{aligned}(2k+2)!\cdot f^{2k+1}(x)\cdot f'(x) &= \sum_{i=0}^{k} C(k,i)\cdot(1-4x)^{i+\frac{1}{2}}\cdot f^{(k+i+2)}(x)\\ &- \sum_{i=0}^{k}(4i+2)\cdot C(k,i)\cdot(1-4x)^{i-\frac{1}{2}}\cdot f^{(k+i+1)}(x)\\ &- \sum_{i=1}^{k} 4i\cdot C(k,i)\cdot(1-4x)^{i-\frac{1}{2}}\cdot f^{(k+i+1)}(x)\\ &+ \sum_{i=1}^{k}(4i)\cdot(4i-2)\cdot C(k,i)\cdot(1-4x)^{i-\frac{3}{2}}\cdot f^{(k+i)}(x)\end{aligned}$$

or note that identity (3) we have

$$\begin{aligned}
(2k+2)! \cdot f^{2k+3}(x) &= \sum_{i=0}^{k} C(k,i) \cdot (1-4x)^{i+1} \cdot f^{(k+i+2)}(x) \\
&- \sum_{i=0}^{k} (4i+2) \cdot C(k,i) \cdot (1-4x)^{i} \cdot f^{(k+i+1)}(x) \\
&- \sum_{i=1}^{k} 4i \cdot C(k,i) \cdot (1-4x)^{i} \cdot f^{(k+i+1)}(x) \\
&+ \sum_{i=1}^{k} (4i) \cdot (4i-2) \cdot C(k,i) \cdot (1-4x)^{i-1} \cdot f^{(k+i)}(x) \\
&= C(k,k) \cdot (1-4x)^{k+1} \cdot f^{(2k+2)}(x) + \sum_{i=1}^{k} C(k,i-1) \cdot (1-4x)^{i} \cdot f^{(k+i+1)}(x) \\
&-2C(k,0) \cdot f^{(k+1)}(x) - \sum_{i=1}^{k} (4i+2) \cdot C(k,i) \cdot (1-4x)^{i} \cdot f^{(k+i+1)}(x) \qquad (6) \\
&- \sum_{i=1}^{k} 4i \cdot C(k,i) \cdot (1-4x)^{i} \cdot f^{(k+i+1)}(x) + 8 \cdot C(k,1) \cdot f^{(k+1)}(x) \\
&+ \sum_{i=1}^{k-1} (4i+4) \cdot (4i+2) \cdot C(k,i+1) \cdot (1-4x)^{i} \cdot f^{(k+i+1)}(x) \\
&= (1-4x)^{k+1} \cdot f^{(2k+2)}(x) + (8 \cdot C(k,1) - 2 \cdot C(k,0)) \cdot f^{(k+1)}(x) \\
&+ (C(k,k-1) - (8k+2) \cdot C(k,k)) \cdot (1-4x)^{k} \cdot f^{(2k+1)}(x) \\
&+ \sum_{i=1}^{k-1} (C(k,i-1) - (8i+2) \cdot C(k,i) + (4i+4)(4i+2) \cdot C(k,i+1)) \\
&\times (1-4x)^{i} \cdot f^{(k+i+1)}(x) \\
&= \sum_{i=0}^{k+1} C(k+1,i) \cdot (1-4x)^{i} \cdot f^{(k+i+1)}(x),
\end{aligned}$$

where we have used the identities $C(k+1,k) = C(k,k-1) - (8k+2) \cdot C(k,k)$, $C(k,k) = 1$, $C(k+1,0) = 8 \cdot C(k,1) - 2 \cdot C(k,0)$ and for all integers $1 \le i \le k-1$,

$$C(k+1,i) = C(k,i-1) - (8i+2) \cdot C(k,i) + (4i+4)(4i+2) \cdot C(k,i+1).$$

It is clear that (6) implies Lemma 1 is true for $h = k+1$.

This proves Lemma 1 by mathematical induction. □

Lemma 2. *For any positive integer h, we have the identity*

$$(2h-1)! \cdot f^{2h}(x) = \sum_{i=0}^{h-1} D(h,i+1) \cdot (1-4x)^{i+\frac{1}{2}} \cdot f^{(h+i)}(x),$$

where $D(h,i)$ are defined as the same as in Theorem 2.

Proof. It is clear that using the methods of proving Lemma 1 we can easily deduce Lemma 2. □

Lemma 3. *Let h be any positive integer. Then for any integer $k \ge 0$, we have the identity*

$$(1-4x)^{k} \cdot f^{(h+k)}(x) = \sum_{n=0}^{\infty} \left(\sum_{i=0}^{\min(n,k)} \frac{C_{n-i+h+k}}{(n-i)!} \binom{k}{i} \cdot (-4)^{i} \right) \cdot x^{n}.$$

Proof. From the binomial theorem we have

$$(1-4x)^k = \sum_{i=0}^{k} \binom{k}{i} \cdot (-4x)^i. \tag{7}$$

On the other hand, from (1) we also have

$$f^{(h+k)}(x) = \sum_{n=0}^{\infty} \frac{(n+h+k)! \cdot C_{n+h+k}}{n!} \cdot x^n. \tag{8}$$

Combining (7) and (8) we have

$$\begin{aligned}
&(1-4x)^k \cdot f^{(h+k)}(x) \\
=& \left(\sum_{i=0}^{k} \binom{k}{i} \cdot (-4x)^i\right) \left(\sum_{n=0}^{\infty} \frac{(n+h+k)! \cdot C_{n+h+k}}{n!} \cdot x^n\right) \\
=& \sum_{n=0}^{\infty} \sum_{i=0}^{k} \frac{(n+h+k)! \cdot C_{n+h+k}}{n!} \cdot \binom{k}{i} \cdot (-4)^i \cdot x^{n+i} \\
=& \sum_{n=0}^{\infty} \left(\sum_{i=0}^{\min(n,k)} \frac{(n-i+h+k)! \cdot C_{n-i+h+k}}{(n-i)!} \binom{k}{i} \cdot (-4)^i\right) \cdot x^n.
\end{aligned}$$

This proves Lemma 3. □

Lemma 4. *Let h be any positive integer. Then for any integer $k \geq 0$, we have the identity*

$$(1-4x)^{k+\frac{1}{2}} \cdot f^{(h+k)}(x) = \sum_{n=0}^{\infty} \left(\sum_{i=0}^{n} \binom{k+\frac{1}{2}}{i} \cdot (-4)^i \cdot \frac{C_{n-i+h+k}}{(n-i)!}\right) \cdot x^n.$$

Proof. From the power series expansion of the function we know that

$$(1-4x)^{k+\frac{1}{2}} = \sum_{n=0}^{\infty} \binom{k+\frac{1}{2}}{n} \cdot (-4)^n \cdot x^n. \tag{9}$$

Applying (8) and (9) we have

$$\begin{aligned}
&(1-4x)^{k+\frac{1}{2}} \cdot f^{(h+k)}(x) \\
=& \left(\sum_{n=0}^{\infty} \binom{k+\frac{1}{2}}{n} \cdot (-4)^n \cdot x^n\right) \left(\sum_{n=0}^{\infty} \frac{(n+h+k)! \cdot C_{n+h+k}}{n!} \cdot x^n\right) \\
=& \sum_{n=0}^{\infty} \left(\sum_{i=0}^{n} \binom{k+\frac{1}{2}}{i} \cdot (-4)^i \cdot \frac{(n-i+h+k)! \cdot C_{n-i+h+k}}{(n-i)!}\right) \cdot x^n.
\end{aligned}$$

This proves Lemma 4. □

3. Proofs of the Theorems

In this section, we shall complete the proofs of our theorems. First we prove Theorem 1. From (1) and the multiplicative properties of the power series we have

$$(2h)! \cdot f^{2h+1}(x) = (2h)! \sum_{n=0}^{\infty} \left(\sum_{a_1+a_2+\cdots+a_{2h+1}=n} C_{a_1} \cdot C_{a_2} \cdots C_{a_{2h+1}}\right) \cdot x^n. \tag{10}$$

On the other hand, from Lemma 1 and Lemma 3 we also have

$$\begin{aligned}(2h)! \cdot f^{2h+1}(x) &= \sum_{i=0}^{h} C(h,i) \cdot (1-4x)^i \cdot f^{(h+i)}(x) \\ &= \sum_{n=0}^{\infty} \left(\sum_{i=0}^{h} C(h,i) \sum_{j=0}^{\min(n,i)} \frac{(n-j+h+i)! C_{n-j+h+i}}{(n-j)!} \binom{i}{j} (-4)^j \right) x^n.\end{aligned} \tag{11}$$

Combining (10) and (11) we may immediately deduce the identity

$$\begin{aligned}&\sum_{a_1+a_2+\cdots+a_{2h+1}=n} C_{a_1} \cdot C_{a_2} \cdot C_{a_3} \cdots C_{a_{2h+1}} \\ &= \frac{1}{(2h)!} \sum_{i=0}^{h} C(h,i) \sum_{j=0}^{\min(n,i)} \frac{(n-j+h+i)! \cdot C_{n-j+h+i}}{(n-j)!} \cdot \binom{i}{j} \cdot (-4)^j.\end{aligned}$$

This proves Theorem 1.

Now we prove Theorem 2. For any positive integer h, from (1) we have

$$f^{2h}(x) = \sum_{n=0}^{\infty} \left(\sum_{a_1+a_2+\cdots+a_{2h}=n} C_{a_1} \cdot C_{a_2} \cdots C_{a_{2h}} \right) \cdot x^n. \tag{12}$$

On the other hand, from Lemma 2 and Lemma 4 we also have

$$\begin{aligned}(2h-1)! \cdot f^{2h}(x) &= \sum_{i=0}^{h-1} D(h,i+1) \cdot (1-4x)^{i+\frac{1}{2}} \cdot f^{(h+i)}(x) \\ &= \sum_{i=0}^{h-1} D(h,i+1) \sum_{n=0}^{\infty} \left(\sum_{j=0}^{n} \binom{i+\frac{1}{2}}{j} (-4)^j \frac{(n-j+h+i)! \cdot C_{n-j+h+i}}{(n-j)!} \right) x^n \\ &= \sum_{n=0}^{\infty} \sum_{i=0}^{h-1} \sum_{j=0}^{n} D(h,i+1) \binom{i+\frac{1}{2}}{j} (-4)^j \frac{(n-j+h+i)! C_{n-j+h+i}}{(n-j)!} x^n.\end{aligned} \tag{13}$$

From (12), (13), and Lemma 2 we may immediately deduce the identity

$$\begin{aligned}&\sum_{a_1+a_2+\cdots+a_{2h}=n} C_{a_1} \cdot C_{a_2} \cdot C_{a_3} \cdots C_{a_{2h}} \\ &= \frac{1}{(2h-1)!} \sum_{i=0}^{h-1} \sum_{j=0}^{n} D(h,i+1) \cdot \binom{i+\frac{1}{2}}{j} \cdot (-4)^j \cdot \frac{(n-j+h+i)! \cdot C_{n-j+h+i}}{(n-j)!}.\end{aligned}$$

This completes the proof of Theorem 2.

4. Conclusions

The main results of this paper are Theorem 1 and Theorem 2. They gave two special expressions for convolution (2). In addition, Corollary 1 gives a close relationship between $C(h,i)$ and C_{h+i}. Corollary 2 gives a close relationship between $D(h,i)$ and D_{h+i-1}. Corollary 3 and Corollary 4 give two exact representations for the special cases of Theorem 1 with $h = 1$ and $h = 2$.

About the new sequences $C(h,i)$ and $D(h,i)$, we proposed two interesting conjectures related to congruence mod p, where p is an odd prime. We believe that these conjectures are correct, but at the moment we cannot prove them. We also believe that these two conjectures will certainly attract the interest of many readers, thus further promoting the study of the properties of $C(h,i)$ and C_{h+i}.

Author Contributions: All authors have equally contributed to this work. All authors read and approved the final manuscript.

Acknowledgments: The authors would like to thank the Editor and referees for their very helpful and detailed comments, which have significantly improved the presentation of this paper.

References

1. Stanley, R.P. *Enumerative Combinatorics (Vol. 2)*; Cambridge Studieds in Advanced Mathematics; Cambridge University Press: Cambridge, UK, 1997; p. 49.
2. Chu, W.C. Further identities on Catalan numbers. *Discret. Math.* **2018**, *341*, 3159–3164. [CrossRef]
3. Joseph, A.; Lamprou, P. A new interpretation of the Catalan numbers arising in the theory of crystals. *J. Algebra* **2018**, *504*, 85–128. [CrossRef]
4. Allen, E.; Gheorghiciuc, I. A weighted interpretation for the super Catalan numbers. *J. Integer Seq.* **2014**, *9*, 17.
5. Tauraso, R. qq-Analogs of some congruences involving Catalan numbers. *Adv. Appl. Math.* **2012**, *48*, 603–614. [CrossRef]
6. Liu, J.C. Congruences on sums of super Catalan numbers. *Results Math.* **2018**, *73*, 73–140. [CrossRef]
7. Qi, F.; Shi, X.T.; Liu, F.F. An integral representation, complete monotonicity, and inequalities of the Catalan numbers. *Filomat* **2018**, *32*, 575–587. [CrossRef]
8. Qi, F.; Shi, X.T.; Mahmoud, M. The Catalan numbers: A generalization, an exponential representation, and some properties. *J. Comput. Anal. Appl.* **2017**, *23*, 937–944.
9. Kim, D.S.; Kim, T. A new approach to Catalan numbers using differential equations. *Russ. J. Math. Phys.* **2017**, *24*, 465–475. [CrossRef]
10. Kim, T.; Kim, D.S. Some identities of Catalan-Daehee polynomials arising from umbral calculus. *Appl. Comput. Math.* **2017**, *16*, 177–189.
11. Kim, T.; Kim, D.S. Differential equations associated with Catalan-Daehee numbers and their applications. *Revista de la Real Academia de Ciencias Exactas, Físicas y Naturales. Serie A. Matemáticas* **2017**, *111*, 1071–1081. [CrossRef]
12. Kim, D.S.; Kim, T. Triple symmetric identities for w-Catalan polynomials. *J. Korean Math. Soc.* **2017**, *54*, 1243–1264.
13. Kim, T.; Kwon, H.-I. Revisit symmetric identities for the λ-Catalan polynomials under the symmetry group of degree n. *Proc. Jangjeon Math. Soc.* **2016**, *19*, 711–716.
14. Kim, T. A note on Catalan numbers associated with p-adic integral on Zp. *Proc. Jangjeon Math. Soc.* **2016**, *19*, 493–501.
15. Basic, B. On quotients of values of Euler's function on the Catalan numbers. *J. Number Theory* **2016**, *169*, 160–173. [CrossRef]
16. Aker, K.; Gursoy, A.E. A new combinatorial identity for Catalan numbers. *Ars Comb.* **2017**, *135*, 391–398.
17. Qi, F.; Guo, B.N. Integral representations of the Catalan numbers and their applications. *Mathematics* **2017**, *5*, 40. [CrossRef]
18. Hein, N.; Huang, J. Modular Catalan numbers. *Eur. J. Combin.* **2017**, *61*, 197–218. [CrossRef]
19. Dilworth, S.J.; Mane, S.R. Applications of Fuss-Catalan numbers to success runs of Bernoulli trials. *J. Probab. Stat.* **2016**, *2016*, 2071582. [CrossRef]
20. Dolgy, D.V.; Jang, G.-W.; Kim, D.S.; Kim, T. Explicit expressions for Catalan-Daehee numbers. *Proc. Jangjeon Math. Soc.* **2017**, *20*, 1–9.
21. Zhang, Y.X.; Chen, Z.Y. A new identity involving the Chebyshev polynomials. *Mathematics* **2018**, *6*, 244. [CrossRef]
22. Zhao, J.H.; Chen, Z.Y. Some symmetric identities involving Fubini polynomials and Euler numbers. *Symmetry* **2018**, *10*, 359.
23. Zhang, W.P. Some identities involving the Fibonacci numbers and Lucas numbers. *Fibonacci Q.* **2004**, *42*, 149–154.

14

Some Convolution Formulae Related to the Second-Order Linear Recurrence Sequence

Zhuoyu Chen [1] **and Lan Qi** [2,*]

1 School of Mathematics, Northwest University, Xi'an 710127, China; chenzymath@stumail.nwu.edu.cn
2 School of Mathematics and Statistics, Yulin University, Yulin 719000, China
* Correspondence: qilanmail@163.com

Abstract: The main aim of this paper is that for any second-order linear recurrence sequence, the generating function of which is $f(t) = \frac{1}{1+at+bt^2}$, we can give the exact coefficient expression of the power series expansion of $f^x(t)$ for $x \in \mathbf{R}$ with elementary methods and symmetry properties. On the other hand, if we take some special values for a and b, not only can we obtain the convolution formula of some important polynomials, but also we can establish the relationship between polynomials and themselves. For example, we can find relationship between the Chebyshev polynomials and Legendre polynomials.

Keywords: Fibonacci numbers; Lucas numbers; Chebyshev polynomials; Legendre polynomials; Jacobi polynomials; Gegenbauer polynomials; convolution formula

MSC: 11B83

---

1. Introduction

For any integer $n \geq 1$ and any real number y, the Fibonacci polynomials $F_n(y)$ and the Lucas polynomials $L_n(y)$ are defined by the second-order linear recurrence sequence

$$F_{n+1}(y) = yF_n(y) + F_{n-1}(y)$$

and

$$L_{n+1}(y) = yL_n(y) + L_{n-1}(y),$$

where the first two terms are $F_0(y) = 0$, $F_1(y) = 1$, $L_0(y) = 2$ and $L_1(y) = y$.

If we take $\alpha = \frac{y+\sqrt{y^2+4}}{2}$, $\beta = \frac{y-\sqrt{y^2+4}}{2}$, according to the properties of the second-order linear recurrence sequence, we have

$$F_n(y) = \frac{\alpha^n - \beta^n}{\alpha - \beta}$$

and

$$L_n(y) = \alpha^n + \beta^n.$$

For any integer $n \geq 0$, the Fibonacci numbers $F_n = F_n(1)$ can be defined by the generating function

$$\frac{1}{1-t-t^2} = \sum_{n=0}^{\infty} F_n t^n.$$

For any integer $n \geq 0$, the first and the second kind Chebyshev polynomials $T_n(y)$ and $U_n(y)$ are defined by the second-order linear recurrence sequence

$$T_{n+2}(y) = 2yT_{n+1}(y) - T_n(y)$$

and

$$U_{n+2}(y) = 2yU_{n+1}(y) - U_n(y),$$

where the first two terms are $T_0(y) = 1$, $T_1(y) = y$, $U_0(y) = 1$ and $U_1(y) = 2y$.

If we take $\alpha = y + \sqrt{y^2 - 1}$, $\beta = y - \sqrt{y^2 - 1}$, according to the properties of the second-order linear recurrence sequence, we have

$$T_n(y) = \frac{\alpha^n + \beta^n}{2}$$

and

$$U_n(y) = \frac{\alpha^{n+1} - \beta^{n+1}}{\alpha - \beta}.$$

On the other hand, the second kind Chebyshev polynomials $U_n(y)$ can be also defined by the generating function

$$\frac{1}{1 - 2yt + t^2} = \sum_{n=0}^{\infty} U_n(y)t^n.$$

Besides Fibonacci polynomials, Lucas polynomials and Chebyshev polynomials, other orthogonal polynomials have also been studied by interested scholars.

For example, the Legendre polynomials $P_n(y)$ are defined by the generating function

$$\left(\frac{1}{1 - 2yt + t^2}\right)^{\frac{1}{2}} = \sum_{n=0}^{\infty} P_n(y)t^n.$$

The Jacobi polynomials $\{P_n^{(\alpha,\beta)}(y)\}_{0 \leq n < \infty}$ are defined by the generating function

$$\left[R(1 + R - t)^{\alpha}(1 + R + t)^{\beta}\right]^{-1} = \sum_{k=0}^{\infty} 2^{-\alpha-\beta} P_n^{(\alpha,\beta)}(y)t^n,$$

where $R = \sqrt{1 - 2yt + t^2}$, $|t| < 1$, $\alpha, \beta > -1$.

The Gegenbauer polynomials $\{C_n^{\lambda}(y)\}_{0 \leq n < \infty}$ are defined by the generating function

$$\left(\frac{1}{1 - 2yt + t^2}\right)^{\lambda} = \sum_{n=0}^{\infty} C_n^{\lambda}(y)t^n, \left(\lambda > -\frac{1}{2}\right).$$

It is well know that polynomials and sequence occupy indispensable positions in the research of number theory. Especially, Fibonacci and Lucas numbers, Chebyshev and Legendre polynomials and others. These polynomials and numbers are closely related and there are a variety of meaningful results which have been researched by interested scholars until now. For example, the identities of Chebyshev polynomials can be found in [1–9], and the contents about Fibonacci and Lucas numbers in [10,11]. Some authors have a research which connects Chebyshev polynomials and Fibonacci or Lucas polynomials (see [12–14]).

In particular, we can find many significant results in the aspect of studying the calculating problem of one kind sums of some important polynomials. For example, Yuankui Ma and Wenpeng Zhang have calculated one kind sums of Fibonacci Polynomials (see [15]) as follows.

Let h be a positive integer, for any integer $n \geq 0$, they proved

$$\sum_{a_1+a_2+\cdots+a_{h+1}=n} F_{a_1}(x)F_{a_2}(x)\cdots F_{a_{h+1}}(x) = \frac{1}{h!}\cdot\sum_{j=1}^{h}\frac{(-1)^{h-j}\cdot S(h,j)}{x^{2h-j}}$$
$$\times\left(\sum_{i=0}^{n}\frac{(n-i+j)!}{(n-i)!}\cdot\binom{2h+i-j-1}{i}\cdot\frac{(-1)^i\cdot 2^i\cdot F_{n-i+j}(x)}{x^i}\right),$$

where the summation is over all $h+1$-tuples with non-negative integer coordinates $(a_1, a_2, \cdots, a_{h+1}$ such that $a_1 + a_2 + \cdots + a_{h+1} = n$, and $S(h,i)$ is a second order non-linear recurrence sequence defined by $S(h,0) = 0$, $S(h,h) = 1$, and $S(h+1,i+1) = 2\cdot(2h-1-i)\cdot S(h,i+1) + S(h,i)$ for all positive integers $1 \leq i \leq h-1$.

Yixue Zhang and Zhuoyu Chen have researched the calculating problem of one kind sums of the second kind Chebyshev polynomials (see [16]) as follows.

Let h be a positive integer, for any integer $n \geq 0$, they proved

$$\sum_{a_1+a_2+\cdots+a_{h+1}=n} U_{a_1}(x)U_{a_2}(x)\cdots U_{a_{h+1}}(x)$$
$$= \frac{1}{2^h\cdot h!}\cdot\sum_{j=1}^{h}\frac{C(h,j)}{x^{2h-j}}\sum_{i=0}^{n}\frac{(n-i+j)!}{(n-i)!}\cdot\binom{2h+i-j-1}{i}\cdot\frac{U_{n-i+j}(x)}{x^i},$$

where $C(h,i)$ is a second order non-linear recurrence sequence defined by $C(h,0) = 0$, $C(h,h) = 1$, $C(h+1,1) = 1\cdot 3\cdot 5\cdots(2h-1) = (2h-1)!!$ and $C(h+1,i+1) = (2h-1-i)\cdot C(h,i+1) + C(h,i)$ for all $1 \leq i \leq h-1$.

Shimeng Shen and Li Chen have studied the calculating problem of one kind sums of Legendre Polynomials (see [17]) as follows.

For any positive integer k and integer $n \geq 0$, they proved

$$(2k-1)!!\sum_{a_1+a_2+\cdots+a_{2k+1}=n} P_{a_1}(x)P_{a_2}(x)\cdots P_{a_k}(x)$$
$$= \sum_{j=1}^{k} C(k,j)\sum_{i=0}^{n}\frac{(n+k+1-i-j)!}{(n-i)!}\cdot\frac{\binom{i+j+k-2}{i}}{x^{k-1+i+j}}\cdot P_{n+k+1-i-j}(x)$$

where$(2k-1)!! = 1\times 3\times 5\cdots(2k-1) = 2^k(\frac{1}{2})_k$, and $C(k,i)$ is a recurrence sequence defined by $C(k,1) = 1$, $C(k+1,k+1) = (2k-1)!!$ and $C(k+1,i+1) = C(k,i+1) + (k-1+i)\cdot C(k,i)$ for all $1 \leq i \leq k-1$.

They have converted the complex sums of $F_n(x)$ into a simple combination of $F_n(x)$, the complex sums of $U_n(x)$ into a simple combination of $U_n(x)$, and the complex sums of $P_n(x)$ into a simple combination of $P_n(x)$.

Very recently, Taekyun Kim and other people researched the properties of Fibonacci numbers through introducing the convolved Fibonacci numbers $p_n(x)$ by generating function as follows (see [18]):

$$\left(\frac{1}{1-t-t^2}\right)^x = \sum_{n=0}^{\infty} p_n(x)\frac{t^n}{n!}, (x \in \mathbf{R}).$$

They researched some new and explicit identities of the convolved Fibonacci numbers for $x \in \mathbf{N}$. For example, for $n \geq 0$ and $r \in \mathbf{N}$, they have proved the recurrence relationship of $p_n(x)$ (see [18]):

$$p_n(x) = \sum_{l=0}^{n} p_l(r)p_{n-l}(x-r) = \sum_{l=0}^{n} p_{n-l}(r)p_l(x-r).$$

The convolved Fibonacci numbers $p_n(x)$ seems to be only connected with the simple power square. In fact, it can establish the relationship between polynomials and themselves, so the further research of $p_n(x)$ is very significant. They have provided us a new perspective to study the properties of some vital polynomials. For example, Taekyun Kim and other people have proved the relationship between $p_n(x)$ and the combination sums about Fibonacci numbers:

$$\frac{p_n(r+1)}{n!} = \sum_{l_1=0}^{n} \sum_{l_2=0}^{n-l_1} \cdots \sum_{l_r=0}^{n-l_1-\cdots-l_{r-1}} F_{l_1} F_{l_2} \cdots F_{l_r} F_{n-l_1-l_2-\cdots-l_r}.$$

They have converted the complex sums of $F_n(x)$ into a calculation problem of $p_n(x)$ and the calculation method is easier and the expression is simpler.

Inspired by this article, in this paper, for any second-order linear recurrence sequence, the generating function of which is $f(t) = \frac{1}{1+at+bt^2}$, we can define

$$\left(\frac{1}{1+at+bt^2}\right)^x = \sum_{n=0}^{\infty} p_n(x)\frac{t^n}{n!}, \ (a,b,x \in \mathbf{R}). \tag{1}$$

Firstly, we give a specific computational formula of $p_n(x)$ for $x \in \mathbf{R}$ using the elementary methods. After that for any polynomial or sequence, the generating function of which is $f(t) = \frac{1}{1+at+bt^2}$, we can obtain its convolved formula easily and directly.

Secondly, if we take some special values for a, b in $f(t)$ and x in $p_n(x)$, we can find some relationship between special polynomials and themselves. For example, we will establish the relationship between the convolved Fibonacci numbers and Lucas numbers, the relationship between the convolved formula of the second kind Chebyshev polynomials and the first kind Chebyshev polynomials, and the relationship between Legendre polynomials and the first kind Chebyshev polynomials and others.

At last, through the computational formula of $p_n(x)$, especially for $x \in \mathbf{N}$, we can also convert the complex sums of F_n into a liner combination of L_n; and express the complex sums of $U_n(y)$ as a liner combination of $T_n(y)$. More importantly, the forms are more common and the calculations are easier than previous results.

We will prove the main results as follows:

Theorem 1. *Let $f(t) = \frac{1}{1-t-t^2}$, for any integer $n \geq 0$ and $x \in \mathbf{R}$, we can obtain*

$$p_n(x) = \frac{1}{2}\sum_{i=0}^{n}(-1)^i \binom{n}{i} \langle x\rangle_i \langle x\rangle_{n-i} L_{n-2i},$$

where $\langle x\rangle_n = x(x+1)(x+2)\cdots(x+n-1)$ and $(x)_0 = 1$.

Theorem 2. *Let $f(t) = \frac{1}{1-2yt+t^2}$, for any integer $n \geq 0$ and $x, y \in \mathbf{R}$, we can obtain*

$$p_n(x;y) = \sum_{i=0}^{n} \binom{n}{i} \langle x\rangle_i \langle x\rangle_{n-i} T_{n-2i}(y).$$

From Theorem 1 we can deduce the following:

Corollary 1. *For any positive integer k, we have the identity*

$$\sum_{a_1+a_2+\cdots+a_k=n} F_{a_1}F_{a_2}\cdots F_{a_k} = \frac{1}{2((k-1)!)^2}\sum_{i=0}^{n}(-1)^i\frac{(k+i-1)!(k+n-i-1)!}{i!(n-i)!}\cdot L_{n-2i}.$$

From Theorem 2 we can deduce the following:

Corollary 2. *For any positive integer k, we have the identity*

$$\sum_{a_1+a_2+\cdots+a_k=n} U_{a_1}(y)\cdot U_{a_2}(y)\cdots U_{a_k}(y) = \frac{1}{((k-1)!)^2}\sum_{i=0}^{n}\frac{(k+i-1)!(k+n-i-1)!}{i!(n-i)!}\cdot T_{n-2i}(y).$$

Corollary 3. *If $x=\frac{1}{2}$, we have the identity*

$$P_n(y) = \frac{1}{2^n}\sum_{i=0}^{n}\frac{(2i-1)!!(2n-2i-1)!!}{i!(n-i)!}\cdot T_{n-2i}(y).$$

Corollary 4. *If $x=-\frac{1}{2}$, we have the identity*

$$R = \sum_{n=0}^{\infty}\frac{1}{2^n}\sum_{i=0}^{n}\frac{(2i-3)!!(2n-2i-3)!!}{i!(n-i)!}\cdot T_{n-2i}(y)\cdot t^n.$$

Corollary 5. *If $x=\lambda>-\frac{1}{2}$, we have the identity*

$$C_n^{\lambda}(y) = \frac{1}{n!}\sum_{i=0}^{n}\binom{n}{i}\langle\lambda\rangle_i\langle\lambda\rangle_{n-i}T_{n-2i}(y).$$

Theorems 1 and 2 give the computational formula of $p_n(x)$ of some famous polynomials. Especially, we know that polynomials are closely connected and they can be converted to each other. According to these theorems, we can obtain the relationship between the polynomials easily. It cannot only extend the application of orthogonal polynomials, but also make replacement calculations according to its complexity. For example, if we make a calculation involving the Gegenbauer polynomials, for simple calculations, we can convert it into Chebyshev polynomials according to Corollary 5.

2. A Simple Lemma

In order to prove our theorems, we are going to introduce a simple lemma.

Lemma 1. *For any integer $n\geq 0$ and $a,b,x\in\mathbf{R}$, we can obtain the equation*

$$p_n(x) = \frac{1}{2}\sum_{i=0}^{n} b^i\binom{n}{i}\langle x\rangle_i\langle x\rangle_{n-i}\left(\left(\frac{-a+\sqrt{a^2-4b}}{2}\right)^{n-2i}+\left(\frac{-a-\sqrt{a^2-4b}}{2}\right)^{n-2i}\right).$$

Proof. Firstly, according Equation (1), we have

$$\sum_{n=0}^{\infty} p_n(x)\frac{t^n}{n!} = \left(\frac{1}{1+at+bt^2}\right)^x = (1-\alpha t)^{-x}(1-\beta t)^{-x}. \tag{2}$$

We can easily know that $\alpha+\beta=-a$, $\alpha\beta=b$ and $\alpha=\frac{-a+\sqrt{a^2-4b}}{2}$, $\beta=\frac{-a-\sqrt{a^2-4b}}{2}$ are two roots of $1+at+bt^2=0$.

Then, applying the properties of power series, we obtain

$$(1-\alpha t)^{-x}=\sum_{n=0}^{\infty}\binom{-x}{n}(-1)^n(\alpha t)^n=\sum_{n=0}^{\infty}\frac{(-x)_n}{n!}(-1)^n\alpha^n t^n \tag{3}$$

and

$$(1-\beta t)^{-x}=\sum_{n=0}^{\infty}\binom{-x}{n}(-1)^n(\beta t)^n=\sum_{n=0}^{\infty}\frac{(-x)_n}{n!}(-1)^n\beta^n t^n, \tag{4}$$

where $(x)_n=x(x-1)(x-2)\cdots(x-n+1)$ and $(x)_0=1$.

Combining Equations (2)–(4), we get

$$\begin{aligned}\sum_{n=0}^{\infty}p_n(x)\frac{t^n}{n!} &= \left(\sum_{n=0}^{\infty}\frac{(-x)_n}{n!}(-1)^n\alpha^n t^n\right)\left(\sum_{n=0}^{\infty}\frac{(-x)_n}{n!}(-1)^n\beta^n t^n\right)\\ &= \sum_{n=0}^{\infty}\left(\sum_{i=0}^{n}\frac{(-x)_i(-1)^i\alpha^i t^i}{i!}\cdot\frac{(-x)_{n-i}(-1)^{n-i}\beta^{n-i}t^{n-i}}{(n-i)!}\right)\\ &= \sum_{n=0}^{\infty}\frac{(-1)^n}{n!}\left(\sum_{i=0}^{n}\binom{n}{i}(-x)_i(-x)_{n-i}\alpha^i\beta^{n-i}\right)t^n. \end{aligned}\tag{5}$$

Similarly, according the symmetry of α and β, we can easily obtain

$$\sum_{n=0}^{\infty}p_n(x)\frac{t^n}{n!}=\sum_{n=0}^{\infty}\frac{(-1)^n}{n!}\left(\sum_{i=0}^{n}\binom{n}{i}(-x)_i(-x)_{n-i}\beta^i\alpha^{n-i}\right)t^n. \tag{6}$$

Then, combining Equations (5) and (6), we know that

$$\begin{aligned}\sum_{n=0}^{\infty}p_n(x)\frac{t^n}{n!} &= \frac{1}{2}\sum_{n=0}^{\infty}\frac{(-1)^n}{n!}\left(\sum_{i=0}^{n}\binom{n}{i}(-x)_i(-x)_{n-i}(\alpha\beta)^i\left(\beta^{n-2i}+\alpha^{n-2i}\right)\right)t^n\\ &= \frac{1}{2}\sum_{n=0}^{\infty}\frac{1}{n!}\left(\sum_{i=0}^{n}b^i\binom{n}{i}\langle x\rangle_i\langle x\rangle_{n-i}\left(\beta^{n-2i}+\alpha^{n-2i}\right)\right)t^n. \end{aligned}\tag{7}$$

Comparing the coefficients of t^n in Equation (7), we get

$$\begin{aligned}p_n(x) &= \frac{1}{2}\sum_{i=0}^{n}b^i\binom{n}{i}\langle x\rangle_i\langle x\rangle_{n-i}\left(\alpha^{n-2i}+\beta^{n-2i}\right)\\ &= \frac{1}{2}\sum_{i=0}^{n}b^i\binom{n}{i}\langle x\rangle_i\langle x\rangle_{n-i}\left(\left(\frac{-a+\sqrt{a^2-4b}}{2}\right)^{n-2i}+\left(\frac{-a-\sqrt{a^2-4b}}{2}\right)^{n-2i}\right). \end{aligned}$$

Now we have completed the proof of the Lemma 1. □

3. Proof of the Theorem

Proof of Theorem 1. If we take $a=-1$ and $b=-1$ in Equation (1), we know that $f(t)$ is the generating function of Fibonacci number. That is,

$$f(t)=\frac{1}{1-t-t^2}=\sum_{n=0}^{\infty}F_n t^n.$$

The convolved Fibonacci numbers $p_n(x)$ are defined by the generating function as [18]

$$f^x(t) = \left(\frac{1}{1-t-t^2}\right)^x = \sum_{n=0}^{\infty} p_n(x)\frac{t^n}{n!}. \tag{8}$$

In this time, $\alpha = \frac{1+\sqrt{5}}{2}, \beta = \frac{1-\sqrt{5}}{2}$.
According to the Lemma 1 and $L_n = \alpha^n + \beta^n$, we can get

$$\begin{aligned} p_n(x) &= \frac{1}{2}\sum_{i=0}^{n} b^i \binom{n}{i} \langle x\rangle_i \langle x\rangle_{n-i} \left(\left(\frac{-a+\sqrt{a^2-4b}}{2}\right)^{n-2i} + \left(\frac{-a-\sqrt{a^2-4b}}{2}\right)^{n-2i}\right) \\ &= \frac{1}{2}\sum_{i=0}^{n} b^i \binom{n}{i} \langle x\rangle_i \langle x\rangle_{n-i} \left(\left(\frac{1+\sqrt{5}}{2}\right)^{n-2i} + \left(\frac{1-\sqrt{5}}{2}\right)^{n-2i}\right) \\ &= \frac{1}{2}\sum_{i=0}^{n} b^i \binom{n}{i} \langle x\rangle_i \langle x\rangle_{n-i} \left(\alpha^{n-2i} + \beta^{n-2i}\right) \\ &= \frac{1}{2}\sum_{i=0}^{n} (-1)^i \binom{n}{i} \langle x\rangle_i \langle x\rangle_{n-i} L_{n-2i}. \end{aligned} \tag{9}$$

In this equation, $p_n(x)$ is expressed as a combined forms of Lucas number. The Proof of Theorem 1 has finished. □

About the convolved Fibonacci numbers $p_n(x)$, Taekyun Kim and others have obtained its some-recurrence formulae in reference [18]. Based on [18], we have given an exact computational formula of $p_n(x)$ for any arbitrary x in Theorem 1. Compared with the results in [18], Theorem 1 is more general and easier.

If we take $x = k \in \mathbf{N}$ in Equation (8), we get

$$\begin{aligned} \sum_{n=0}^{\infty} p_n(k)\frac{t^n}{n!} &= \left(\frac{1}{1-t-t^2}\right)^k = \left(\sum_{n=0}^{\infty} F_n t^n\right)^k \\ &= \left(\sum_{a_1=0}^{\infty} F_{a_1}\cdot t^{a_1}\right)\left(\sum_{a_2=0}^{\infty} F_{a_2}\cdot t^{a_2}\right)\cdots\left(\sum_{a_k=0}^{\infty} F_{a_k}\cdot t^{a_k}\right) \\ &= \left(\sum_{a_1=0}^{\infty}\sum_{a_2=0}^{\infty}\cdots\sum_{a_k=0}^{\infty} F_{a_1}\cdot F_{a_2}\cdots F_{a_k}\cdot t^{a_1+a_2\cdots+a_k}\right) \\ &= \sum_{n=0}^{\infty}\left(\sum_{a_1+a_2+\cdots+a_k=n} F_{a_1}\cdot F_{a_2}\cdots F_{a_k}\right)\cdot t^n, \end{aligned}$$

and then combining Equation (9), we can obtain

$$\begin{aligned} &\sum_{a_1+a_2+\cdots+a_k=n} F_{a_1}\cdot F_{a_2}\cdots F_{a_k} \\ = &\frac{1}{2n!}\sum_{i=0}^{n}(-1)^i\binom{n}{i}\langle k\rangle_i\langle k\rangle_{n-i}L_{n-2i} \\ = &\frac{1}{2((k-1)!)^2}\sum_{i=0}^{n}(-1)^i\frac{(k+i-1)!(k+n-i-1)!}{i!(n-i)!}L_{n-2i}. \end{aligned}$$

The proof of Corollary 1 has finished.
For every $F_{a_l} (1 \le l \le k)$, $\sum_{a_1+a_2+\cdots+a_k=n} F_{a_1}\cdot F_{a_2}\cdots F_{a_k}$ is symmetry.

Proof of Theorem 2. If we take $a = -2y$ and $b = 1$ in Equation (1), we all know $f(t;y)$ is the generating function of the second-kind Chebyshev polynomials $U_n(y)$

$$f(t;y) = \frac{1}{1-2yt+t^2} = \sum_{n=0}^{\infty} U_n(y)t^n.$$

The convolved second-kind Chebyshev polynomials $p_n(x;y)$ are defined by the generating function as [18]

$$f^x(t;y) = \left(\frac{1}{1-2yt+t^2}\right)^x = \sum_{n=0}^{\infty} p_n(x;y)\frac{t^n}{n!}. \tag{10}$$

In this time, $\alpha = y + \sqrt{y^2-1}$, $\beta = y - \sqrt{y^2-1}$.
According to the Lemma 1 and $T_n(y) = \frac{1}{2}(\alpha^n + \beta^n)$, we can get

$$p_n(x;y) = \sum_{i=0}^{n} \binom{n}{i} \langle x\rangle_i \langle x\rangle_{n-i} T_{n-2i}(y). \tag{11}$$

In this equation, $p_n(x;y)$ is expressed as a combined form of the first-kind Chebyshev polynomials $T_n(x)$. □

If we take $x = k \in \mathbf{N}$ in Equation (10), and combining Equation (11) we can easily prove the Corollary 2.

Take $x = \frac{1}{2}$ in Equation (10), we know $f^{\frac{1}{2}}(t;y)$ is the generating function of the Legendre polynomials $P_n(x)$ as follows:

$$f^{\frac{1}{2}}(t) = \left(\frac{1}{1-2yt+t^2}\right)^{\frac{1}{2}} = \sum_{n=0}^{\infty} P_n(y)t^n = \sum_{n=0}^{\infty} p_n\left(\frac{1}{2};y\right)\frac{t^n}{n!}.$$

According to Theorem 2, we can easily obtain

$$\begin{aligned} p_n\left(\frac{1}{2};y\right) &= \sum_{i=0}^{n}\binom{n}{i}\left\langle\frac{1}{2}\right\rangle_i\left\langle\frac{1}{2}\right\rangle_{n-i} T_{n-2i}(y) \\ &= \frac{1}{2^{2n}}\sum_{i=0}^{n}\frac{n!(2i)!(2(n-i))!}{(i!)^2((n-i)!)^2}\cdot T_{n-2i}(y) \\ &= \frac{n!}{2^n}\sum_{i=0}^{n}\frac{(2i-1)!!(2n-2i-1)!!}{i!(n-i)!}\cdot T_{n-2i}(y). \end{aligned}$$

In a word, we know the Legendre polynomials $P_n(x)$ can be expressed as combined forms of the first kind Chebyshev polynomials $T_n(x)$ as follows:

$$P_n(y) = \frac{1}{2^n}\sum_{i=0}^{n}\frac{(2i-1)!!(2n-2i-1)!!}{i!(n-i)!}\cdot T_{n-2i}(y).$$

The proof of Corollary 3 has finished.
If we take $x = -\frac{1}{2}$ in (10), then we can easily obtain

$$\begin{aligned} R &= \sum_{n=0}^{\infty} p(-\frac{1}{2};y)\frac{t^n}{n!} = \sum_{n=0}^{\infty}\frac{1}{2^{2n-2}}\sum_{i=0}^{n}\frac{(2(i-1))!(2(n-i-1))!}{i!(i-1)!(n-i)!(n-i-1)!}\cdot T_{n-2i}(y)\cdot t^n \\ &= \sum_{n=0}^{\infty}\frac{1}{2^n}\sum_{i=0}^{n}\frac{(2i-3)!!(2n-2i-3)!!}{i!(n-i)!}\cdot T_{n-2i}(y)\cdot t^n. \end{aligned}$$

The proof of Corollary 4 has finished.

Taking $x = \lambda > -\frac{1}{2}$ in Equation (10), we know $f^{\lambda}(t;y)$ is the generating function of the Gegenbauer polynomials $\{C_n^{\lambda}(y)\}_{0\leq n<\infty}$ as follows:

$$\left(\frac{1}{1-2yt+t^2}\right)^{\lambda} = \sum_{n=0}^{\infty} C_n^{\lambda}(y)t^n = f^{\lambda}(t;y) = \sum_{n=0}^{\infty} p_n(\lambda;y)\frac{t^n}{n!}.$$

According to Theorem 2, we can easily obtain

$$p_n(\lambda;y) = \sum_{i=0}^{n} \binom{n}{i} \langle\lambda\rangle_i \langle\lambda\rangle_{n-i} T_{n-2i}(y).$$

The proof of Corollary 5 has finished.

Author Contributions: Writing—original draft: Z.C.; Writing—review and editing: L.Q.

Acknowledgments: The authors would like to thank the referees for their very helpful and detailed comments, which have significantly improved the presentation of this paper.

References

1. Chen, L.; Zhang, W. Chebyshev polynomials and their some interesting applications. *Adv. Differ. Equ.* **2017**, *2017*, 303.
2. Wang, S. Some new identities of Chebyshev polynomials and their applications. *Adv. Differ. Equ.* **2015**, *2015*, 335.
3. Li, X. Some identities involving Chebyshev polynomials. *Math. Probl. Eng.* **2015**, *2015*, 950695. [CrossRef]
4. Ma, Y.; Lv, X. Several identities involving the reciprocal sums of Chebyshev polynomials. *Math. Probl. Eng.* **2017**, *2017*, 4194579. [CrossRef]
5. Cesarano, C. Identities and generating functions on Chebyshev polynomials. *Georgian Math. J.* **2012**, *19*, 427–440. [CrossRef]
6. Cesarano, C. Integral representations and new generating functions of Chebyshev polynomials. *Hacet. J. Math. Stat.* **2015**, *44*, 535–546. [CrossRef]
7. Lee, C.; Wong, K. On Chebyshev's Polynomials and Certain Combinatorial Identities. *Bull. Malays. Math. Sci. Soc.* **2011**, *34*, 279–286.
8. Wang, T.; Zhang, H. Some identities involving the derivative of the first kind Chebyshev polynomials. *Math. Probl. Eng.* **2015**, *2015*, 146313. [CrossRef]
9. Wang, T.; Zhang, W. Two identities involving the integral of the first kind Chebyshev polynomials. *Bull. Math. Soc. Sci. Math. Roum.* **2017**, *108*, 91–98.
10. Zhang, W. Some identities involving the Fibonacci numbers and Lucas numbers. *Fibonacci Q.* **2004**, *42*, 149–154.
11. Ma, R.; Zhang, W. Several identities involving the Fibonacci numbers and Lucas numbers. *Fibonacci Q.* **2007**, *45*, 164–170.
12. Kim, T.; Kim, D.; Dolgy, D.; Park, J. Sums of finite products of Chebyshev polynomials of the second kind and of Fibonacci polynomials. *J. Inequal. Appl.* **2018**, *2018*, 148. [CrossRef] [PubMed]
13. Kim, T.; Kim, D.; Kwon, J.; Dolgy, D. Expressing Sums of Finite Products of Chebyshev Polynomials of the Second Kind and of Fibonacci Polynomials by Several Orthogonal Polynomials. *Mathematics* **2018**, *6*, 210. [CrossRef]
14. Kim, T.; Kim, D.; Dolgy, D.; Kwon, J. Representing Sums of Finite Products of Chebyshev Polynomials of the First Kind and Lucas Polynomials by Chebyshev Polynomials. *Mathematics* **2019**, *7*, 26. [CrossRef]
15. Ma, Y.; Zhang, W. Some Identities Involving Fibonacci Polynomials and Fibonacci Numbers. *Mathematics* **2018**, *6*, 334. [CrossRef]

16. Zhang, Y.; Chen, Z. A New Identity Involving the Chebyshev Polynomials. *Mathematics* **2018**, *6*, 244. [CrossRef]
17. Shen, S.; Chen, L. Some Types of Identities Involving the Legendre Polynomials. *Mathematics* **2019**, *7*, 114. [CrossRef]
18. Kim, T.; Dolgy, D.; Kim, D.; Seo, J. Convolved Fibonacci Numbers and Their Applications. *ARS Comb.* **2017**, *135*, 119–131.

A Note on Modified Degenerate Gamma and Laplace Transformation

YunJae Kim [1], Byung Moon Kim [2], Lee-Chae Jang [3] and Jongkyum Kwon [4,*]

1 Department of Mathematics, Dong-A University, Busan 49315, Korea; kimholzi@gmail.com
2 Department of Mechanical System Engineering, Dongguk University, Gyungju-si, Gyeongsangbukdo 38066, Korea; kbm713@dongguk.ac.kr
3 Graduate School of Education, Konkuk University, Seoul 139-701, Korea; lcjang@konkuk.ac.kr
4 Department of Mathematics Education and ERI, Gyeongsang National University, Jinju, Gyeongsangnamdo 52828, Korea
* Correspondence: mathkjk26@gnu.ac.kr

Abstract: Kim-Kim studied some properties of the degenerate gamma and degenerate Laplace transformation and obtained their properties. In this paper, we define modified degenerate gamma and modified degenerate Laplace transformation and investigate some properties and formulas related to them.

Keywords: the degenerate gamma function; the modified degenerate gamma function; the degenerate Laplace transform; the modified degenerate Laplace transform

---

1. Introduction

It is well known that gamma function is defied by

$$\Gamma(s) = \int_0^\infty e^{-t} t^{s-1} dt, \text{ where } s \in \mathbf{C} \text{ with } Re(s) > 0, \tag{1}$$

(see [1,2]). From (1), we note that

$$\Gamma(s+1) = s\Gamma(s), \text{ and } \Gamma(n+1) = n!, \text{ where } n \in \mathbf{N}. \tag{2}$$

Let $f(t)$ be a function defined for $t \geq 0$. Then, the integral

$$L(f(t)) = \int_0^\infty e^{-st} f(t) dt, \tag{3}$$

(see [1–4]), is said to be the Laplace transform of f, provided that the integral converges. For $\lambda \in (0, \infty)$. Kim-Kim [2] introduced the degenerate gamma function for the complex variable $s$with $0 < Re(s) < \frac{1}{\lambda}$ as follows:

$$\Gamma_\lambda(s) = \int_0^\infty (1+\lambda t)^{-\frac{1}{\lambda}} t^{s-1} dt, \tag{4}$$

(see [2]) and degenerate Laplace transformation which was defined by

$$L_\lambda(f(t)) = \int_0^\infty (1+\lambda t)^{-\frac{s}{\lambda}} f(t) dt, \tag{5}$$

(see [2,5]), if the integral converges. The authors obtained some properties and interesting formulas related to the degenerate gamma function. For examples, For $\lambda \in (0,1)$ and $0 < Re(s) < \frac{1-\lambda}{\lambda}$,

$$\Gamma_\lambda(s+1) = \frac{s}{(1-\lambda)^{s-1}}\Gamma_{\frac{\lambda}{1-\lambda}}(s), \tag{6}$$

and $\lambda \in (0, \frac{1}{k+s})$ with $k \in \mathbf{N}$ and $0 < Re(s) < \frac{1-\lambda}{\lambda}$,

$$\Gamma_\lambda(s+1) = \frac{s(s-1)\cdots(s-(k+1)+1)}{(1-\lambda)(1-2\lambda)\cdots(1-k\lambda)(1-(k+1)\lambda)}\Gamma_{\frac{\lambda}{1-(k+1)\lambda}}(s-k), \tag{7}$$

and for $k \in \mathbf{N}$ and $\lambda \in (0, \frac{1}{k})$,

$$\Gamma_\lambda(k) = \frac{(k-1)!}{(1-\lambda)(1-2\lambda)\cdots(1-k\lambda)}. \tag{8}$$

The authors obtained some formulas related to the degenerate Laplace transformation. For examples,

$$L_\lambda(1) = \frac{1}{s-\lambda}, \text{ if } s > \lambda, \tag{9}$$

and

$$L_\lambda((1+\lambda t)^{-\frac{a}{\lambda}}) = \frac{1}{s+a-\lambda}, \text{ if } s > -a+\lambda, \tag{10}$$

and

$$L_\lambda(\cos_\lambda(at)) = \frac{s-\lambda}{(s-\lambda)^2+a^2}, \tag{11}$$

and

$$L_\lambda(\sin_\lambda(at)) = \frac{a}{(s-\lambda)^2+a^2}, \tag{12}$$

where $\cos_\lambda(t) = \frac{1}{2}\left((1+\lambda t)^{\frac{it}{\lambda}} + (1+\lambda t)^{-\frac{it}{\lambda}}\right)$ and $\sin_\lambda(t) = \frac{1}{2i}\left((1+\lambda t)^{\frac{it}{\lambda}} - (1+\lambda t)^{-\frac{it}{\lambda}}\right)$.

Furthermore, the authors obtained that

$$L_\lambda(t^n) = \frac{n!}{(s-\lambda)(s-2\lambda)\cdots(s-n\lambda)(s-(n+1)\lambda)}, \tag{13}$$

for $n \in \mathbf{N}$ and $s > (n+1)\lambda$, and

$$\begin{aligned} L_\lambda(f^{(n)}(t)) &= s(s+\lambda)(s+2\lambda)\cdots(s+(n-1)\lambda)L_\lambda((1+\lambda t)^{-n}f(t)) \\ &- \sum_{i=0}^{n-1} f^{(i)}(0)\left(\prod_{t=1}^{n-i-1} s+(l-1)\lambda\right). \end{aligned} \tag{14}$$

where $f, f^{(1)}, \cdots, f^{(n-1)}$ are continuous on $(0,\infty)$ and are of degenerate exponential order and $f^{(n)}(t)$ is piecewise continuous on $(0,\infty)$, and

$$L_\lambda((\log(1+\lambda t))^n f(t)) = (-1)^n\lambda^n\left(\frac{d}{ds}\right)^n L_\lambda(s), \tag{15}$$

for $n \in \mathbf{N}$.

At first, L. Carlitz introduced the degenerate special polynomials (see [6,7]). The recently works which can be cited in this and researchers have studied the degenerate special polynomials and numbers (see [2,8–19]). Recently, the concept of degenerate gamma function and degenerate Laplace transformation was introduced by Kim-Kim [2]. They studied some properties of the degenerate gamma and degenerate Laplace transformation and obtained their properties. We observe whether or not that holds. Thus, we consider the modified degenerate Laplace transform which are satisfied (16). The degenerate gamma and degenerate Laplace transformation applied to engineer's mathematical toolbox as they make solving linear ODEs and related initial value problems. This paper consists of two sections. The first section contains the modified degenerate gamma function and investigate the properties of the modified gamma function. The second part of the paper provide the modified degenerate Laplace transformation and investigate interesting results of the modified degenerate Laplace transformation.

$$L_\lambda(f*g) = L_\lambda(f)L_\lambda(g) \tag{16}$$

2. Modified Degenerate Gamma Function

In this section, we will define modified degenerate gamma functions which are different to degenerate gamma functions. For each $\lambda \in (0,\infty)$, we define modified degenerate gamma function for the complex variable s with $0 < Re(s)$ as follows:

$$\Gamma^*_\lambda(s) = \int_0^\infty (1+\lambda)^{-\frac{t}{\lambda}} t^{s-1} dt. \tag{17}$$

Let $\lambda \in (0,1)$. Then, for $0 < Re(s)$, we have

$$\begin{aligned}\Gamma^*_\lambda(s+1) &= \int_0^\infty (1+\lambda)^{-\frac{t}{\lambda}} t^s dt \\ &= \frac{1}{(\log(1+\lambda)^{-\frac{1}{\lambda}}}(1+\lambda)^{-\frac{t}{\lambda}} t^s \mid_0^\infty + \frac{\lambda}{\log(1+\lambda)}\int_0^\infty s(1+\lambda)^{-\frac{t}{\lambda}} t^{s-1} dt \\ &= \frac{\lambda}{\log(1+\lambda)} s\Gamma^*_\lambda(s).\end{aligned} \tag{18}$$

Therefore, by (18), we obtain the following theorem.

Theorem 1. *Let $\lambda \in (0,1)$. Then, for $0 < Re(s)$, we have*

$$\Gamma^*_\lambda(s+1) = \frac{\lambda s}{\log(1+\lambda)}\Gamma^*_\lambda(s). \tag{19}$$

Then, for $0 < Re(s)$ and $\lambda \in (0,1)$, repeatly we calculate

$$\Gamma^*_\lambda(s+1) = \frac{\lambda s}{\log(1+\lambda)}\Gamma^*_\lambda(s) = \frac{\lambda^2(s-1)}{(\log(1+\lambda))^2}\Gamma^*_\lambda(s-1). \tag{20}$$

Thus, continuing this process, for $0 < Re(s)$ and $\lambda \in (0,1)$, we have

$$\Gamma^*_\lambda(s+1) = \frac{\lambda^k(s-1)\cdots(s-k+1)}{(\log(1+\lambda))^k}\Gamma^*_\lambda(s-k). \tag{21}$$

Therefore, by (21), we obtain the following theorem.

Theorem 2. *Let $\lambda \in (0,1)$. Then, for $0 < Re(s)$, we have*

$$\Gamma^*_\lambda(s+1) = \tfrac{\lambda^k(s-1)\cdots(s-k+1)}{(\log(1+\lambda))^k}\Gamma^*_\lambda(s-k). \tag{22}$$

Let us take $s = k+1$. Then, by Theorem 2, we get

$$\begin{aligned}\Gamma^*_\lambda(k+2) &= \frac{\lambda^{k+1}k\cdots 2}{(\log(1+\lambda))^{k+1}}\Gamma^*_\lambda(1) \\ &= \frac{\lambda^{k+1}k!}{(\log(1+\lambda))^{k+1}}\Gamma^*_\lambda(1)\end{aligned} \tag{23}$$

and

$$\begin{aligned}\Gamma_{\lambda}^{*}(1) &= \int_{0}^{\infty}(1+\lambda)^{-\frac{t}{\lambda}}dt \\ &= -\frac{\lambda}{(\log(1+\lambda)}(1+\lambda)^{-\frac{t}{\lambda}}\mid_{0}^{\infty} \\ &= \frac{\lambda}{(\log(1+\lambda)}.\end{aligned} \tag{24}$$

Therefore, by (23) and (24), we obtain the following theorem.

Theorem 3. *For $k \in \mathbf{N}$ and $\lambda \in (0,1)$, we have*

$$\Gamma_{\lambda}^{*}(k+1) = \frac{\lambda^{k+1}k!}{(\log(1+\lambda))^{k+1}}. \tag{25}$$

3. Modified Degenerate Laplace Transformation

In this section, we will define modified Laplace transformation which are different to degenerate Laplace transformation. Let $\lambda \in (0,\infty)$ and let $f(t)$ be a function defined for $t \geq 0$. Then the integral

$$\mathcal{L}_{\lambda}^{*}(f(t)) = \int_{0}^{\infty}(1+\lambda s)^{-\frac{t}{\lambda}}f(t)dt. \tag{26}$$

is said to be the modified degenerate Laplace transformation of f if the integral converges which is also defined by $\mathcal{L}_{\lambda}^{*}(f(t)) = F_{\lambda}(s)$.

From (26), we get

$$\mathcal{L}_{\lambda}^{*}(\alpha f(t)+\beta g(t)) = \alpha\mathcal{L}_{\lambda}^{*}(f(t))+\beta\mathcal{L}_{\lambda}^{*}(g(t)), \tag{27}$$

where α and β are constant real numbers.

First, we observe that for $n \in \mathbf{N}$,

$$\begin{aligned}\mathcal{L}_{\lambda}^{*}(t^{n}) &= \int_{0}^{\infty}(1+\lambda s)^{-\frac{t}{\lambda}}t^{n}dt \\ &= -\frac{\lambda}{\log(1+\lambda s)}(1+\lambda s)^{-\frac{t}{\lambda}}t^{n}\mid_{0}^{\infty}+\frac{\lambda n}{\log(1+\lambda s)}\int_{0}^{\infty}(1+\lambda s)^{-\frac{t}{\lambda}}t^{n-1}dt \\ &= \frac{\lambda n}{\log(1+\lambda s)}\mathcal{L}_{\lambda}^{*}(t^{n-1}) \\ &= \frac{\lambda n}{\log(1+\lambda s)}\left(-\frac{\lambda}{\log(1+\lambda s)}(1+\lambda s)^{-\frac{t}{\lambda}}t^{n-1}\mid_{0}^{\infty}+\frac{\lambda(n-1)}{\log(1+\lambda s)}\int_{0}^{\infty}(1+\lambda s)^{-\frac{t}{\lambda}}t^{n-2}dt\right) \\ &= \left(\frac{\lambda}{\log(1+\lambda s)}\right)^{2}n(n-1)\mathcal{L}_{\lambda}^{*}(t^{n-2}) \\ &= \cdots \\ &= \left(\frac{\lambda}{\log(1+\lambda s)}\right)^{n}n!\mathcal{L}_{\lambda}^{*}(1) \\ &= \left(\frac{\lambda}{\log(1+\lambda s)}\right)^{n+1}n!.\end{aligned} \tag{28}$$

Therefore, by (28), we obtain the following theorem.

Theorem 4. *For $k \in \mathbf{N}$ and $\lambda \in (0,1)$, we have*

$$\mathcal{L}_{\lambda}^{*}(t^{n}) = \left(\frac{\lambda}{\log(1+\lambda s)}\right)^{n+1}n!. \tag{29}$$

Secondly, we note that if f is a periodic function with a period T.

$$\begin{aligned}\mathcal{L}_{\lambda}^{*}(f(t)) &= \int_{0}^{\infty}(1+\lambda s)^{-\frac{t}{\lambda}}f(t)dt \\ &= \int_{0}^{T}(1+\lambda s)^{-\frac{t}{\lambda}}f(t)dt + \int_{T}^{\infty}(1+\lambda s)^{-\frac{t}{\lambda}}f(t)dt \\ &= \int_{0}^{T}(1+\lambda s)^{-\frac{t}{\lambda}}f(t)dt + \int_{0}^{\infty}(1+\lambda s)^{-\frac{t+T}{\lambda}}f(t+T)dt \\ &= \int_{0}^{T}(1+\lambda s)^{-\frac{t}{\lambda}}f(t)dt + (1+\lambda s)^{-\frac{T}{\lambda}}\int_{0}^{\infty}(1+\lambda s)^{-\frac{t}{\lambda}}f(t)dt\end{aligned} \tag{30}$$

By (30), we get

$$\left(1-(1+\lambda s)^{-\frac{T}{\lambda}}\right)\mathcal{L}_{\lambda}^{*}(f(t)) = \int_{0}^{T}(1+\lambda s)^{-\frac{t}{\lambda}}f(t)dt. \tag{31}$$

Thus, by (31), we get

$$\mathcal{L}_{\lambda}^{*}(f(t)) = \frac{1}{\left(1-(1+\lambda s)^{-\frac{T}{\lambda}}\right)}\int_{0}^{T}(1+\lambda s)^{-\frac{t}{\lambda}}f(t)dt. \tag{32}$$

We recall that the degenerate Bernoulli numbers are introduced as

$$\frac{t}{(1+\lambda)^{-\frac{t}{\lambda}}} = \sum_{n=0}^{\infty}B_{n,\lambda}\frac{t^{n}}{n!}, \tag{33}$$

Thus, by (32) and (33), we have

$$\begin{aligned}\frac{1}{1-(1+\lambda S)^{-\frac{T}{\lambda}}} &= -\frac{1}{TS}\frac{ST}{(1+\lambda s)^{-\frac{TS}{\lambda S}}-1} \\ &= -\frac{1}{TS}\sum_{n=0}^{\infty}B_{n,\lambda S}(-1)^{n}S^{n}\frac{T^{n}}{n!}.\end{aligned} \tag{34}$$

Therefore, by (33) and (34), we obtain the following theorem.

Theorem 5. *If f is a function defined $t \geq 0$ and $\mathcal{L}_{\lambda}^{*}(f(t))$ exists, then we have*

$$\begin{aligned}\mathcal{L}_{\lambda}^{*}(f(t)) &= -\frac{1}{TS}\sum_{n=0}^{\infty}B_{n,\lambda S}(-1)^{n}S^{n}\int_{0}^{T}(1+\lambda s)^{-\frac{t}{\lambda}}f(t)dt\frac{T^{n}}{n!} \\ &= -\frac{1}{TS}\sum_{n=0}^{\infty}B_{n,\lambda S}(-1)^{n}S^{n}\mathcal{L}_{\lambda}^{*}(U(t-T)f(t)),\end{aligned} \tag{35}$$

where $U(t-a) = \begin{cases}0, \text{for } 0 \geq t \geq a, \\ 1, \text{for } t \leq a.\end{cases}$ *is the Heaviside function.*

Thirdly, we observe the modified degenerate Laplace transformation of $f(t-a)U(t-a)$ as follows:

$$\begin{aligned}\mathcal{L}_{\lambda}^{*}(f(t-a)U(t-a)) &= \int_{0}^{\infty}(1+\lambda s)^{-\frac{t}{\lambda}}f(t-a)U(t-a)dt \\ &= \int_{a}^{\infty}(1+\lambda s)^{-\frac{t}{\lambda}}f(t-a)dt \\ &= \int_{0}^{\infty}(1+\lambda s)^{-\frac{t+a}{\lambda}}f(t)dt \\ &= (1+\lambda s)^{-\frac{a}{\lambda}}\int_{0}^{\infty}(1+\lambda s)^{-\frac{t}{\lambda}}f(t)dt \\ &= (1+\lambda s)^{-\frac{a}{\lambda}}\mathcal{L}_{\lambda}^{*}(f(t)).\end{aligned} \tag{36}$$

Therefore, by (36), we obtain the following theorem.

Theorem 6. *For $\lambda \in (0,1)$ and $a \in (0,\infty)$ we have*

$$\mathcal{L}_{\lambda}^{*}(f(t-a)U(t-a)) = (1+\lambda s)^{-\frac{a}{\lambda}}\mathcal{L}_{\lambda}^{*}(f(t)), \tag{37}$$

where $U(t-a)$ is the Heviside function.

Fourthly, we observe the modified degenerate Laplace transformation of the convolution $f * g$ of two function f, g as follows:

$$\begin{aligned}\mathcal{L}_{\lambda}^{*}(f)\mathcal{L}_{\lambda}^{*}(g) &= \left(\int_{0}^{\infty}(1+\lambda s)^{-\frac{t}{\lambda}}f(t)dt\right)\left(\int_{0}^{\infty}(1+\lambda s)^{-\frac{\tau}{\lambda}}g(\tau)d\tau\right)\\ &= \int_{0}^{\infty}\int_{0}^{\infty}(1+\lambda s)^{-\frac{t+\tau}{\lambda}}f(t)g(\tau)dtd\tau\\ &= \int_{0}^{\infty}f(t)\int_{\tau}^{\infty}(1+\lambda s)^{-\frac{\mu}{\lambda}}g(\mu-\tau)d\mu d\tau\\ &= \int_{0}^{\infty}\int_{\tau}^{\infty}f(t)(1+\lambda s)^{-\frac{\mu}{\lambda}}g(\mu-\tau)d\mu d\tau\\ &= \int_{0}^{\infty}(f*g)\,(1+\lambda s)^{-\frac{\mu}{\lambda}}d\mu\\ &= \mathcal{L}_{\lambda}(f*g).\end{aligned} \tag{38}$$

Therefore, by (38), we obtain the following theorem.

Theorem 7. *For $\lambda \in (0,1]$, we have*

$$\mathcal{L}_{\lambda}^{*}(f*g) = \mathcal{L}_{\lambda}^{*}(f)\mathcal{L}_{\lambda}^{*}(g). \tag{39}$$

We note that

$$\begin{aligned}\mathcal{L}_{\lambda}^{*}(1) &= \int_{0}^{\infty}(1+\lambda s)^{-\frac{t}{\lambda}}1dt\\ &= -\frac{\lambda}{\log(1+\lambda s)}(1+\lambda s)^{-\frac{t}{\lambda}}\Big|_{0}^{\infty}\\ &= \frac{\lambda}{\log(1+\lambda s)}.\end{aligned} \tag{40}$$

By (40), we have

$$L_{\lambda}^{*}(f*1) = L_{\lambda}^{*}(f)L_{\lambda}^{*}(1) = L_{\lambda}^{*}(f)\frac{\lambda}{\log(1+\lambda s)}. \tag{41}$$

Therefore, by (41), we obtain the following theorem.

Theorem 8. *For $\lambda \in (0,1]$, we have*

$$\mathcal{L}_{\lambda}^{*-1}(L_{\lambda}^{*}(f)\frac{\lambda}{\log(1+\lambda s)}) = f*1(t) = \int_{0}^{t}f(t)dt. \tag{42}$$

Fifthly, we observe that the modified degenerate Laplace transformation of derivative of f which is $f(t) = 0((1+\lambda s)^{-\frac{t}{\lambda}})$, where $f(t) = 0(u(t))$ means

$$\begin{aligned}\mathcal{L}_\lambda^*(f') &= \int_0^\infty (1+\lambda s)^{-\frac{t}{\lambda}} f' dt \\ &= (1+\lambda s)^{-\frac{t}{\lambda}} f(t) \mid_0^\infty + \int_0^\infty \frac{\log(1+\lambda s)}{\lambda} (1+\lambda s)^{-\frac{t}{\lambda}} f(t) dt \\ &= -f(0) + \frac{\log(1+\lambda s)}{\lambda} \mathcal{L}_\lambda^*(f).\end{aligned} \tag{43}$$

and

$$\begin{aligned}\mathcal{L}_\lambda^*(f^{(2)}) &= \int_0^\infty (1+\lambda s)^{-\frac{t}{\lambda}} f^{(2)} dt \\ &= (1+\lambda s)^{-\frac{t}{\lambda}} f'(t) \mid_0^\infty + \frac{\log(1+\lambda s)}{\lambda} \int_0^\infty (1+\lambda s)^{-\frac{t}{\lambda}} f'(t) dt \\ &= -f(0) + \frac{\log(1+\lambda s)}{\lambda} \left(-f(0) + \frac{\log(1+\lambda s)}{\lambda} \mathcal{L}_\lambda^*(f)\right) \\ &= \left(\frac{\log(1+\lambda s)}{\lambda}\right)^2 \mathcal{L}_\lambda^*(f) - \frac{\log(1+\lambda s)}{\lambda} f(0) - f'(0).\end{aligned} \tag{44}$$

By using mathematical induction, we obtain the following theorem.

Theorem 9. *For $\lambda \in (0,1]$, we have*

$$\mathcal{L}_\lambda^*(f^{(n)}) = \left(\frac{\log(1+\lambda s)}{\lambda}\right)^n \mathcal{L}_\lambda^*(f) - \sum_{i=0}^{n-1} \left(\frac{\log(1+\lambda s)}{\lambda}\right)^{n-1-i} f^{(i)}(0). \tag{45}$$

Finally, we observe

$$\begin{aligned}\frac{dF_\lambda^*}{ds} &= \int_0^\infty \frac{\lambda}{1+\lambda s} \left(-\frac{t}{\lambda}\right) (1+\lambda s)^{-\frac{t}{\lambda}} f(t) dt \\ &= -\frac{1}{1+\lambda s} \int_0^\infty (1+\lambda s)^{-\frac{t}{\lambda}} t f(t) dt \\ &= -\frac{1}{1+\lambda s} \mathcal{L}_\lambda^*(t f(t)).\end{aligned} \tag{46}$$

By (46), we obtain the following theorem.

Theorem 10. *For $\lambda \in (0,1]$ and $0 < Re(s)$, we have*

$$\frac{dF_\lambda^*}{ds} = -\frac{1}{1+\lambda s} \mathcal{L}_\lambda^*(t f(t)). \tag{47}$$

4. Conclusions

Kim-Kim ([9]) defined a degenerate gamma function and a degenerate Laplace transformation. The motivation of this paper is to define modified degenerate gamma functions and modified degenerate Laplace transformations which are different to degenerate gamma function and degenerate Laplace transformation and to obtain more useful results which are Theorems 7 and 8 for the modified degenerate Laplace transformation. We do not obtain these result from the degenerate Laplace transformation. Also, we investigated some results which are Theorems 1 and 3 for modified degenerate gamma functions. Furthermore, Theorems 6 and 9 are some interesting properties which are applied to differential equations in engineering mathematics.

Author Contributions: All authors contributed equally to this work. All authors read and approved the final manuscript.

References

1. Kreyszig, E.; Kreyszig, H.; Norminton, E.J. *Advanced Engineering Mathematics*; John Wiley & Sons Inc.: New Jersey, NJ, USA, 2011.
2. Kim, T.; Kim, D.S. Degenerate Laplace transform and degenerate Gamma function. *Russ. J. Math. Phys.* **2017**, *24*, 241–248. [CrossRef]
3. Chung, W.S.; Kim, T.; Kwon, H.I. On the q-analog of the Laplace transform. *Russ. J. Math. Phys.* **2014**, *21*, 156–168. [CrossRef]
4. Spiegel, M.R. *Laplace Transforms (Schaum's Outlines)*; McGraw Hill: New York, NY, USA, 1965.
5. Upadhyaya, L.M. On the degenerate Laplace transform. *Int. J. Eng. Sci. Res.* **2018**, *6*, 198–209. [CrossRef]
6. Carlitz, L. A degenerate Staudt-Clausen Theorem. *Arch. Math. (Basel)* **1956**, *7*, 28–33. [CrossRef]
7. Carlitz, L. Degenerate Stirling, Bernoulli and Eulerian numbers. *Util. Math.* **1979**, *15*, 51–88.
8. Dolgy, D.V.; Kim, T.; Seo, J.-J. On the symmetric identities of modified degenerate Bernoulli polynomials. *Proc. Jangjeon Math. Soc.* **2016**, *19*, 301–308.
9. Kim, D.S.; Kim, T. Some identities of degenerate Euler polynomials arising from p-adic fermionic integral on $\mathbb{Z}_p$. *Integral Transf. Spec. Funct.* **2015**, *26*, 295–302. [CrossRef]
10. Kim, T. Degenerate Euler Zeta function. *Russ. J. Math. Phys.* **2015**, *22*, 469–472. [CrossRef]
11. Kim, T. On the degenerate q-Bernoulli polynomials. *Bull. Korean Math. Soc.* **2016**, *53*, 1149–1156. [CrossRef]
12. Kim, T.; Kim, D.S.; Dolgy, D.V. Degenerate q-Euler polynomials. *Adv. Difference Equ.* **2015**, *2015*, 13662. [CrossRef]
13. Kim, T.; Dolgy, D.V.; Kim, D.S. Symmetric identities for degenerate generalized Bernoulli polynomials. *J. Nonlinear Sci. Appl.* **2016**, *9*, 677–683. [CrossRef]
14. Kim, T.; Kim, D.S.; Seo, J.J. Differential equations associated with degenerate Bell polynomials. *Int. J. Pure Appl. Math.* **2016**, *108*, 551–559.
15. Kwon, H.I.; Kim, T.; Seo, J.-J. A note on degenerate Changhee numbers and polynomials. *Proc. Jangjeon Math. Soc.* **2015**, *18*, 295–305.
16. Kim, T.; Jang, G.-W. A note on degenerate gamma function and degenerate Stirling number of the second kind. *Adv. Stud. Contemp. Math. (Kyungshang)* **2018**, *28*, 207–214.
17. Kim, T.; Yao, Y.; Kim, D.S.; Jang, G.-W. Degenerate r-Stirling numbers and r-Bell polynomials. *Russ. J. Math. Phys.* **2018**, *25*, 44–58. [CrossRef]
18. Kim, T.; Kim, D.S. A new approach to Catalan numbers using differential equations. *Russ. J. Math. Phys.* **2017**, *24*, 465–475. [CrossRef]
19. Jang, G.-W.; Kim, T.; Kwon, H.I. On the extension of degenerate Stirling polynomials of the second kind and degenerate Bell polynomials. *Adv. Stud. Contemp. Math. (Kyungshang)* **2018**, *28*, 305–316.

16

Bernoulli Polynomials and their Some New Congruence Properties

Ran Duan and Shimeng Shen *

School of Mathematics, Northwest University, Xi'an 710127, China; duan.ran.stumail@stumail.nwu.edu.cn
* Correspondence: millieshen28@163.com

Abstract: The aim of this article is to use the fundamental modus and the properties of the Euler polynomials and Bernoulli polynomials to prove some new congruences related to Bernoulli polynomials. One of them is that for any integer h or any non-negative integer n, we obtain the congruence $B_{2n+1}(2h) \equiv 0 \bmod (2n+1)$, where $B_n(x)$ are Bernoulli polynomials.

Keywords: Euler polynomials; Bernoulli polynomials; elementary method; identity; congruence

MSC: 11B68; 11A07

---

1. Introduction

As usual, for the real number x, if $m \geq 0$ denotes any integer, the famous Bernoulli polynomials $B_m(x)$ (see [1–4]) and Euler polynomials $E_m(x)$ (see [2–5]) are decided by the coefficients of the series of powers:

$$\frac{z \cdot e^{zx}}{e^z - 1} = \sum_{m=0}^{\infty} \frac{B_m(x)}{m!} \cdot z^m \tag{1}$$

and:

$$\frac{2e^{zx}}{e^z + 1} = \sum_{m=0}^{\infty} \frac{E_m(x)}{m!} \cdot z^m. \tag{2}$$

If $x = 0$, then $E_m = E_m(0)$ and $B_m = B_m(0)$ are known as the mth Euler numbers and m^{th} Bernoulli numbers, respectively. For example, some values of B_m and E_m are $B_0 = 1$, $B_1 = -\frac{1}{2}$, $B_2 = \frac{1}{6}$, $B_3 = 0$, $B_4 = -\frac{1}{30}$, $B_5 = 0$, $B_6 = \frac{1}{42}$ and $E_0 = 1$, $E_1 = -\frac{1}{2}$, $E_2 = 0$, $E_3 = \frac{1}{4}$, $E_4 = 0$, $E_5 = -\frac{1}{2}$, $E_6 = 0$, etc. These polynomials and numbers occupy a very important position in number theory and combinatorics; this is not only because Bernoulli and Euler polynomials are well known, but also because they have a wide range of theoretical and applied values. Because of this, many scholars have studied the properties of these polynomials and numbers, and they also have obtained some valuable research conclusions. For instance, Zhang Wenpeng [6] studied a few combinational identities. As a continuation of the conclusion in [6], he showed that if p is a prime, one can obtain the congruence expression:

$$(-1)^{\frac{p-1}{2}} \cdot 2^{p-1} \cdot E_{p-1}\left(\frac{1}{2}\right) \equiv \begin{cases} 0 \mod p & \text{if } p \equiv 1 \bmod 4; \\ -2 \mod p & \text{if } p \equiv 3 \bmod 4. \end{cases}$$

Hou Yiwei and Shen Shimeng [3] proved the identity:

$$E_{2n-1} = -\frac{(2^{2n} - 1)}{n} \cdot B_{2n}.$$

As some corollaries of [3], Hou Yiwei and Shen Shimeng obtained several interesting congruences. For example, for p in an odd prime, one can obtain the expression:

$$E_{\frac{p-3}{2}} \equiv 0 \,(\bmod\ p), \text{ if } p \equiv 1 \bmod 8.$$

Zhao Jianhong and Chen Zhuoyu [7] obtained the following deduction: if m is a positive integer, $k \geq 2$, one obtains the equation:

$$\sum_{a_1+a_2+\cdots+a_k=m} \frac{E_{a_1}}{(a_1)!} \cdot \frac{E_{a_2}}{(a_2)!} \cdots \frac{E_{a_k}}{(a_k)!} = \frac{2^{k-1}}{(k-1)!} \cdot \frac{1}{m!} \sum_{i=0}^{k-1} C(k-1,i) E_{m+k-1-i},$$

for which the summation is taken over all k-dimensional nonnegative integer coordinates $(a_1, a_2, \cdots, a_k)$ such that the equation $a_1 + a_2 + \cdots + a_k = m$, and the sequence $\{C(k,i)\}$ is decided as follows: for any integers $0 \leq i \leq k$, $C(k,k) = k!$, $C(k,0) = 1$,

$$C(k+1, i+1) = C(k, i+1) + (k+1)C(k,i), \text{ for all } 0 \leq i < k,$$

providing $C(k,i) = 0$, if $i > k$, and k is a positive integer.

T.Kim et al. did a good deal of research work and obtained a series of significant results; see [5,8–14]. Specifically, in [5], T. Kim found many valuable results involving Euler numbers and polynomials connected with zeta functions. Other papers in regard to the Bernoulli polynomials and Euler polynomials can be found in [15–19]; we will not go into detail here.

Here, we will make use of the properties of the Euler numbers, Euler polynomials, Bernoulli numbers, and Bernoulli polynomials to verify a special relationship between the Bernoulli polynomials and Euler polynomials. As some of the applications of our conclusions, we also deduce two unusual congruences involving the Bernoulli polynomials.

Theorem 1. *For any positive integers m and h, the following identity should be obtained, that is:*

$$2 \cdot B_{2m+1}(2h) = (2m+1) \cdot \left(E_{2m}(2h) + 2 \sum_{i=0}^{2h-1} E_{2m}(i) \right).$$

Theorem 2. *For any positive integers m and h, we derive the identity as below:*

$$B_{2m}(2h) - B_{2n} + m\left(E_{2m-1}(2h) - E_{2m-1}\right) = (2m) \cdot \sum_{i=1}^{2h} E_{2m-1}(i).$$

From these deductions, the following several corollaries can be inferred:

Corollary 1. *Let m be a non-negative integer. Thus, for any integer h, we obtain the congruence:*

$$B_{2m+1}(2h) \equiv 0 \bmod (2m+1),$$

where $\frac{a}{b} \equiv 0 \bmod k$ implies $(a,b) = 1$ and $k \mid a$ for any integers $b (b \neq 0)$ and a.

Corollary 2. *For any positive integer m and integer h, $2^{2m-1} \cdot (B_{2m}(2h) - B_{2m})$ must be an integer, and:*

$$2^{2m-1} \cdot (B_{2m}(2h) - B_{2m}) \equiv 0 \bmod m.$$

Corollary 3. *For any integer h, let p be an odd prime; as a result, we have:*

$$B_p(2h) \equiv 0 \bmod p \text{ and } B_{2p}(2h) \equiv B_{2p} \bmod p.$$

Corollary 4. *Let p be an odd prime. In this way, there exits an integer N with $N \equiv 1 \bmod p$ such that the polynomial congruence:*

$$N \cdot B_p(x) \equiv (x-2)(x-1)x \cdot (x-p+1) \equiv x \cdot \left(x^{p-1}-1\right) \bmod p.$$

Some notes: It is well known that congruences regarding Bernoulli numbers have interesting applications in number theory; in particular, for studying the class numbers of class-groups of number fields. Therefore, our corollaries will promote the further development of research in this field. Some important results in this field can also be found in [20–23]. Here, we will not list them one by one.

2. Several Lemmas

In this part, we will provide three straightforward lemmas. Henceforth, we will handle certain mathematical analysis knowledge and the properties of the Euler polynomials and Bernoulli polynomials, all of which can be discovered from [1–3]. Thus, they will not be repeated here.

Lemma 1. *If $m \geq 0$ is an integer, polynomial $2^m \cdot E_m(x)$ denotes the integral coefficient polynomial of x.*

Proof. First, from Definition 2 of the Euler polynomials $E_m(x)$, we have:

$$2e^{xz} = (e^z+1) \cdot \frac{2e^{xz}}{e^z+1} = \left(1 + \sum_{m=0}^{\infty} \frac{1}{n!} \cdot z^m\right)\left(\sum_{m=0}^{\infty} \frac{E_m(x)}{m!} \cdot z^m\right). \tag{3}$$

On the other hand, we also have:

$$2e^{xz} = 2 \cdot \sum_{m=0}^{\infty} \frac{x^m}{m!} \cdot z^m. \tag{4}$$

uniting (3) and (4), then comparing the coefficients of the power series, we obtain that:

$$2x^m = E_m(x) + \sum_{k=0}^{m} \binom{m}{k} E_k(x)$$

or identity:

$$2E_m(x) = 2x^m - \sum_{k=0}^{m-1} \binom{m}{k} E_k(x). \tag{5}$$

Note that $E_0(x) = 1$, $E_1(x) = x - \frac{1}{2}$, so from (5) and mathematical induction, we may immediately deduce that $2^m \cdot E_m(x)$ is an integral coefficient polynomial of x. □

Lemma 2. *If m is a positive integer, the following equation can be obtained:*

$$2^m \cdot B_m(x) = B_m(2x) - \frac{1}{2} \cdot m \cdot E_{m-1}(2x).$$

Proof. From Definitions 1 and 2 of the Euler polynomials and Bernoulli polynomials, we discover the identity as below:

$$\begin{aligned} \frac{2ze^{2xz}}{e^{2z}-1} &= \sum_{m=0}^{\infty} \frac{2^m \cdot B_m(x)}{m!} \cdot z^m = \left(\frac{z \cdot e^{2xz}}{e^z-1} - \frac{z \cdot e^{2xz}}{e^z+1}\right) \\ &= \sum_{m=0}^{\infty} \frac{B_m(2x)}{m!} \cdot z^m - \frac{1}{2} \sum_{m=0}^{\infty} \frac{E_m(2x)}{m!} \cdot z^{m+1}. \end{aligned} \tag{6}$$

Relating the coefficients of the power series in (6), we obtain:

$$2^m \cdot B_m(x) = B_m(2x) - \frac{m}{2} \cdot E_{m-1}(2x).$$

This proves Lemma 2. □

Lemma 3. *If m is a positive integer, then for any positive integer M, we will be able to obtain the identities:*

$$2^m \cdot \left(B_m\left(M\right) - B_m\right) = m \cdot \sum_{i=0}^{2M-1} E_{m-1}(i).$$

Proof. On the basis of Definition 2 of the Euler polynomials, we obtain:

$$\sum_{i=0}^{N-1} \frac{2ze^{iz}}{e^z+1} = \sum_{m=0}^{\infty} \frac{1}{n!} \left(\sum_{i=0}^{N-1} E_m(i) \right) \cdot z^{m+1}. \tag{7}$$

In another aspect, we also obtain:

$$\sum_{i=0}^{N-1} \frac{2ze^{iz}}{e^z+1} = \frac{2z\left(e^{Nz}-1\right)}{\left(e^z+1\right)\left(e^z-1\right)} = \frac{2ze^{Nz}-2z}{e^{2z}-1}$$
$$= \sum_{m=0}^{\infty} \frac{2^m \cdot B_m\left(\frac{N}{2}\right)}{m!} \cdot z^m - \sum_{m=0}^{\infty} \frac{2^m \cdot B_m}{m!} \cdot z^m. \tag{8}$$

Combining (7) and (8), then comparing the coefficients of the power series, we will obtain:

$$2^m \cdot \left(B_m\left(\frac{N}{2}\right) - B_m \right) = m \cdot \sum_{i=0}^{N-1} E_{m-1}(i). \tag{9}$$

Now, Lemma 3 follows from (9) with $N = 2M$. □

3. Proofs of the Theorems

Applying three simple lemmas in Section 2, we can easily finish the proofs of our theorems. Above all, we study Theorem 1. For any positive integer m, from Lemma 2, we have:

$$2^{2m+1} \cdot B_{2m+1}(M) = B_{2m+1}(2M) - \frac{2m+1}{2} \cdot E_{2m}(2M). \tag{10}$$

Note that $B_{2m+1} = 0$. From Lemma 3, we also have:

$$2^{2m+1} \cdot B_{2m+1}\left(M\right) = (2m+1) \cdot \sum_{i=0}^{2M-1} E_{2m}(i). \tag{11}$$

Combining (10) and (11), we have:

$$B_{2m+1}(2M) = \frac{2m+1}{2} \cdot E_{2m}(2M) + (2m+1) \cdot \sum_{i=0}^{2M-1} E_{2m}(i).$$

Afterwards, we prove Theorem 2. According to Lemma 2 with $x = M$ and $x = 0$, we have:

$$2^{2m} \cdot B_{2m}(M) = B_{2m}(2M) - m \cdot E_{2m-1}(2M) \tag{12}$$

and:

$$2^{2m} \cdot B_{2m} = B_{2m} - m \cdot E_{2m-1}. \tag{13}$$

Applying Lemma 3, we also have:

$$2^{2m} \cdot (B_{2m}(M) - B_{2m}) = (2m) \cdot \sum_{i=0}^{2M-1} E_{2m-1}(i). \tag{14}$$

Combining (12), (13), and (14), we have the identity:

$$B_{2m}(2M) - B_{2m} = m \cdot E_{2m-1}(2M) - m \cdot E_{2m-1} + 2m \cdot \sum_{i=0}^{2M-1} E_{2m-1}(i).$$

This proves Theorem 2.

From Lemma 1, we know that all $2^{2m} \cdot E_{2m}(i)$ $(i = 0, 1, \cdots, 2M)$ are integers, and $(2^{2m}, 2m+1) = 1$, so on the basis of Theorem 1, we may directly deduce the congruence:

$$B_{2m+1}(2M) \equiv 0 \bmod (2m+1). \tag{15}$$

Since $B_{2m+1}(x)$ is an odd function (that is, $B_{2m+1}(-x) = -B_{2m+1}(x)$), and $B_{2m+1} = 0$, so (15) also holds for any integer M and non-negative integer m.

This completes the proof of Corollary 1.

Now, we study Corollary 2. On the basis of Lemma 1, we know that $2^{2m-1} \cdot E_{2m-1}(i)$ is an integer for all $1 \leq i \leq 2M$, so from Theorem 1, we know that $2^{2m-1} \cdot (B_{2m}(2M) - B_{2m})$ must be an integer, and it can be divided by m, that is,

$$2^{2m-1} \cdot (B_{2m}(2M) - B_{2m}) \equiv 0 \bmod m. \tag{16}$$

Note that $B_{2m}(x)$ is an even function, and if $M = 0$, after that, the left-hand side of (16) becomes zero; thus, the congruence (16) is correct for all integers M.

This completes the proof of Corollary 2.

Corollary 3 is a special case of Corollary 1 with $2m + 1 = p$ and Corollary 2 with $2m = 2p$.

Now, we prove Corollary 4. Since $B_p(x)$ is a pth rational coefficient polynomial of x and its first item is x^p, from Lemma 3, we know that the congruence equation $B_p(2x) \equiv 0 \bmod p$ has exactly p different solutions $x = 0, 1, 2, \cdots p-1$, so there exits an integer N with $N \equiv 1 \bmod p$ satisfied with $N \cdot B_p(x)$, an integral coefficient polynomial of x. From [1] (see Theorem 5.23), we have the congruence:

$$N \cdot B_p(x) \equiv x(x-1)(x-2) \cdot (x-p+1) \bmod p.$$

This completes the proofs of our all results.

4. Conclusions

As we all know, the congruences of Bernoulli numbers have important applications in number theory; in particular, for studying the class numbers of class-groups of number fields. The main results of this paper are two theorems involving Bernoulli and Euler polynomials and numbers and four corollaries (or congruences). Two theorems gave some new equations regarding Bernoulli polynomials and Euler polynomials. As some applications of these theorems, we gave four interesting congruences involving Bernoulli polynomials. Especially, Corollaries 1 and 4 are very simple and beautiful. It is clear that Corollary 4 is a good reference for further research on Bernoulli polynomials.

Author Contributions: All authors have equally contributed to this work. All authors read and approved the final manuscript.

Acknowledgments: The authors would like to thank the Editor and referee for their very helpful and detailed comments, which have significantly improved the presentation of this paper.

References

1. Apostol, T.M. *Introduction to Analytic Number Theory*; Springer: New York, NY, USA, 1976.
2. Knuth, D.E.; Buckholtz, T.J. Computation of Tangent, Euler, and Bernoulli numbers. *Math. Comput.* **1967**, *21*, 663–688. [CrossRef]
3. Hou, Y.W.; Shen, S.M. The Euler numbers and recursive properties of Dirichlet L-functions. *Adv. Differ. Equ.* **2018**, *2018, 397.* [CrossRef]
4. Liu, G.D. Identities and congruences involving higher-order Euler-Bernoulli numbers and polynonials. *Fibonacci Q.* **2001**, *39*, 279–284.
5. Kim, T. Euler numbers and polynomials associated with zeta functions. *Abstr. Appl. Anal.* **2008**, *2018*, 581582. [CrossRef]
6. Zhang, W.P. Some identities involving the Euler and the central factorial numbers. *Fibonacci Q.* **1998**, *36*, 154–157.
7. Zhao, J.H.; Chen, Z.Y. Some symmetric identities involving Fubini polynomials and Euler numbers. *Symmetry* **2018**, *10*, 303.
8. Kim, D.S.; Kim, T. Some symmetric identities for the higher-order q-Euler polynomials related to symmetry group S_3 arising from p-Adic q-fermionic integrals on $\mathbb{Z}_p$. *Filomat* **2016**, *30*, 1717–1721. [CrossRef]
9. Kim, T. Symmetry of power sum polynomials and multivariate fermionic p-Adic invariant integral on $\mathbb{Z}_p$. *Russ. J. Math. Phys.* **2009**, *16*, 93–96. [CrossRef]
10. Kim, T.; Kim, D.S.; Jang, G.W. A note on degenerate Fubini polynomials. *Proc. Jiangjeon Math. Soc.* **2017**, *20*, 521–531.
11. Kim, D.S.; Park, K.H. Identities of symmetry for Bernoulli polynomials arising from quotients of Volkenborn integrals invariant under S_3. *Appl. Math. Comput.* **2013**, *219*, 5096–5104. [CrossRef]
12. Kim, T.; Kim, D.S. An identity of symmetry for the degernerate Frobenius-Euler polynomials. *Math. Slovaca* **2018**, *68*, 239–243. [CrossRef]
13. Kim, T.; Kim, S.D.; Jang, G.W.; Kwon, J. Symmetric identities for Fubini polynomials. *Symmetry* **2018**, *10*, 219. [CrossRef]
14. Kim, D.S.; Rim, S.-H.; Kim, T. Some identities on Bernoulli and Euler polynomials arising from orthogonality of Legendre polynomials. *J. Inequal. Appl.* **2012**, *2012*, 227. [CrossRef]
15. Simsek, Y. Identities on the Changhee numbers and Apostol-type Daehee polynomials. *Adv. Stud. Contemp. Math.* **2017**, *27*, 199–212.
16. Guy, R.K. *Unsolved Problems in Number Theory*, 2nd ed.; Springer: New York, NY, USA, 1994.
17. Liu, G.D. The solution of problem for Euler numbers. *Acta Math. Sin.* **2004**, *47*, 825–828.
18. Zhang, W.P.; Xu, Z.F. On a conjecture of the Euler numbers. *J. Number Theory* **2007**, *127*, 283–291. [CrossRef]
19. Cho, B.; Park, H. Evaluating binomial convolution sums of divisor functions in terms of Euler and Bernoulli polynomials. *Int. J. Number Theory* **2018**, *14*, 509–525. [CrossRef]
20. Wagstaff, S.S., Jr. Prime divisors of the Bernoulli and Euler Numbers. *Number Theory Millenn.* **2002**, *3*, 357–374.
21. Bayad, A.; Aygunes, A. Hecke operators and generalized Bernoulli-Euler polynomials. *J. Algebra Number Theory Adv. Appl.* **2010**, *3*, 111–122.
22. Kim, D.S.; Kim, T. Some p-Adic integrals on $\mathbb{Z}_p$ Associated with trigonometric functions. *Russ. J. Math. Phys.* **2018**, *25*, 300–308. [CrossRef]
23. Powell, B.J. Advanced problem 6325. *Am. Math. Mon.* **1980**, *87*, 836.

Some Identities and Inequalities Involving Symmetry Sums of Legendre Polynomials

Tingting Wang * and Liang Qiao

College of Science, Northwest A&F University, Yangling 712100, China; 18309225762@163.com
* Correspondence: ttwang@nwsuaf.edu.cn

Abstract: By using the analysis methods and the properties of Chebyshev polynomials of the first kind, this paper studies certain symmetry sums of the Legendre polynomials, and gives some new and interesting identities and inequalities for them, thus improving certain existing results.

Keywords: Legendre polynomials; Chebyshev polynomials of the first kind; power series; symmetry sums; polynomial identities; polynomial inequalities

1. Introduction

For any integer $n \geq 0$, the Legendre polynomials $\{P_n(x)\}$ are defined as follows:

$$P_n(x) = \frac{2n-1}{n} x P_{n-1}(x) - \frac{n-1}{n} P_{n-2}(x)$$

for all $n \geq 2$, with $P_0(x) = 1$ and $P_1(x) = x$, see [1,2] for more information.

The first few terms of $P_n(x)$ are $P_2(x) = \frac{1}{2}(3x^2-1)$, $P_3(x) = \frac{1}{2}(5x^3-3x)$, $P_4(x) = \frac{1}{8}(35x^4-30x^2+3)$, $P_5(x) = \frac{1}{8}(63x^5-70x^3+15x), \cdots$.

In fact, the general term of $P_n(x)$ is given by the formula

$$P_n(x) = \frac{1}{2^n} \cdot \sum_{k=0}^{\left[\frac{n}{2}\right]} \frac{(-1)^k \cdot (2n-2k)!}{k! \cdot (n-k)! \cdot (n-2k)!} \cdot x^{n-2k},$$

where $[y]$ denotes the greatest integer less than or equal to y.

It is clear that $P_n(x)$ is an orthogonal polynomial (see [1,2]). That is,

$$\int_{-1}^{1} P_m(x) P_n(x) dx = \begin{cases} 0, & \text{if } m \neq n; \\ \dfrac{2}{2n+1}, & \text{if } m = n. \end{cases}$$

The generating function of $P_n(x)$ is

$$\frac{1}{\sqrt{1-2xt+t^2}} = \sum_{n=0}^{\infty} P_n(x) \cdot t^n, \ |x| \leq 1, \ |t| < 1. \tag{1}$$

These polynomials play a vital role in the study of function orthogonality and approximation theory, as a result, some scholars have dedicated themselves to studying their various natures and obtained a series of meaningful research results. The studies that are concerned with this content can be found in [1–20]. Recently, Shen Shimeng and Chen Li [3] give certain symmetry sums of $P_n(x)$, and proved the following result:

For any positive integer k and integer $n \geq 0$, one has the identity

$$\begin{aligned}&(2k-1)!! \sum_{a_1+a_2+\cdots+a_{2k+1}=n} P_{a_1}(x)P_{a_2}(x)\cdots P_{a_{2k+1}}(x)\\ = &\sum_{j=1}^{k} C(k,j) \sum_{i=0}^{n} \frac{(n+k+1-i-j)!}{(n-i)!} \cdot \frac{\binom{i+j+k-2}{i}}{x^{k-1+i+j}} \cdot P_{n+k+1-i-j}(x),\end{aligned}$$

where $(2k-1)!! = (2k-1)\cdot(2k-3)\cdots 3\cdot 1$, and $C(k,i)$ is a recurrence sequence defined by $C(k,1) = 1$, $C(k+1,k+1) = (2k-1)!!$ and $C(k+1,i+1) = C(k,i+1) + (k-1+i)\cdot C(k,i)$ for all $1 \leq i \leq k-1$.

The calculation formula for the sum of Legendre polynomials given above is virtually a linear combination of some $P_n(x)$, and the coefficients $C(k,i)$ are very regular. However, the result is in the form of a recursive formula, in other words, especially when k is relatively large, the formula is not actually easy to use for calculating specific values.

In an early paper, Zhou Yalan and Wang Xia [4] obtained some special cases with $k = 3$ and $k = 5$. It is even harder to calculate their exact values for the general positive integer k, especially if k is large enough.

Naturally, we want to ask a question: Is there a more concise and specific formula for the calculation of the above problems? This is the starting point of this paper. We used the different methods to come up with additional simpler identities. It is equal to saying that we have used the analysis method and the properties of the first kind of Chebyshev polynomials, thereby establishing the symmetry of the Legendre polynomial and symmetry relationship with the first kind of Chebyshev polynomial, and proved the following three results:

Theorem 1. *For any integers $k \geq 1$ and $n \geq 0$, we have the identity*

$$\sum_{a_1+a_2+\cdots+a_k=n} P_{a_1}(x)\cdot P_{a_2}(x)\cdots P_{a_k}(x) = \sum_{i=0}^{n} \frac{<\frac{k}{2}>_i}{i!}\cdot\frac{<\frac{k}{2}>_{n-i}}{(n-i)!}\cdot T_{n-2i}(x),$$

where $<x>_0 = 1$, $<x>_k = x(x+1)(x+2)\cdots(x+k-1)$ *for all integers* $k \geq 1$, *and* $T_n(x) = T_{-n}(x) = \frac{1}{2}\left(\left(x+\sqrt{x^2-1}\right)^n + \left(x+\sqrt{x^2-1}\right)^n\right)$ *denotes Chebyshev polynomials of the first kind.*

Theorem 2. *Let $q > 1$ is an integer, χ is any primitive character* mod q. *Then for any integers $k \geq 1$ and $n \geq 0$, we have the inequality*

$$\begin{aligned}&\left|\sum_{a_1+a_2+\cdots+a_k=n}\sum_{a=1}^{q}\chi(a)P_{a_1}\left(\cos\frac{2\pi a}{q}\right)\cdot P_{a_2}\left(\cos\frac{2\pi a}{q}\right)\cdots P_{a_k}\left(\cos\frac{2\pi a}{q}\right)\right|\\ &\leq \sqrt{q}\cdot\binom{n+k-1}{k-1}.\end{aligned}$$

Theorem 3. *For any integer $n \geq 0$ with $2 \nmid n$, we have the identity*

$$\begin{aligned}&\int_{-\frac{\pi}{2}}^{\frac{\pi}{2}}\left(\sum_{a_1+a_2+\cdots+a_k=n} P_{a_1}(\sin\theta)\cdot P_{a_2}(\sin\theta)\cdots P_{a_k}(\sin\theta)\right)^2 d\theta\\ = \;& 2\pi\cdot\sum_{i=0}^{\left[\frac{n}{2}\right]}\left(\frac{<\frac{k}{2}>_i}{i!}\cdot\frac{<\frac{k}{2}>_{n-i}}{(n-i)!}\right)^2;\end{aligned}$$

If $n = 2m$, then we have

$$\int_{-\frac{\pi}{2}}^{\frac{\pi}{2}} \left(\sum_{a_1+a_2+\cdots+a_k=n} P_{a_1}(\sin\theta) \cdot P_{a_2}(\sin\theta) \cdots P_{a_k}(\sin\theta) \right)^2 d\theta$$
$$= 2\pi \cdot \sum_{i=0}^{m} \left(\frac{<\frac{k}{2}>_i}{i!} \cdot \frac{<\frac{k}{2}>_{2m-i}}{(2m-i)!} \right)^2 - \pi \cdot \left(\frac{<\frac{k}{2}>_m}{m!} \right)^4.$$

Essentially, the main result of this paper is Theorem 1, which not only reveals the profound properties of Legendre polynomials and Chebyshev polynomials, but also greatly simplifies the calculation of the symmetry sum of Legendre polynomials in practice. We can replace the calculation of the symmetric sum of the Legendre polynomial with the first single Chebyshev polynomial calculation, which can greatly simplify the calculation of the symmetric sum.

Theorem 2 gives an upper bound estimate of the character sum of Legendre polynomials. Theorem 3 reveals the orthogonality of the symmetry sum of Legendre polynomials, which is a generalization of the orthogonality of functions. Of course, Theorems 2 and 3 can also be seen as the direct application of Theorem 1 in analytical number theory and the orthogonality of functions. This is of great significance in analytic number theory, and it has also made new contributions to the study of Gaussian sums.

In fact if we taking $k = 1$, and note that the identity $\frac{<\frac{1}{2}>_h}{h!} = \frac{1}{4^h} \cdot \binom{2h}{h}$, then from our theorems we may immediately deduce the following three corollaries.

Corollary 1. *For any integer $n \geq 0$, we have the identity*

$$P_n(x) = \frac{1}{4^n} \sum_{i=0}^{n} \binom{2i}{i} \binom{2n-2i}{n-i} \cdot T_{n-2i}(x),$$

where $T_n(x)$ denotes Chebyshev polynomials of the first kind.

Corollary 2. *Let $q > 1$ is an integer, χ is any primitive character* mod q. *Then for any integer $n \geq 0$, we have the inequality*

$$\left| \sum_{a=1}^{q} \chi(a) P_n\left(\cos \frac{2\pi a}{q}\right) \right| \leq \sqrt{q}.$$

Corollary 3. *For any integer $n \geq 0$ with $2 \nmid n$, we have the identity*

$$\int_{-\frac{\pi}{2}}^{\frac{\pi}{2}} P_n^2(\sin\theta)\, d\theta = \frac{2\pi}{4^{2n}} \sum_{i=0}^{[\frac{n}{2}]} \binom{2i}{i}^2 \binom{2n-2i}{n-i}^2;$$

If $n = 2m$, then we have the identity

$$\int_{-\frac{\pi}{2}}^{\frac{\pi}{2}} P_n^2(\sin\theta)\, d\theta = \frac{2\pi}{4^{2n}} \sum_{i=0}^{[\frac{n}{2}]} \binom{2i}{i}^2 \binom{2n-2i}{n-i}^2 - \frac{\pi}{4^{2n}} \cdot \binom{2m}{m}^4.$$

2. Proofs of the Theorems

In this section, we will directly prove the main results in this paper by by means of the properties of characteristic roots.

Proof of Theorem 1. First we prove Theorem 1. Let $\alpha = x + \sqrt{x^2-1}$ and $\beta = x - \sqrt{x^2-1}$ be two characteristic roots of the characteristic equation $\lambda^2 - 2x\lambda + 1 = 0$. Then from the definition and properties of Chebyshev polynomials $T_n(x)$ of the first kind, we have

$$T_n(x) = T_{-n}(x) = \frac{1}{2}\left(\alpha^n + \beta^n\right), \ n \geq 0.$$

For any positive integer k, combining properties of power series and Formula (1) we have the identity

$$\left(\frac{1}{\sqrt{1-2xt+t^2}}\right)^k = \frac{1}{(1-2xt+t^2)^{\frac{k}{2}}} = \frac{1}{(1-\alpha t)^{\frac{k}{2}}(1-\beta t)^{\frac{k}{2}}}$$
$$= \sum_{n=0}^{\infty}\left(\sum_{a_1+a_2+\cdots+a_k=n} P_{a_1}(x)\cdot P_{a_2}(x)\cdot P_{a_3}(x)\cdots P_{a_k}(x)\right)\cdot t^n. \tag{2}$$

At the same time, we focus on the power series

$$\frac{1}{(1-x)^{\frac{k}{2}}} = \sum_{n=0}^{\infty}\frac{<\frac{k}{2}>_n}{n!}\cdot x^n, \ |x|<1, \tag{3}$$

where $< x >_0 = 1, < x >_h = x(x+1)(x+2)\cdots(x+h-1)$ for all integers $h \geq 1$.

So for any positive integer k, note that $\alpha\cdot\beta = 1$, from (3) and the symmetry properties of α and β we have

$$\left(\frac{1}{\sqrt{1-2xt+t^2}}\right)^k = \frac{1}{(1-\alpha t)^{\frac{k}{2}}(1-\beta t)^{\frac{k}{2}}}$$
$$= \left(\sum_{n=0}^{\infty}\frac{<\frac{k}{2}>_n}{n!}\cdot\alpha^n\cdot t^n\right)\left(\sum_{n=0}^{\infty}\frac{<\frac{k}{2}>_n}{n!}\cdot\beta^n\cdot t^n\right)$$
$$= \sum_{n=0}^{\infty}\left(\sum_{i=0}^{n}\frac{<\frac{k}{2}>_i}{i!}\cdot\frac{<\frac{k}{2}>_{n-i}}{(n-i)!}\cdot\alpha^i\cdot\beta^{n-i}\right)\cdot t^n$$
$$= \sum_{n=0}^{\infty}\left(\sum_{i=0}^{n}\frac{<\frac{k}{2}>_i}{i!}\cdot\frac{<\frac{k}{2}>_{n-i}}{(n-i)!}\cdot\beta^{n-2i}\right)\cdot t^n$$
$$= \sum_{n=0}^{\infty}\left(\sum_{i=0}^{n}\frac{<\frac{k}{2}>_i}{i!}\cdot\frac{<\frac{k}{2}>_{n-i}}{(n-i)!}\cdot\alpha^{n-2i}\right)\cdot t^n$$
$$= \sum_{n=0}^{\infty}\left(\sum_{i=0}^{n}\frac{<\frac{k}{2}>_i}{i!}\cdot\frac{<\frac{k}{2}>_{n-i}}{(n-i)!}\cdot\frac{1}{2}\left(\alpha^{n-2i}+\beta^{n-2i}\right)\right)\cdot t^n$$
$$= \sum_{n=0}^{\infty}\left(\sum_{i=0}^{n}\frac{<\frac{k}{2}>_i}{i!}\cdot\frac{<\frac{k}{2}>_{n-i}}{(n-i)!}\cdot T_{n-2i}(x)\right)\cdot t^n. \tag{4}$$

Combining (2) and (4), and then by comparing the coefficients on both sides of the power series, we can find

$$\sum_{a_1+a_2+\cdots+a_k=n} P_{a_1}(x)\cdot P_{a_2}(x)\cdots P_{a_k}(x) = \sum_{i=0}^{n}\frac{<\frac{k}{2}>_i}{i!}\cdot\frac{<\frac{k}{2}>_{n-i}}{(n-i)!}\cdot T_{n-2i}(x).$$

This proves Theorem 1. □

Proof of Theorem 2. The proof of Theorem 2 is next. Let $q > 1$ be any integer, χ denotes any primitive character mod q. Then from Theorem 1 with $x = \cos\left(\frac{2\pi a}{q}\right)$ and the identity $T_n(\cos\theta) = \cos(n\theta)$, we have

$$\begin{aligned}
&\sum_{a_1+a_2+\cdots+a_k=n}\sum_{a=1}^{q}\chi(a)P_{a_1}\left(\cos\frac{2\pi a}{q}\right)\cdot P_{a_2}\left(\cos\frac{2\pi a}{q}\right)\cdots P_{a_k}\left(\cos\frac{2\pi a}{q}\right)\\
=&\sum_{i=0}^{n}\frac{<\frac{k}{2}>_i}{i!}\cdot\frac{<\frac{k}{2}>_{n-i}}{(n-i)!}\cdot\sum_{a=1}^{q}\chi(a)\cdot\cos\left(\frac{2\pi a(n-2i)}{q}\right)\\
=&\frac{1}{2}\sum_{i=0}^{n}\frac{<\frac{k}{2}>_i}{i!}\cdot\frac{<\frac{k}{2}>_{n-i}}{(n-i)!}\cdot\sum_{a=1}^{q}\chi(a)\left(e\left(\frac{a(n-2i)}{q}\right)+e\left(\frac{-a(n-2i)}{q}\right)\right)\\
=&\frac{\tau(\chi)}{2}\sum_{i=0}^{n}\frac{<\frac{k}{2}>_i}{i!}\cdot\frac{<\frac{k}{2}>_{n-i}}{(n-i)!}\cdot\left(\overline{\chi}(n-2i)+\overline{\chi}\left(-(n-2i)\right)\right), \qquad (5)
\end{aligned}$$

where $e(y)=e^{2\pi iy}$, and $\sum_{a=1}^{q}\chi(a)e\left(\frac{na}{q}\right)=\overline{\chi}(n)\tau(\chi)$.

Note that for any primitive character χ mod q, from the properties of Gauss sums, we have $|\tau(\chi)|=\sqrt{q}$, and for any positive integer $k\geq 1$, we have

$$\begin{aligned}
&\frac{1}{(1-x)^k}=\sum_{n=0}^{\infty}\binom{n+k-1}{k-1}\cdot x^n=\frac{1}{(1-x)^{\frac{k}{2}}}\cdot\frac{1}{(1-x)^{\frac{k}{2}}}\\
&=\sum_{n=0}^{\infty}\left(\sum_{i=0}^{n}\frac{<\frac{k}{2}>_i}{i!}\cdot\frac{<\frac{k}{2}>_{n-i}}{(n-i)!}\right)\cdot x^n
\end{aligned}$$

or

$$\binom{n+k-1}{k-1}=\sum_{i=0}^{n}\frac{<\frac{k}{2}>_i}{i!}\cdot\frac{<\frac{k}{2}>_{n-i}}{(n-i)!}. \qquad (6)$$

Combining (5) and (6), there will be an estimation formula immediately deduced

$$\begin{aligned}
&\left|\sum_{a_1+a_2+\cdots+a_k=n}\sum_{a=1}^{q}\chi(a)P_{a_1}\left(\cos\frac{2\pi a}{q}\right)\cdot P_{a_2}\left(\cos\frac{2\pi a}{q}\right)\cdots P_{a_k}\left(\cos\frac{2\pi a}{q}\right)\right|\\
=&\frac{\sqrt{q}}{2}\cdot\left|\sum_{i=0}^{n}\frac{<\frac{k}{2}>_i}{i!}\cdot\frac{<\frac{k}{2}>_{n-i}}{(n-i)!}\cdot\left[\overline{\chi}(n-2i)+\overline{\chi}\left(-(n-2i)\right)\right]\right|\\
\leq&\sqrt{q}\cdot\sum_{i=0}^{n}\frac{<\frac{k}{2}>_i}{i!}\cdot\frac{<\frac{k}{2}>_{n-i}}{(n-i)!}=\sqrt{q}\cdot\binom{n+k-1}{k-1}.
\end{aligned}$$

Theorem 2 is proven completely. □

Proof of Theorem 3. We prove Theorem 3 below. From the orthogonality of Chebyshev polynomials of the first kind we know that

$$\int_{-1}^{1}\frac{T_m(x)T_n(x)}{\sqrt{1-x^2}}dx=\begin{cases}0, & \text{if } m\neq n;\\ \frac{\pi}{2}, & \text{if } m=n>0,\\ \pi, & \text{if } m=n=0.\end{cases} \qquad (7)$$

If integer $n\geq 1$ with $2\nmid n$, then for any integer $0\leq i\leq n$, we have $n-2i\neq 0$, note that $T_n(x)=T_{-n}(x)$, so from (7) and Theorem 1 we have

$$\int_{-1}^{1} \frac{1}{\sqrt{1-x^2}} \left(\sum_{a_1+a_2+\cdots+a_k=n} P_{a_1}(x) \cdot P_{a_2}(x) \cdots P_{a_k}(x) \right)^2 dx$$
$$= \int_{-1}^{1} \frac{1}{\sqrt{1-x^2}} \left(\sum_{i=0}^{n} \frac{<\frac{k}{2}>_i}{i!} \cdot \frac{<\frac{k}{2}>_{n-i}}{(n-i)!} \cdot T_{n-2i}(x) \right)^2 dx$$
$$= 4 \int_{-1}^{1} \frac{1}{\sqrt{1-x^2}} \left(\sum_{i=0}^{[\frac{n}{2}]} \frac{<\frac{k}{2}>_i}{i!} \cdot \frac{<\frac{k}{2}>_{n-i}}{(n-i)!} \cdot T_{n-2i}(x) \right)^2 dx$$
$$= 2\pi \cdot \sum_{i=0}^{[\frac{n}{2}]} \left(\frac{<\frac{k}{2}>_i}{i!} \cdot \frac{<\frac{k}{2}>_{n-i}}{(n-i)!} \right)^2. \tag{8}$$

For $n = 2m$, if $n - 2i = 0$, then $i = m$. So from (7), Theorem 1 and the methods of proving (8) we have

$$\int_{-1}^{1} \frac{1}{\sqrt{1-x^2}} \left(\sum_{a_1+a_2+\cdots+a_k=n} P_{a_1}(x) \cdot P_{a_2}(x) \cdots P_{a_k}(x) \right)^2 dx$$
$$= \int_{-1}^{1} \frac{1}{\sqrt{1-x^2}} \left(\left(\frac{<\frac{k}{2}>_m}{m!} \right)^2 + 2 \sum_{i=0}^{m-1} \frac{<\frac{k}{2}>_i}{i!} \cdot \frac{<\frac{k}{2}>_{2m-i}}{(2m-i)!} \cdot T_{2m-2i}(x) \right)^2 dx$$
$$= \pi \cdot \left(\frac{<\frac{k}{2}>_m}{m!} \right)^4 + 2\pi \cdot \sum_{i=0}^{m-1} \left(\frac{<\frac{k}{2}>_i}{i!} \cdot \frac{<\frac{k}{2}>_{n-i}}{(n-i)!} \right)^2$$
$$= 2\pi \cdot \sum_{i=0}^{m} \left(\frac{<\frac{k}{2}>_i}{i!} \cdot \frac{<\frac{k}{2}>_{2m-i}}{(2m-i)!} \right)^2 - \pi \cdot \left(\frac{<\frac{k}{2}>_m}{m!} \right)^4. \tag{9}$$

Let $x = \sin\theta$, then we have

$$\int_{-1}^{1} \frac{1}{\sqrt{1-x^2}} \left(\sum_{a_1+a_2+\cdots+a_k=n} P_{a_1}(x) \cdot P_{a_2}(x) \cdots P_{a_k}(x) \right)^2 dx$$
$$= \int_{-\frac{\pi}{2}}^{\frac{\pi}{2}} \left(\sum_{a_1+a_2+\cdots+a_k=n} P_{a_1}(\sin\theta) \cdot P_{a_2}(\sin\theta) \cdots P_{a_k}(\sin\theta) \right)^2 d\theta. \tag{10}$$

Now Theorem 3 follows from (8), (9), and (10). □

3. Conclusions

Three theorems and three inferences are the main results in the paper. Theorem 1 gives proof of the symmetry of Legendre polynomials and the symmetry relationship with Chebyshev polynomials of the first kind. This conclusion also improves the early results in [4], and also gives us a different representation for the result in [3]. Theorem 2 obtained an inequality involving Dirichlet characters and Legendre polynomials; this is actually a new contribution to the study of Legendre polynomials and character sums mod q. Theorem 3 established an integral identity involving the symmetry sums of the Legendre polynomials. The three corollaries are some special cases of our three theorems for $k = 1$, and can not only enrich the research content of the Legendre polynomials, but also promote its research development.

Author Contributions: All authors have equally contributed to this work. All authors read and approved the final manuscript.

Acknowledgments: The authors would like to thank the editor and referee for their very helpful and detailed comments, which have significantly improved the presentation of this paper.

References

1. Borwein, P.; Erdèlyi, T. *Polynomials and Polynomial Inequalities*; Springer: New York, NY, USA, 1995.
2. Jackson, D. *Fourier Series and Orthogonal Polynomials*; Dover Publications: Mineola, NY, USA, 2004.
3. Shen, S.M.; Chen, L. Some types of identities involving the Legendre polynomials. *Mathematics* **2019**, *7*, 114. [CrossRef]
4. Zhou, Y.L.; Wang, X. The relationship of Legendre polynomials and Chebyshev polynomials. *Pure Appl. Math.* **1999**, *15*, 75–81.
5. Zhang, Y.X.; Chen, Z.Y. A new identity involving the Chebyshev polynomials. *Mathematics* **2018**, *6*, 244. [CrossRef]
6. Kim, T.; Kim, D.S.; Dolgy, D.V. Sums of finite products of Legendre and Laguerre polynomials. *Adv. Differ. Equ.* **2018**, *2018*, 277. [CrossRef]
7. Wang, S.Y. Some new identities of Chebyshev polynomials and their applications. *Adv. Differ. Equ.* **2015**, *2015*, 355.
8. He, Y. Some results for sums of products of Chebyshev and Legendre polynomials. *Adv. Differ. Equ.* **2019**, *2019*, 357. [CrossRef]
9. Chen, L.; Zhang, W.P. Chebyshev polynomials and their some interesting applications. *Adv. Differ. Equ.* **2017**, *2017*, 303.
10. Li, X.X. Some identities involving Chebyshev polynomials. *Math. Probl. Eng.* **2015**, *2015*, 950695. [CrossRef]
11. Wang, T.T.; Zhang, H. Some identities involving the derivative of the first kind Chebyshev polynomials. *Math. Probl. Eng.* **2015**, *2015*, 146313. [CrossRef]
12. Zhang, W.P.; Wang, T.T. Two identities involving the integral of the first kind Chebyshev polynomials. *Bull. Math. Soc. Sci. Math. Roum.* **2017**, *108*, 91–98.
13. Kim, T.; Dolgy, D.V.; Kim, D.S. Representing sums of finite products of Chebyshev polynomials of the second kind and Fibonacci polynomials in terms of Chebyshev polynomials. *Adv. Stud. Contemp. Math.* **2018**, *28*, 321–336.
14. Cesarano, C. Identities and generating functions on Chebyshev polynomials. *Georgian Math. J.* **2012**, *19*, 427–440. [CrossRef]
15. Cesarano, C. Generalized Chebyshev polynomials. *Hacet. J. Math. Stat.* **2014**, *43*, 731–740.
16. Dattoli, G.; Srivastava, H.M.; Cesarano, C. The Laguerre and Legendre polynomials from an operational point of view. *Appl. Math. Comput.* **2001**, *124*, 117–127. [CrossRef]
17. Islam, S.; Hossain, B. Numerical solutions of eighth order BVP by the Galerkin residual technique with Bernstein and Legendre polynomials. *Appl. Math. Comput.* **2015**, *261*, 48–59. [CrossRef]
18. Khalil, H.; Rahmat, A.K. A new method based on Legendre polynomials for solutions of the fractional two-dimensional heat conduction equation. *Comput. Math. Appl.* **2014**, *67*, 1938–1953. [CrossRef]
19. Wan, J.; Zudilin, W. Generating functions of Legendre polynomials: A tribute to Fred Brafman. *J. Approx. Theory* **2013**, *170*, 198–213. [CrossRef]
20. Nemati, S.; Lima, P. M.; Ordokhani, Y. Numerical solution of a class of two-dimensional nonlinear Volterra integral equations using Legendre polynomials. *J. Comput. Appl. Math.* **2013**, *242*, 53–69. [CrossRef]

18

On p-adic Integral Representation of q-Bernoulli Numbers Arising from Two Variable q-Bernstein Polynomials

Dae San Kim [1], Taekyun Kim [2,3], Cheon Seoung Ryoo [4,*] and Yonghong Yao [2,5]

1. Department of Mathematics, Sogang University, Seoul 121-742, Korea; dskim@sogang.ac.kr
2. Department of Mathematics, Tianjin Polytechnic University, Tianjin 300387, China; tkkim@kw.ac.kr (T.K.); yaoyonghong@aliyun.com (Y.Y.)
3. Department of Mathematics, Kwangwoon University, Seoul 139-701, Korea
4. Department of Mathematics, Hannam University, Daejeon 306-791, Korea
5. Institute of Fundamental and Frontier Sciences, University of Electronic Science and Technology of China, Chengdu 610054, China

* Correspondence: ryoocs@hnu.kr

Abstract: The q-Bernoulli numbers and polynomials can be given by Witt's type formulas as p-adic invariant integrals on $\mathbb{Z}_p$. We investigate some properties for them. In addition, we consider two variable q-Bernstein polynomials and operators and derive several properties for these polynomials and operators. Next, we study the evaluation problem for the double integrals on $\mathbb{Z}_p$ of two variable q-Bernstein polynomials and show that they can be expressed in terms of the q-Bernoulli numbers and some special values of q-Bernoulli polynomials. This is generalized to the problem of evaluating any finite product of two variable q-Bernstein polynomials. Furthermore, some identities for q-Bernoulli numbers are found.

Keywords: q-Bernoulli numbers; q-Bernoulli polynomials; two variable q-Bernstein polynomials; two variable q-Bernstein operators; p-adic integral on $\mathbb{Z}_p$

---

1. Introduction

Let p be a fixed prime number. Throughout this paper, $\mathbb{N}$, $\mathbb{Z}_p$, $\mathbb{Q}_p$, and $\mathbb{C}_p$ will denote the set of natural numbers, the ring of p-adic integers, the field of p-adic rational numbers, and the completion of the algebraic closure of $\mathbb{Q}_p$, respectively. The p-adic norm $|\cdot|_p$ is normalized as $|p|_p = \frac{1}{p}$. Assume that q is an indeterminate in $\mathbb{C}_p$ such that $|1-q|_p < p^{-\frac{1}{p-1}}$.

It is known that the q-number is defined by

$$[x]_q = \frac{1-q^x}{1-q},$$

see [1–20].

Please note that $\lim_{q\to 1}[x]_q = x$. Let $UD(\mathbb{Z}_p)$ be the space of uniformly differentiable functions on $\mathbb{Z}_p$. For $f \in UD(\mathbb{Z}_p)$, the p-adic q-integral on $\mathbb{Z}_p$ is defined by Kim as

$$I_q(f) = \int_{\mathbb{Z}_p} f(x)d\mu_q(x) = \lim_{N\to\infty} \frac{1}{[p^N]_q} \sum_{x=0}^{p^N-1} f(x)q^x, \tag{1}$$

see [9,10].

As $q \to 1$ in (1), we have the p-adic integral on $\mathbb{Z}_p$ which is given by

$$I_1(f) = \lim_{q\to 1} I_q(f) = \int_{\mathbb{Z}_p} f(x)d\mu_1(x) = \lim_{N\to\infty} \frac{1}{p^N} \sum_{x=0}^{p^N-1} f(x), \tag{2}$$

see [7–11,17].

From (2), we note that

$$I_1(f_1) - I_1(f) = f'(0), \tag{3}$$

see [9]. Where $f_1(x) = f(x+1)$ and $f'(0) = \frac{df(x)}{dx}|_{x=0}$.

Thus, by (3), we get

$$\int_{\mathbb{Z}_p} e^{(x+y)t} d\mu_1(y) = \frac{t}{e^t - 1} e^{xt} = \sum_{n=0}^{\infty} B_n(x) \frac{t^n}{n!}, \tag{4}$$

see [6,9], where $B_n(x)$ are the ordinary Bernoulli polynomials.

From (4), we note that

$$\int_{\mathbb{Z}_p} (x+y)^n d\mu_1(y) = B_n(x), (n \geq 0), \tag{5}$$

see [7–11,17,18].

When $x = 0$, $B_n = B_n(0), (n \geq 0)$, are called the ordinary Bernoulli numbers.

The Equation (4) implies the following recurrence relation for Bernoulli numbers:

$$B_0 = 1, \quad (B+1)^n - B_n = \begin{cases} 1 & \text{if } n = 1, \\ 0 & \text{if } n > 1, \end{cases} \tag{6}$$

with the usual convention about replacing B^n by B_n (see [21]).

In [3,4], L. Carlitz introduced the q-Bernoulli numbers given by the recurrence relation

$$\beta_{0,q} = 1, \quad q(q\beta_q + 1)^n - \beta_{n,q} = \begin{cases} 1 & \text{if } n = 1, \\ 0 & \text{if } n > 1, \end{cases} \tag{7}$$

with the usual convention about replacing β_q^n by $\beta_{n,q}$.

He also defined q-Bernoulli polynomials as

$$\beta_{n,q}(x) = (q^x\beta_q + [x]_q)^n = \sum_{l=0}^{n} \binom{n}{l} [x]_q^{n-l} q^{lx} \beta_{l,q}, \tag{8}$$

see [3,4].

In 1999, Kim proved the following formula.

$$\int_{\mathbb{Z}_p} [x+y]_q^n d\mu_q(y) = \beta_{n,q}(x), \ (n \geq 0), \tag{9}$$

see [10].

In the view of (5) and (9), we define the q-Bernoulli polynomials, different from Carlitz's q-Bernoulli polynomials, as

$$B_{n,q}(x) = \int_{\mathbb{Z}_p} [x+y]_q^n d\mu_1(y), \ (n \geq 0), \tag{10}$$

see [8,9].

When $x = 0$, $B_{n,q} = B_{n,q}(0)$ are called the q-Bernoulli numbers.

From (3) and (10), we have

$$B_{0,q}=1,\ (qB_q+1)^n-B_{n,q}=\begin{cases}\frac{\log q}{q-1} & \text{if } n=1,\\ 0 & \text{if } n>1,\end{cases} \tag{11}$$

with the usual convention about replacing B_q^n by $B_{n,q}$.

By (10), we easily get

$$B_{n,q}(x)=(q^xB_q+[x]_q)^n=\sum_{l=0}^{n}\binom{n}{l}[x]_q^{n-l}q^{lx}B_{l,q}, \tag{12}$$

see [9].

As is known, the p-adic q-Bernstein operator is given by

$$\mathbb{B}_{n,q}(f|x)=\sum_{k=0}^{n}\binom{n}{k}f\left(\frac{k}{n}\right)[x]_q^k[1-x]_{q^{-1}}^{n-k}=\sum_{k=0}^{n}f\left(\frac{k}{n}\right)B_{k,n}(x|q),$$

where $n,k\in\mathbb{N}\cup\{0\}$, $x\in\mathbb{Z}_p$, and f is a continuous function on $\mathbb{Z}_p$ (see [7]). Here

$$B_{k,n}(x|q)=\binom{n}{k}[x]_q^k[1-x]_{q^{-1}}^{n-k},\ (n,k\geq 0),$$

are called the p-adic q-Bernstein polynomials of degree n (see [7]). Please note that $\lim_{q\to 1}B_{k,n}(x|q)=B_{k,n}(x)$, where $B_{k,n}$ are the Bernstein polynomials (see [1,2,18–20,22]).

Here we cannot go without mentioning that Phillips (see [16]) introduced earlier in 1997 a different version of q-Bernstein polynomials from Kim's. Let f be a function defined on $[0,1]$, q any positive real number, and let

$$[n]_q!=[1]_q[2]_q\dots[n]_q,(n\geq 1),\quad [0]_q!=1,\quad \begin{bmatrix}n\\k\end{bmatrix}_q=\frac{[n]_q!}{[k]_q![n-k]_q!}.$$

Then Phillips' q-Bernstein polynomial of order n for f is given by

$$\mathbb{B}_n(f,q;x)=\sum_{k=0}^{n}f\left(\frac{[k]_q}{[n]_q}\right)\begin{bmatrix}n\\k\end{bmatrix}_q x^k\prod_{s=0}^{n-1-k}(1-q^sx),$$

Many results of Phillips' q-Bernstein polynomials for $q>1$ were obtained for instance in [14,15], while those for $q\in(0,1)$ were derived for example in [12,13]. However, all of these and other related papers deal only with analytic properties of those q-Bernstein polynomials and some applications of them.

The Volkenborn integral and the fermionic p-adic, the p-adic q-invariant and the fermionic p-adic q-invariant integrals introduced by Kim have been studied for more than twenty years. Numberous results of arithmetic or combinatorial nature have been found by Kim and his colleagues around the world.

The present and related paper (see [5,6]) concern about Kim's q-Bernstein polynomials which have some merits over Phillips'. Indeed, by considering p-adic integrals on $\mathbb{Z}_p$ of them we can easily derive integral representations of q-Bernoulli numbers in the present paper, those of a q-analogue of Euler numbers in [5] and those of q-Euler numbers in [6]. These approaches also yield some identities for q-Bernoulli numbers, q-analogue of Euler numbers and q-Euler numbers. In conclusion, the Phillips' q-Bernstein polynomials are more analytic nature, while the Kim's are more arithmetic and combinatorial nature.

In this paper, we will study q-Bernoulli numbers and polynomials, which is introduced as p-adic invariant integrals on $\mathbb{Z}_p$, and investigate some properties for these numbers and polynomials. Also, we will consider two variable q-Bernstein polynomials and operators and derive several properties for these polynomials and operators. Next, we will consider p-adic integrals on $\mathbb{Z}_p$ of any finite product of two variable q-Bernstein polynomials and show that they can be expressed in terms of the q-Bernoulli numbers and some special values of q-Bernoulli polynomials. Furthermore, some identities for q-Bernoulli numbers will be found.

2. Some Integral Representations of q-Bernoulli Numbers and Polynomials

First, we consider the two variable q-Bernstein operator of order n which is given by

$$\mathbb{B}_{n,q}(f|x_1,x_2) = \sum_{k=0}^{n} \binom{n}{k} f\left(\frac{k}{n}\right) [x_1]_q^k [1-x_2]_{q^{-1}}^{n-k} = \sum_{k=0}^{n} f\left(\frac{k}{n}\right) B_{k,n}(x_1,x_2|q),$$

where $n \in \mathbb{N}$, and $x_1, x_2 \in \mathbb{Z}_p$.

Here, for $n, k \geq 0$,

$$B_{k,n}(x_1,x_2|q) = \binom{n}{k} [x_1]_q^k [1-x_2]_{q^{-1}}^{n-k} \tag{13}$$

are called two variable q-Bernstein polynomials of degree n (see [6,7]). In particular, this implies that $B_{k,n}(x_1,x_2|q) = 0$, for $0 \leq n < k$. In (13), if $x_1 = x_2 = x$, then $B_{k,n}(x,x|q) = B_{k,n}(x|q)$ are the q-Bernstein polynomials. It is not difficult to show that the generating function of $B_{k,n}(x_1,x_2|q)$ is given by

$$F_q^{(k)}(x_1,x_2|t) = \frac{(t[x_1]_q)^k}{k!} e^{(t[1-x_2]_{q^{-1}})} = \sum_{n=k}^{\infty} B_{k,n}(x_1,x_2|q)\frac{t^n}{n!}, \tag{14}$$

where $k \in \mathbb{N} \cup \{0\}$ (see [6,7]).

From (13), we easily get

$$B_{n-k,n}(1-x_2, 1-x_1|q^{-1}) = B_{k,n}(x_1,x_2|q), \ (0 \leq k \leq n).$$

For $1 \leq k \leq n-1$, we have the following properties (see [6,7]):

$$[1-x_2]_{q^{-1}} B_{k,n-1}(x_1,x_2|q) + [x_1]_q B_{k-1,n-1}(x_1,x_2|q) = B_{k,n}(x_1,x_2|q), \tag{15}$$

$$\frac{\partial}{\partial x_1} B_{k,n}(x_1,x_2|q) = \frac{\log q}{q-1} n \left((q-1)[x_1]_q B_{k-1,n-1}(x_1,x_2|q) + B_{k-1,n-1}(x_1,x_2|q)\right), \tag{16}$$

$$\frac{\partial}{\partial x_2} B_{k,n}(x_1,x_2|q) = \frac{\log q}{1-q} n((q-1)[x_2]_q B_{k,n-1}(x_1,x_2|q) + B_{k,n-1}(x_1,x_2|q)). \tag{17}$$

From (13) and q-Bernstein operator, we note that

$$\mathbb{B}_{n,q}(1|x_1,x_2) = \sum_{k=0}^{n} \binom{n}{k} [x_1]_q^k [1-x_2]_{q^{-1}}^{n-k} = (1+[x_1]_q - [x_2]_q)^n, \tag{18}$$

$$\mathbb{B}_{n,q}(t|x_1,x_2) = [x_1]_q (1+[x_1]_q - [x_2]_q)^{n-1},$$

$$\mathbb{B}_{n,q}(t^2|x_1,x_2) = \frac{n-1}{n}[x_1]_q^2 (1+[x_1]_q - [x_2]_q)^{n-2} + \frac{[x_1]_q}{n}(1+[x_1]_q - [x_2]_q)^{n-1},$$

and

$$\mathbb{B}_{n,q}(f|x_1,x_2) = \sum_{l=0}^{n} \binom{n}{l} [x_2]_q^l \sum_{k=0}^{l} \binom{l}{k} (-1)^{l-k} f\left(\frac{k}{n}\right) \left(\frac{[x_1]_q}{[x_2]_q}\right)^k, \tag{19}$$

where $n \in \mathbb{N}$ and f is a continuous function on $\mathbb{Z}_p$.

To see this, we first observe that

$$[1-x_2]_{q^{-1}} = 1-[x_2]_q, \quad \binom{n}{k}\binom{n-k}{l-k} = \binom{n}{l}\binom{l}{k}.$$

Then (19) can be obtained as follows:

$$\begin{aligned}
\mathbb{B}(f|x_1,x_2) &= \sum_{k=0}^{n} f(\frac{k}{n})\binom{n}{k}[x_1]_q^k(1-[x_2]_q)^{n-k} \\
&= \sum_{k=0}^{n} f(\frac{k}{n})\binom{n}{k}[x_1]_q^k \sum_{l=0}^{n-k}\binom{n-k}{l}(-1)^l[x_2]_q^l \\
&= \sum_{k=0}^{n} f(\frac{k}{n})\binom{n}{k}[x_1]_q^k \sum_{l=k}^{n}\binom{n-k}{l-k}(-1)^{l-k}[x_2]_q^{l-k} \\
&= \sum_{l=0}^{n}\binom{n}{l}[x_2]_q^l \sum_{k=0}^{l}\binom{l}{k}(-1)^{l-k} f\left(\frac{k}{n}\right)\left(\frac{[x_1]_q}{[x_2]_q}\right)^k.
\end{aligned}$$

It is easy to show that

$$\frac{1}{(1+[x_1]_q-[x_2]_q)^{n-j}} \sum_{k=j}^{n} \frac{\binom{k}{j}}{\binom{n}{j}} B_{k,n}(x_1,x_2|q) = [x_1]_q^j, \tag{20}$$

where $j \in \mathbb{N}\cup\{0\}$ and $x_1, x_2 \in \mathbb{Z}_p$.

Indeed, by making use of (18), we see that

$$\begin{aligned}
\sum_{k=j}^{n} \frac{\binom{k}{j}}{\binom{n}{j}} B_{k,n}(x_1,x_2|q) &= \sum_{k=j}^{n}\binom{n-j}{k-j}[x_1]_q^k[1-x_2]_{q^{-1}}^{n-k} \\
&= \sum_{k=0}^{n-j}\binom{n-j}{k}[x_1]_q^{k+j}[1-x_2]_{q^{-1}}^{n-j-k} \\
&= [x_1]^j(1+[x_1]_q-[x_2]_q)^{n-j}.
\end{aligned}$$

From (2), we have

$$\int_{\mathbb{Z}_p}[1-x+y]_{q^{-1}}^n d\mu_1(y) = (-1)^n q^n \int_{\mathbb{Z}_p}[x+y]_q^n d\mu_1(y), \ (n \geq 0). \tag{21}$$

By (10) and (21), we get

$$B_{n,q^{-1}}(1-x) = (-1)^n q^n B_{n,q}(x), \ (n \geq 0). \tag{22}$$

Again, from (11) and (12), we can derive the following equation.

$$B_{n,q}(2) = nq\frac{\log q}{q-1} + (qB_q+1)^n = nq\frac{\log q}{q-1} + B_{n,q}, \ (n > 1). \tag{23}$$

Thus, by (23), we obtain the following lemma.

Lemma 1. *For $n \in \mathbb{N}$ with $n > 1$, we have*

$$B^{n,q}(2) = nq\frac{\log q}{q-1} + B^{n,q}.$$

By (2), (10) and (22), we get

$$\begin{aligned}\int_{\mathbb{Z}_p}[1-x]_{q^{-1}}^n d\mu_1(x) &= (-1)^n q^n \int_{\mathbb{Z}_p}[x-1]_q^n d\mu_1(x)\\ &= (-1)^n q^n B_{n,q}(-1)\\ &= B_{n,q^{-1}}(2),\ (n \geq 0).\end{aligned} \tag{24}$$

For $n \in \mathbb{N}$ with $n > 1$, by (21), Lemma 1, and (24), we have

$$\begin{aligned}\int_{\mathbb{Z}_p}[1-x]_{q^{-1}}^n d\mu_1(x) &= \int_{\mathbb{Z}_p}[x+2]_{q^{-1}}^n d\mu_1(x)\\ &= (-1)^n q^n \int_{\mathbb{Z}_p}[x-1]_q^n d\mu_1(x)\\ &= n\frac{\log q}{q-1} + \int_{\mathbb{Z}_p}[x]_{q^{-1}}^n d\mu_1(x)\\ &= \frac{n\log q}{q-1} + B_{n,q^{-1}}.\end{aligned} \tag{25}$$

Let us take the double p-adic integral on $\mathbb{Z}_p$ for the two variable q-Bernstein polynomials. Then we have

$$\begin{aligned}&\int_{\mathbb{Z}_p}\int_{\mathbb{Z}_p} B_{k,n}(x_1,x_2|q)d\mu_1(x_1)d\mu_1(x_2)\\ &= \binom{n}{k}\int_{\mathbb{Z}_p}\int_{\mathbb{Z}_p}[x_1]_q^k[1-x_2]_{q^{-1}}^{n-k}d\mu_1(x_1)d\mu_1(x_2)\\ &= \binom{n}{k}B_{k,q}\int_{\mathbb{Z}_p}[1-x_2]_{q^{-1}}^{n-k}d\mu_1(x_2)\\ &= \begin{cases}\binom{n}{k}B_{k,q}\left(B_{n-k,q^{-1}} + \frac{\log q}{q-1}(n-k)\right), & \text{if } n > k+1,\\ (k+1)B_{k,q}B_{1,q^{-1}}(2), & \text{if } n = k+1,\\ B_{k,q}, & \text{if } n = k,\\ 0, & \text{if } 0 \leq n < k.\end{cases}\end{aligned} \tag{26}$$

Therefore, we obtain the following theorem.

Theorem 1. *For $n, k \in \mathbb{N} \cup \{0\}$, we have*

$$\begin{aligned}&\int_{\mathbb{Z}_p}\int_{\mathbb{Z}_p} B_{k,n}(x_1,x_2|q)d\mu_1(x_1)d\mu_1(x_2)\\ &= \begin{cases}\binom{n}{k}B_{k,q}\left(B_{n-k,q^{-1}} + \frac{\log q}{q-1}(n-k)\right), & \text{if } n > k+1,\\ (k+1)B_{k,q}B_{1,q^{-1}}(2), & \text{if } n = k+1,\\ B_{k,q} & \text{if } n = k,\\ 0, & \text{if } 0 \leq n < k.\end{cases}\end{aligned}$$

For $n, k \in \mathbb{N} \cup \{0\}$, we have

$$\begin{aligned}
&\int_{\mathbb{Z}_p}\int_{\mathbb{Z}_p} B_{k,n}(x_1, x_2|q)d\mu_1(x_1)d\mu_1(x_2) \\
&= \int_{\mathbb{Z}_p}\int_{\mathbb{Z}_p} \binom{n}{k}[x_1]_q^k[1-x_2]_{q^{-1}}^{n-k}d\mu_1(x_1)d\mu_1(x_2) \\
&= \binom{n}{k}\int_{\mathbb{Z}_p}\int_{\mathbb{Z}_p}(1-[1-x_1]_{q^{-1}})^k[1-x_2]_{q^{-1}}^{n-k}d\mu_1(x_1)d\mu_1(x_2) \\
&= \binom{n}{k}\sum_{l=0}^{k}\binom{k}{l}(-1)^{k-l}\int_{\mathbb{Z}_p}\int_{\mathbb{Z}_p}[1-x_1]_{q^{-1}}^{k-l}[1-x_2]_{q^{-1}}^{n-k}d\mu_1(x_1)d\mu_1(x_2) \\
&= \binom{n}{k}\int_{\mathbb{Z}_p}[1-x_2]_{q^{-1}}^{n-k}d\mu_1(x_2) \\
&\times\left(1-k\int_{\mathbb{Z}_p}[1-x_1]_{q^{-1}}d\mu_1(x_1)+\sum_{l=0}^{k-2}\binom{k}{l}(-1)^{k-l}\left((k-l)\frac{\log q}{q-1}+B_{k-l,q^{-1}}\right)\right).
\end{aligned} \tag{27}$$

Thus, by (27), we get

$$\begin{aligned}
&\binom{n}{k}^{-1}\frac{\int_{\mathbb{Z}_p}\int_{\mathbb{Z}_p} B_{k,n}(x_1, x_2|q)d\mu_1(x_1)d\mu_1(x_2)}{\int_{\mathbb{Z}_p}[1-x_2]_{q^{-1}}^{n-k}d\mu_1(x_2)} \\
&= 1-k\int_{\mathbb{Z}_p}[1-x_1]_{q^{-1}}d\mu_1(x_1)+\sum_{l=0}^{k-2}\binom{k}{l}(-1)^{k-l}\left((k-l)\frac{\log q}{q-1}+B_{k-l,q^{-1}}\right) \\
&= 1-k\Big(1-\frac{\log q-q+1}{(q-1)^2}\Big) \\
&\quad +k\sum_{l=0}^{k-2}\binom{k-1}{l}(-1)^{k-l}\frac{\log q}{q-1}+\sum_{l=0}^{k-2}(-1)^{k-l}\binom{k}{l}B_{k-l,q^{-1}}
\end{aligned} \tag{28}$$

$$\begin{aligned}
&= 1-k\Big(1-\frac{\log q-q+1}{(q-1)^2}\Big)+k\frac{\log q}{q-1}+\sum_{l=0}^{k-2}(-1)^{k-l}\binom{k}{l}B_{k-l,q^{-1}} \\
&= 1-kq\left(\frac{q-\log q-1}{(q-1)^2}\right)+\sum_{l=0}^{k-2}(-1)^{k-l}\binom{k}{l}B_{k-l,q^{-1}}.
\end{aligned} \tag{29}$$

Therefore, by (28), we obtain the following theorem.

Theorem 2. *For $n, k \in \mathbb{N} \cup \{0\}$ with $k > 1$, we have*

$$\begin{aligned}
&\frac{\int_{\mathbb{Z}_p}\int_{\mathbb{Z}_p} B_{k,n}(x_1, x_2|q)d\mu_1(x_1)d\mu_1(x_2)}{\binom{n}{k}\int_{\mathbb{Z}_p}[1-x_2]_{q^{-1}}^{n-k}d\mu_1(x_2)} \\
&= \binom{n}{k}\left(1-kq\left(\frac{q-\log q-1}{(q-1)^2}\right)+\sum_{l=0}^{k-2}(-1)^{k-l}\binom{k}{l}B_{k-l,q^{-1}}\right).
\end{aligned}$$

Therefore, by Theorems 1 and 2, we obtain the following corollary.

Corollary 1. *For $k \in \mathbb{N}$ with $k > 1$, we have*

$$B_{k,q} = 1-kq\left(\frac{q-\log q-1}{(q-1)^2}\right)+\sum_{l=0}^{k-2}(-1)^{k-l}\binom{k}{l}B_{k-l,q^{-1}}.$$

For $m, n \in \mathbb{N} \cup \{0\}$, we have

$$\begin{aligned}&\int_{\mathbb{Z}_p}\int_{\mathbb{Z}_p} B_{k,n}(x_1, x_2|q)B_{k,m}(x_1, x_2|q)d\mu_1(x_1)d\mu_1(x_2)\\&=\binom{n}{k}\binom{m}{k}\int_{\mathbb{Z}_p}[x_1]_q^{2k}d\mu_1(x_1)\int_{\mathbb{Z}_p}[1-x_2]_{q^{-1}}^{n+m-2k}d\mu_1(x_2).\end{aligned} \tag{30}$$

Thus, by (29), we get

$$\begin{aligned}&\frac{\int_{\mathbb{Z}_p}\int_{\mathbb{Z}_p} B_{k,n}(x_1, x_2|q)B_{k,m}(x_1, x_2|q)d\mu_1(x_1)d\mu_1(x_2)}{\int_{\mathbb{Z}_p}[1-x_2]_{q^{-1}}^{n+m-2k}d\mu_1(x_2)}\\&=\binom{n}{k}\binom{m}{k}B_{2k,q}.\end{aligned}$$

Hence, we have the following proposition.

Proposition 1. *For $m, n, k \in \mathbb{N} \cup \{0\}$, we have*

$$\begin{aligned}&\frac{\int_{\mathbb{Z}_p}\int_{\mathbb{Z}_p} B_{k,n}(x_1, x_2|q)B_{k,m}(x_1, x_2|q)d\mu_1(x_1)d\mu_1(x_2)}{\int_{\mathbb{Z}_p}[1-x_2]_{q^{-1}}^{n+m-2k}d\mu_1(x_2)}\\&=\binom{n}{k}\binom{m}{k}B_{2k,q}.\end{aligned}$$

Let $m, n, k \in \mathbb{N} \cup \{0\}$. Then we get

$$\begin{aligned}&\int_{\mathbb{Z}_p}\int_{\mathbb{Z}_p} B_{k,n}(x_1, x_2|q)B_{k,m}(x_1, x_2|q)d\mu_1(x_1)d\mu_1(x_2)\\&=\sum_{l=0}^{2k}\binom{n}{k}\binom{m}{k}\binom{2k}{l}(-1)^{2k-l}\\&\quad\times\int_{\mathbb{Z}_p}\int_{\mathbb{Z}_p}[1-x_1]_{q^{-1}}^{2k-l}[1-x_2]_{q^{-1}}^{n+m-2k}d\mu_1(x_1)d\mu_1(x_2)\end{aligned} \tag{31}$$

Thus, from (30), we have

$$\begin{aligned}&\binom{n}{k}^{-1}\binom{m}{k}^{-1}\frac{\int_{\mathbb{Z}_p}\int_{\mathbb{Z}_p} B_{k,n}(x_1, x_2|q)B_{k,m}(x_1, x_2|q)d\mu_1(x_1)d\mu_1(x_2)}{\int_{\mathbb{Z}_p}[1-x_2]_{q^{-1}}^{n+m-2k}d\mu_1(x_2)}\\&=1-2k\int_{\mathbb{Z}_p}[1-x_1]_{q^{-1}}d\mu_1(x_1)+\sum_{l=0}^{2k-2}\binom{2k}{l}(-1)^{2k-l}\int_{\mathbb{Z}_p}[1-x_1]_{q^{-1}}^{2k-l}d\mu_1(x_1)\\&=1-2k\Big(1-\frac{\log q-q+1}{(q-1)^2}\Big)+2k\frac{\log q}{q-1}+\sum_{l=0}^{2k-2}\binom{2k}{l}(-1)^{2k-l}B_{2k-l,q^{-1}}\\&=1-2kq\left(\frac{q-\log q-1}{(q-1)^2}\right)+\sum_{l=0}^{2k-2}\binom{2k}{l}(-1)^{2k-l}B_{2k-l,q^{-1}}.\end{aligned} \tag{32}$$

By (31), we have the following proposition.

Proposition 2. *For $m, n, k \in \mathbb{N} \cup \{0\}$, with $k \geq 1$, we have*

$$\frac{\int_{\mathbb{Z}_p}\int_{\mathbb{Z}_p} B_{k,n}(x_1,x_2|q)B_{k,m}(x_1,x_2|q)d\mu_1(x_1)d\mu_1(x_2)}{\int_{\mathbb{Z}_p}[1-x_2]_{q^{-1}}^{n+m-2k}d\mu_1(x_2)} = \binom{n}{k}\binom{m}{k}\left(1-2kq\left(\frac{q-\log q-1}{(q-1)^2}\right)+\sum_{l=0}^{2k-2}\binom{2k}{l}(-1)^{2k-l}B_{2k-l,q^{-1}}\right).$$

Therefore, by Propositions 1 and 2, we obtain the following corollary.

Corollary 2. *For $k \in \mathbb{N}$, we have*

$$B_{2k,q} = 1-2k\left(\frac{q^2-q-\log q}{(q-1)^2}\right) + \sum_{l=0}^{2k-2}\binom{2k}{l}(-1)^{2k-l}\left((2k-l)\frac{\log q}{q-1}+B_{2k-l,q^{-1}}\right).$$

For $m \in \mathbb{N}$, let $n_1, n_2, \cdots, n_m, k \in \mathbb{N} \cup \{0\}$. Then we note that

$$\begin{aligned}&\int_{\mathbb{Z}_p}\int_{\mathbb{Z}_p}\left(\prod_{i=1}^{m} B_{k,n_i}(x_1,x_2|q)\right)d\mu_1(x_1)d\mu_1(x_2)\\ &=\sum_{l=0}^{mk}\left(\prod_{i=1}^{m}\binom{n_i}{k}\right)\binom{mk}{l}(-1)^{mk-l}\\ &\times\int_{\mathbb{Z}_p}\int_{\mathbb{Z}_p}[1-x_1]_{q^{-1}}^{mk-l}[1-x_2]_{q^{-1}}^{n_1+n_2+\cdots+n_m-mk}d\mu_1(x_1)d\mu_1(x_2)\end{aligned} \tag{33}$$

Thus, by (32), we have

$$\begin{aligned}&\prod_{i=1}^{m}\binom{n_i}{k}^{-1}\frac{\int_{\mathbb{Z}_p}\int_{\mathbb{Z}_p}\left(\prod_{i=1}^{m} B_{k,n_i}(x_1,x_2|q)\right)d\mu_1(x_1)d\mu_1(x_2)}{\int_{\mathbb{Z}_p}[1-x_2]_{q^{-1}}^{n_1+n_2+\cdots+n_m-mk}d\mu_1(x_2)}\\ &=\sum_{l=0}^{mk}\binom{mk}{l}(-1)^{mk-l}\int_{\mathbb{Z}_p}[1-x_1]_{q^{-1}}^{mk-l}d\mu_1(x_1)\\ &=1-mk\int_{\mathbb{Z}_p}[1-x_1]_{q^{-1}}d\mu_1(x_1)\\ &\quad+\sum_{l=0}^{mk-2}\binom{mk}{l}(-1)^{mk-l}\int_{\mathbb{Z}_p}[1-x_1]_{q^{-1}}^{mk-l}d\mu_1(x_1)\end{aligned}$$

Therefore we obtain the following theorem.

Theorem 3. *For $n_1, n_2, \cdots, n_m \in \mathbb{N} \cup \{0\}$, and $k, m \in \mathbb{N}$ with $mk > 1$, we have*

$$\begin{aligned}&\prod_{i=1}^{m}\binom{n_i}{k}^{-1}\frac{\int_{\mathbb{Z}_p}\int_{\mathbb{Z}_p}\left(\prod_{i=1}^{m} B_{k,n_i}(x_1,x_2|q)\right)d\mu_1(x_1)d\mu_1(x_2)}{\int_{\mathbb{Z}_p}[1-x_2]_{q^{-1}}^{n_1+n_2+\cdots+n_m-mk}d\mu_1(x_2)}\\ &=1-mk\left(\frac{q^2-q-\log q}{(q-1)^2}\right)\\ &\quad+\sum_{l=0}^{mk-2}\binom{mk}{l}(-1)^{mk-l}\left((mk-l)\frac{\log q}{q-1}+B_{mk-l,q^{-1}}\right).\end{aligned}$$

On the other hand, we easily get

$$\frac{\int_{\mathbb{Z}_p}\int_{\mathbb{Z}_p}\left(\prod_{i=1}^{m}B_{k,n_i}(x_1,x_2|q)\right)d\mu_1(x_1)d\mu_1(x_2)}{\int_{\mathbb{Z}_p}[1-x_2]_{q^{-1}}^{n_1+n_2+\cdots+n_m-mk}d\mu_1(x_2)}=\prod_{i=1}^{m}\binom{n_i}{k}B_{mk,q}. \tag{34}$$

Therefore, by Theorem 3 and (33), we obtain the following corollary.

Corollary 3. *For $m,k\in\mathbb{N}$ with $mk>1$, we have*

$$\begin{aligned}B_{mk,q}&=1-mk\left(\frac{q^2-q-\log q}{(q-1)^2}\right)\\&+\sum_{l=0}^{mk-2}\binom{mk}{l}(-1)^{mk-l}\left((mk-l)\frac{\log q}{q-1}+B_{mk-l,q^{-1}}\right).\end{aligned}$$

3. Conclusions

Here we studied q-Bernoulli numbers and polynomials which are different from the classical Carlitz q-Bernoulli numbers $\beta_{n,q}$ and polynomials $\beta_{n,q}(x)$, and arise naturally from some p-adic invariant integrals on $\mathbb{Z}_p$, as was shown in (10). After investigating some of their properties, we turned our attention to two variable q-Bernstein polynomials and operators, which was introduced by Kim and generalizes the single variable q-Bernstein polynomials and operators in [6]. As a preparation, we derived several properties of these polynomials and operators.

Next, we considered the evaluation problem for the double integrals on $\mathbb{Z}_p$ of two variable q-Bernstein polynomials and showed that they can be expressed in terms of the q-Bernoulli numbers and some special values of q-Bernoulli polynomials. This was further generalized to the problem of evaluating the product of two and that of an arbitrary number of two variable q-Bernstein polynomials. It was shown again that they can be expressed in terms of the q-Bernoulli numbers. Also, some identities for q-Bernoulli numbers were found along the way.

Finally, we would like to mention that, along the same line, in [5] we studied some properties of a q-analogue of Euler numbers and polynomials arising from the p-adic fermionic integrals on $\mathbb{Z}_p$. Then we considered p-adic fermionic integrals on $\mathbb{Z}_p$ of the two variable q-Bernstein polynomials and of products of the two variable q-Bernstein polynomials, and showed that they can be expressed in terms of the q-analogues of Euler numbers.

Author Contributions: T.K. and D.S.K. conceived the framework and structured the whole paper; T.K. wrote the paper; C.S.R. and Y.Y. checked the results of the paper; D.S.K., C.S.R., Y.Y. and T.K. completed the revision of the article.

Acknowledgments: The authors would like to thank the referees for their valuable comments which improved the original manuscript in its present form.

References

1. Bayad, A.; Kim, T. Identities involving values of Bernstein q-Bernoulli, and q-Euler polynomials. *Russ. J. Math. Phys.* **2011**, *18*, 133–143. [CrossRef]
2. Carlitz, L. Expansions of q-Bernoulli numbers. *Duke Math. J.* **1958**, *25*, 355–364. [CrossRef]
3. Carlitz, L. q-Bernoulli and Eulerian numbers. *Trans. Am. Math. Soc.* **1954**, *76*, 332–350.
4. Kim, T. Some identities on the q-integral representation of the product of several q-Bernstein-type polynomials. *Abstr. Appl. Anal.* **2011**, *2011*, 634675. [CrossRef]
5. Jang, L.C.; Kim, T.; Kim, D.S.; Dolgy, D.V. On p-adic fermionic integrals of q-Bernstein polynomials associated with q-Euler numbers and polynomials. *Symmetry* **2018**, *10*, 311. [CrossRef]

6. Kim, T. A note on q-Bernstein polynomials. *Russ. J. Math. Phys.* **2011**, *18*, 73–82. [CrossRef]
7. Kim, T. A study on the q-Euler numbers and the fermionic q-integral of the product of several type q-Bernstein polynomials on $\mathbb{Z}_p$. *Adv. Stud. Contemp. Math.* **2013**, *23*, 5–11.
8. Kim, T. On p-adic q-Bernoulli numbers. *J. Korean Math. Soc.* **2000**, *37*, 21–30.
9. Kim, T.; Kim, H.S. Remark on the p-adic q-Bernoulli numbers. Algerbraic number theory (Hapcheon/Saga, 1996). *Adv. Stud. Contemp. Math.* **1999**, *1*, 127–136.
10. Kim, T. q-Volkenborn integration. *Russ. J. Math. Phys.* **2002**, *9*, 288–299.
11. Kurt, V. Some relation between the Bernstein polynomials and second kind Bernpolli polynomials. *Adv. Stud. Contemp. Math.* **2013**, *23*, 43–48.
12. Oruç, H.; Phillips, G.M. A generalization of Bernstein polynomials. *Proc. Edinb. Math. Soc.* **1999**, *42*, 403–413. [CrossRef]
13. Oruç, H.; Tuncer, N. On the convergence and iterates of q-Bernstein polynomials. *J. Approx. Theory* **2002**, *117*, 301–313. [CrossRef]
14. Ostrovska, S. On the q-Bernstein polynomials. *Adv. Stud. Contemp. Math.* **2015**, *11*, 193–204.
15. Ostrovska, S. On the q-Bernstein polynomials of the logarithmic function in the case $q > 1$. *Math. Slovaca* **2016**, *66*, 73–78. [CrossRef]
16. Phillips, G.M. Bernstein polynomials based on the q-integers. *Ann. Numer. Math.* **1997**, *4*, 511–518.
17. Rim, S.-H.; Joung, J.; Jin, J.-H.; Lee, S.-J. A note on the weighted Carlitz's type q-Euler numbers and q-Bernstein polynomials. *Proc. Jangjeon Math. Soc.* **2012**, *15*, 195–201.
18. Kim, D.S.; Kim, T. Some p-adic integrals on $\mathbb{Z}_p$ associated with trigonometric functions. *Russ. J. Math. Phys.* **2018**, *25*, 300–308. [CrossRef]
19. Siddiqui, M.A.; Agrawal, R.R.; Gupta, N. On a class of modified new Bernstein operators. *Adv. Stud. Contemp. Math.* **2014**, *24*, 97–107.
20. Simsek, Y. On parametrization of the q-Bernstein basis functions and their applications. *J. Inequal. Spec. Funct.* **2017**, *8*, 158–169.
21. Kim, T.; Kim, D.S. Identities for degenerate Bernoulli polynomials and Korobov polynomials of the first kind. *Sci. China Math.* **2018**. [CrossRef]
22. Taberski, R. Approximation properties of the integral Bernstein operators and their derivatives in some classes of locally integral functions. *Funct. Approx. Comment. Math.* **1992**, *21*, 85–96.

Some Identities on Type 2 Degenerate Bernoulli Polynomials of the Second Kind

Taekyun Kim [1,2], Lee-Chae Jang [3,*], Dae San Kim [4] and Han Young Kim [2]

1 School of Sciences, Xian Technological University, Xi'an 710021, China; tkkim@kw.ac.kr
2 Department of Mathematics, Kwangwoon University, Seoul 139-701, Korea; gksdud213@kw.ac.kr
3 Graduate School of Education, Konkuk University, Seoul 05029, Korea
4 Department of Mathematics, Sogang University, Seoul 121-742, Korea; dskim@sogang.ac.kr
* Correspondence: Lcjang@konkuk.ac.kr

Abstract: In recent years, many mathematicians studied various degenerate versions of some special polynomials for which quite a few interesting results were discovered. In this paper, we introduce the type 2 degenerate Bernoulli polynomials of the second kind and their higher-order analogues, and study some identities and expressions for these polynomials. Specifically, we obtain a relation between the type 2 degenerate Bernoulli polynomials of the second and the degenerate Bernoulli polynomials of the second, an identity involving higher-order analogues of those polynomials and the degenerate Stirling numbers of the second kind, and an expression of higher-order analogues of those polynomials in terms of the higher-order type 2 degenerate Bernoulli polynomials and the degenerate Stirling numbers of the first kind.

Keywords: type 2 degenerate Bernoulli polynomials of the second kind; degenerate central factorial numbers of the second kind

---

1. Introduction

In [1,2], Carlitz initiated study of the degenerate Bernoulli and Euler polynomials and obtained some arithmetic and combinatorial results on them. In recent years, many mathematicians have drawn their attention to various degenerate versions of some old and new polynomials and numbers, namely some degenerate versions of Bernoulli numbers and polynomials of the second kind, Changhee numbers of the second kind, Daehee numbers of the second kind, Bernstein polynomials, central Bell numbers and polynomials, central factorial numbers of the second kind, Cauchy numbers, Eulerian numbers and polynomials, Fubini polynomials, Stirling numbers of the first kind, Stirling polynomials of the second kind, central complete Bell polynomials, Bell numbers and polynomials, type 2 Bernoulli numbers and polynomials, type 2 Bernoulli polynomials of the second kind, poly-Bernoulli numbers and polynomials, poly-Cauchy polynomials, and of Frobenius–Euler polynomials, to name a few [3–10] and the references therein.

They have studied those polynomials and numbers with their interest not only in combinatorial and arithmetic properties but also in differential equations and certain symmetric identities [7,9] and references therein, and found many interesting results related to them [3–6,8,10]. It is remarkable that studying degenerate versions is not only limited to polynomials but also extended to transcendental functions. Indeed, the degenerate gamma functions were introduced in connection with degenerate Laplace transforms [11,12].

The motivation for this research is to introduce the type 2 degenerate Bernoulli polynomials of the second kind defined by

$$\frac{(1+t)-(1+t)^{-1}}{\log_{\lambda}(1+t)}(1+t)^{x}=\sum_{n=0}^{\infty}b_{n,\lambda}^{*}(x)\frac{t^{n}}{n!},$$

and investigate its arithmetic and combinatorial properties. The facts in Section 1 are some known definitions and results that are needed throughout this paper. However, all of the results in Section 2 are new.

We will spend the rest of this section in recalling some necessary stuffs for the next section.

As is known, the type 2 Bernoulli polynomials are defined by the generating function [5,13]

$$\frac{t}{e^{t}-e^{-t}}e^{xt}=\sum_{n=0}^{\infty}B_{n}^{*}(x)\frac{t^{n}}{n!}. \tag{1}$$

From (1), we note that

$$B_{n}^{*}(x)=2^{n-1}B_{n}\left(\frac{x+1}{2}\right),\quad (n\geq 0), \tag{2}$$

where $B_n(x)$ are the ordinary Bernoulli polynomials given by

$$\frac{t}{e^{t}-1}e^{xt}=\sum_{n=0}^{\infty}B_{n}(x)\frac{t^{n}}{n!}.$$

Also, the type 2 Euler polynomials are given by [5,13]

$$e^{xt}\mathrm{sech}t=\frac{2}{e^{t}+e^{-t}}e^{xt}=\sum_{n=0}^{\infty}E_{n}^{*}(x)\frac{t^{n}}{n!}. \tag{3}$$

Note that

$$E_{n}^{*}(x)=2^{n}E_{n}\left(\frac{x+1}{2}\right),\quad (n\geq 0), \tag{4}$$

where $E_n(x)$ are the ordinary Euler polynomials given by [14,15]

$$\frac{2}{e^{t}+1}e^{xt}=\sum_{n=0}^{\infty}E_{n}(x)\frac{t^{n}}{n!}.$$

The central factorial numbers of the second kind are defined as [5,8]

$$x^{n}=\sum_{k=0}^{n}T(n,k)x^{[k]}, \tag{5}$$

or equivalently as

$$\frac{1}{k!}(e^{\frac{t}{2}}-e^{-\frac{t}{2}})^{k}=\sum_{n=k}^{\infty}T(n,k)\frac{t^{n}}{n!}, \tag{6}$$

where $x^{[0]}=1$, $x^{[n]}=x\left(x+\frac{n}{2}-1\right)\left(x+\frac{n}{2}-2\right)\cdots\left(x-\frac{n}{2}+1\right),\ (n\geq 1)$.

It is well known that the Daehee polynomials are defined by [16,17]

$$\frac{\log(1+t)}{t}(1+t)^{x}=\sum_{k=0}^{n}D_{n}(x)\frac{t^{n}}{n!}. \tag{7}$$

When $x=0$, $D_n=D_n(0)$ are called the Daehee numbers.

The Bernoulli polynomials of the second kind of order r are defined by [15]

$$\left(\frac{t}{\log(1+t)}\right)^r (1+t)^x = \sum_{k=0}^{n} b_n^{(r)}(x)\frac{t^n}{n!}. \tag{8}$$

Note that $b_n^{(r)}(x) = B_n^{(n-r+1)}(x+1),\ (n \geq 0)$. Here $B_n^{(r)}(x)$ are the ordinary Bernoulli polynomials of order r given by [8,15–18]

$$\left(\frac{t}{e^t - 1}\right)^r e^{xt} = \sum_{k=0}^{n} B_n^{(r)}(x)\frac{t^n}{n!}. \tag{9}$$

It is known that the Stirling numbers of the second kind are defined by [8]

$$\frac{1}{k!}\left(e^t - 1\right)^k = \sum_{n=k}^{\infty} S_2(n,k)\frac{t^n}{n!}, \tag{10}$$

and the Stirling numbers of the first kind by [8]

$$\frac{1}{k!}\log^k(1+t) = \sum_{n=k}^{\infty} S_1(n,k)\frac{t^n}{n!}. \tag{11}$$

For any nonzero $\lambda \in \mathbb{R}$, the degenerate exponential function is defined by [11,12]

$$e_\lambda^x(t) = (1+\lambda t)^{\frac{x}{\lambda}} = \sum_{n=0}^{\infty} (x)_{n,\lambda}\frac{t^n}{n!}, \tag{12}$$

where $(x)_{0,\lambda} = 1$, $(x)_{n,\lambda} = x(x-\lambda)\cdots(x-(n-1)\lambda),\ (n \geq 1)$.

In particular, we let

$$e_\lambda(t) = e_\lambda^1(t) = (1+\lambda t)^{\frac{1}{\lambda}}. \tag{13}$$

In [1,2], Carlitz introduced the degenerate Bernoulli polynomials which are given by the generating function

$$\frac{t}{e_\lambda(t) - 1}e_\lambda^x(t) = \sum_{n=0}^{\infty} \beta_{n,\lambda}(x)\frac{t^n}{n!}. \tag{14}$$

Also, he considered the degenerate Euler polynomials given by [1,2]

$$\frac{2}{e_\lambda(t) + 1}e_\lambda^x(t) = \sum_{n=0}^{\infty} \mathcal{E}_{n,\lambda}(x)\frac{t^n}{n!}. \tag{15}$$

Recently, Kim-Kim considered the degenerate central factorial numbers of the second kind given by [8,13]

$$\frac{1}{k!}\left(e_\lambda^{\frac{1}{2}}(t) - e_\lambda^{-\frac{1}{2}}(t)\right)^k = \sum_{n=k}^{\infty} T_\lambda(n,k)\frac{t^n}{n!}. \tag{16}$$

Note that $\lim_{\lambda\to 0} T_\lambda(n,k) = T(n,k)$.

2. Type 2 Degenerate Bernoulli Polynomials of the Second Kind

Let $\log_\lambda t$ be the compositional inverse of $e_\lambda(t)$ in (13). Then we have

$$\log_\lambda t = \frac{1}{\lambda}\left(t^\lambda - 1\right). \tag{17}$$

Note that $\lim_{\lambda\to 0}\log_\lambda t = \log t$. Now, we define the *degenerate Daehee polynomials* by

$$\frac{\log_\lambda(1+t)}{t}(1+t)^x = \sum_{n=0}^{\infty} D_{n,\lambda}(x)\frac{t^n}{n!}. \tag{18}$$

Note that $\lim_{\lambda\to 0} D_{n,\lambda}(x) = D_n(x)$, $(n \geq 0)$. In view of (8), we also consider the degenerate Bernoulli polynomials of the second kind of order α given by

$$\left(\frac{t}{\log_\lambda(1+t)}\right)^\alpha (1+t)^x = \sum_{n=0}^{\infty} b_{n,\lambda}^{(\alpha)}(x)\frac{t^n}{n!}. \tag{19}$$

Note that $\lim_{\lambda\to 0} b_{n,\lambda}^{(\alpha)}(x) = b_n^{(\alpha)}(x)$, $(n \geq 0)$. From (19), we have

$$\left(\frac{\lambda t}{(1+t)^{\frac{\lambda}{2}} - (1+t)^{-\frac{\lambda}{2}}}\right)^\alpha (1+t)^{x-\frac{\lambda\alpha}{2}} = \sum_{n=0}^{\infty} b_{n,\lambda}^{(\alpha)}(x)\frac{t^n}{n!}. \tag{20}$$

For $\alpha = r \in \mathbb{N}$, and replacing t by $e^{2t} - 1$ in (20), we get

$$\begin{aligned}
\sum_{m=0}^{\infty} b_{m,\lambda}^{(r)}(x)\frac{1}{m!}(e^{2t}-1)^m &= \left(\frac{\lambda t}{e^{t\lambda} - e^{-t\lambda}}\right)^r \frac{1}{t^r}(e^{2t}-1)^r e^{(2x-\lambda r)t} \\
&= \sum_{k=0}^{\infty} B_k^*\left(\frac{2x}{\lambda} - r\right)\frac{\lambda^k t^k}{k!}\sum_{m=0}^{\infty} S_2(m+r,r)2^{m+r}\frac{1}{\binom{m+r}{r}}\frac{t^m}{m!} \\
&= \sum_{n=0}^{\infty}\left(\sum_{m=0}^{n}\binom{n}{m}B_{n-m}^*\left(\frac{2x}{\lambda} - r\right)\lambda^{n-m}\frac{S_2(m+r,r)}{\binom{m+r}{r}}2^{m+r}\right)\frac{t^n}{n!}.
\end{aligned} \tag{21}$$

On the other hand,

$$\begin{aligned}
\sum_{m=0}^{\infty} b_{m,\lambda}^{(r)}(x)\frac{1}{m!}(e^{2t}-1)^m &= \sum_{m=0}^{\infty} b_{m,\lambda}^{(r)}(x)\sum_{n=m}^{\infty} S_2(n,m)2^n\frac{t^n}{n!} \\
&= \sum_{n=0}^{\infty}\left(\sum_{m=0}^{n} b_{m,\lambda}^{(r)}(x)2^n S_2(n,m)\right)\frac{t^n}{n!}.
\end{aligned} \tag{22}$$

From (21) and (22), we have

$$\sum_{m=0}^{n} b_{m,\lambda}^{(r)}(x)S_2(n,m) = \sum_{m=0}^{n}\binom{n}{m}B_{n-m}^*\left(\frac{2x}{\lambda} - r\right)\lambda^{n-m}\frac{S_2(m+r,r)}{\binom{m+r}{r}}2^{m+r-n}. \tag{23}$$

Now, we define the *type* 2 *degenerate Bernoulli polynomials of the second kind* by

$$\frac{(1+t)-(1+t)^{-1}}{\log_\lambda(1+t)}(1+t)^x = \sum_{n=0}^{\infty} b_{n,\lambda}^*(x)\frac{t^n}{n!}. \tag{24}$$

When $x = 0$, $b_{n,\lambda}^* = b_{n,\lambda}^*(0)$ are called the type 2 degenerate Bernoulli numbers of the second kind. Note that $\lim_{\lambda\to 0} b_{n,\lambda}^*(x) = b_n^*(x)$, where $b_n^*(x)$ are the type 2 Bernoulli polynomials of the second kind given by

$$\frac{(1+t)-(1+t)^{-1}}{\log(1+t)}(1+t)^x = \sum_{n=0}^{\infty} b_n^*(x)\frac{t^n}{n!}.$$

From (19) and (24), we note that

$$\begin{aligned}\frac{(1+t)-(1+t)^{-1}}{\log_\lambda(1+t)}(1+t)^x &= \frac{t}{\log_\lambda(1+t)}(1+t)^x\left(1+\frac{1}{1+t}\right)\\ &= \frac{t}{\log_\lambda(1+t)}(1+t)^x + \frac{t}{\log_\lambda(1+t)}(1+t)^{x-1}\\ &= \sum_{n=0}^{\infty}\left(b_{n,\lambda}^{(1)}(x)+b_{n,\lambda}^{(1)}(x-1)\right)\frac{t^n}{n!}.\end{aligned} \tag{25}$$

Therefore, we obtain the following theorem.

Theorem 1. *For $n \geq 0$, we have*

$$b_{n,\lambda}^{*}(x) = b_{n,\lambda}^{(1)}(x) + b_{n,\lambda}^{(1)}(x-1).$$

Moreover,

$$\sum_{m=0}^{n} b_{m,\lambda}^{(r)}(x)S_2(n,m) = \sum_{m=0}^{n}\binom{n}{m}B_{n-m}^{*}\left(\frac{2x}{\lambda}-r\right)\lambda^{n-m}\frac{S_2(m+r,r)}{\binom{m+r}{r}}2^{m+r-n},$$

where r is a positive integer.

Now, we observe that

$$\begin{aligned}\frac{(1+t)-(1+t)^{-1}}{\log_\lambda(1+t)}(1+t)^x &= \sum_{l=0}^{\infty} b_{l,\lambda}^{*}\frac{t^l}{l!}\sum_{m=0}^{\infty}(x)_m\frac{t^m}{m!}\\ &= \sum_{n=0}^{\infty}\left(\sum_{l=0}^{n}\binom{n}{l}b_{l,\lambda}^{*}(x)_{n-l}\right)\frac{t^n}{n!},\end{aligned} \tag{26}$$

where $(x)_0 = 1$, $(x)_n = x(x-1)\cdots(x-n+1)$, $(n \geq 1)$. From (24) and (26), we get

$$b_{n,\lambda}^{*}(x) = \sum_{l=0}^{n}\binom{n}{l}b_{l,\lambda}^{*}(x)_{n-l}, \quad (n \geq 0). \tag{27}$$

For $\alpha \in \mathbb{R}$, let us define the *type 2 degenerate Bernoulli polynomials of the second kind of order* α by

$$\left(\frac{(1+t)-(1+t)^{-1}}{\log_\lambda(1+t)}\right)^{\alpha}(1+t)^x = \sum_{n=0}^{\infty} b_{n,\lambda}^{*(\alpha)}(x)\frac{t^n}{n!} \tag{28}$$

When $x = 0$, $b_{n,\lambda}^{*(\alpha)} = b_{n,\lambda}^{*(\alpha)}(0)$ are called the type 2 degenerate Bernoulli numbers of the second kind of order α.

Let $\alpha = k \in \mathbb{N}$. Then we have

$$\sum_{n=0}^{\infty} b_{n,\lambda}^{*(k)}(x)\frac{t^n}{n!} = \left(\frac{(1+t)-(1+t)^{-1}}{\log_\lambda(1+t)}\right)^{k}(1+t)^x. \tag{29}$$

By replacing t by $e_\lambda(t)-1$ in (29), we get

$$\begin{aligned}\frac{k!}{t^k}\frac{1}{k!}\left(e_\lambda(t)-e_\lambda^{-1}(t)\right)^k e_\lambda^x(t) &= \sum_{l=0}^{\infty} b_{l,\lambda}^{*(k)}(x)\frac{1}{l!}(e_\lambda(t)-1)^l \\ &= \sum_{l=0}^{\infty} b_{l,\lambda}^{*(k)}(x)\sum_{n=l}^{\infty} S_{2,\lambda}(n,l)\frac{t^n}{n!} \\ &= \sum_{n=0}^{\infty}\left(\sum_{l=0}^{n} b_{l,\lambda}^{*(k)}(x)S_{2,\lambda}(n,l)\right)\frac{t^n}{n!},\end{aligned} \tag{30}$$

where $S_{2,\lambda}(n,l)$ are the degenerate Stirling numbers of the second kind given by [6]

$$\frac{1}{k!}(e_\lambda(t)-1)^k = \sum_{n=k}^{\infty} S_{2,\lambda}(n,k)\frac{t^n}{n!}. \tag{31}$$

On the other hand, we also have

$$\begin{aligned}\frac{k!}{t^k}\frac{1}{k!}\left(e_\lambda(t)-e_\lambda^{-1}(t)\right)^k e_\lambda^x(t) &= \frac{k!}{t^k}\frac{1}{k!}\left(e_\lambda^2(t)-1\right)^k e_\lambda^{x-k}(t) \\ &= \frac{k!}{t^k}\frac{1}{k!}\left(e_{\frac{\lambda}{2}}(2t)-1\right)^k e_\lambda^{x-k}(t) \\ &= \sum_{m=0}^{\infty} S_{2,\frac{\lambda}{2}}(m+k,k)\frac{2^{m+k}}{\binom{m+k}{k}}\frac{t^m}{m!}\sum_{l=0}^{\infty}(x-k)_{l,\lambda}\frac{t^l}{l!} \\ &= \sum_{n=0}^{\infty}\left(\sum_{m=0}^{n}\frac{\binom{n}{m}2^{m+k}}{\binom{m+k}{k}}S_{2,\frac{\lambda}{2}}(m+k,k)(x-k)_{n-m,\lambda}\right)\frac{t^n}{n!}.\end{aligned} \tag{32}$$

Therefore, by (30) and (32), we obtain the following theorem.

Theorem 2. *For $n \geq 0$, we have*

$$\sum_{l=0}^{n} b_{l,\lambda}^{*(k)}(x)S_{2,\lambda}(n,l) = \sum_{l=0}^{n}\frac{\binom{n}{l}2^{l+k}}{\binom{l+k}{k}}S_{2,\frac{\lambda}{2}}(l+k,k)(x-k)_{n-l,\lambda}.$$

In particular,

$$2^{n+k}S_{2,\frac{\lambda}{2}}(n+k,k) = \binom{n+k}{k}\sum_{l=0}^{n} b_{l,\lambda}^{*(k)}(k)S_{2,\lambda}(n,l).$$

For $\alpha \in \mathbb{R}$, we recall that the type 2 degenerate Bernoulli polynomials of order α are defined by [5,13]

$$\left(\frac{t}{e_\lambda(t)-e_\lambda^{-1}(t)}\right)^\alpha e_\lambda^x(t) = \sum_{n=0}^{\infty}\beta_{n,\lambda}^{*(\alpha)}(x)\frac{t^n}{n!}. \tag{33}$$

For $k \in \mathbb{N}$, let us take $\alpha = -k$ and replace t by $\log_\lambda(1+t)$ in (33). Then we have

$$\begin{aligned}\left(\frac{(1+t)-(1+t)^{-1}}{\log_\lambda(1+t)}\right)^k(1+t)^x &= \sum_{l=0}^{\infty}\beta_{l,\lambda}^{*(-k)}(x)\frac{1}{l\,!}\left(\log_\lambda(1+t)\right)^l \\ &= \sum_{l=0}^{\infty}\beta_{l,\lambda}^{*(-k)}(x)\sum_{n=l}^{\infty}S_{1,\lambda}(n.l)\frac{t^n}{n!} \\ &= \sum_{n=0}^{\infty}\left(\sum_{l=0}^{n}\beta_{l,\lambda}^{*(-k)}S_{1,\lambda}(n.l)\right)\frac{t^n}{n!},\end{aligned} \tag{34}$$

where $S_{1,\lambda}(n,l)$ are the degenerate Stirling numbers of the first kind given by

$$\frac{1}{k!}(\log_\lambda(1+t))^k = \sum_{n=k}^{\infty} S_{1,\lambda}(n,k)\frac{t^n}{n!}. \tag{35}$$

Note here that $\lim_{\lambda\to 0} S_{1,\lambda}(n,l) = S_1(n,l)$. Therefore, by (26) and (34), we obtain the following theorem.

Theorem 3. *For $n \geq 0$ and $k \in \mathbb{N}$, we have*

$$b_{n,\lambda}^{*(k)}(x) = \sum_{l=0}^{n} \beta_{l,\lambda}^{*(-k)}(x) S_{1,\lambda}(n,l).$$

We observe that

$$\begin{aligned}
\frac{1}{k!}t^k &= \frac{1}{k!}\left((1+t)^{\frac{1}{2}} - (1+t)^{-\frac{1}{2}}\right)^k (1+t)^{\frac{k}{2}} \\
&= \frac{1}{k!}\left(e_\lambda^{\frac{1}{2}}(\log_\lambda(1+t)) - e_\lambda^{-\frac{1}{2}}(\log_\lambda(1+t))\right)^k (1+t)^{\frac{k}{2}} \\
&= \sum_{l=k}^{\infty} T_\lambda(l,k)\frac{1}{l!}(\log_\lambda(1+t))^l \sum_{r=0}^{\infty} \binom{\frac{k}{2}}{r}_{\!r} \frac{t^r}{r!} \\
&= \sum_{l=k}^{\infty} T_\lambda(l,k) \sum_{m=l}^{\infty} S_{1,\lambda}(m,l)\frac{t^m}{m!} \sum_{r=0}^{\infty} \left(\frac{k}{2}\right)_r \frac{t^r}{r!} \\
&= \sum_{m=k}^{\infty}\sum_{l=k}^{m} T_\lambda(l,k) S_{1,\lambda}(m,l)\frac{t^m}{m!} \sum_{r=0}^{\infty} \left(\frac{k}{2}\right)_r \frac{t^r}{r!} \\
&= \sum_{n=k}^{\infty}\left(\sum_{m=k}^{n}\sum_{l=k}^{m} T_\lambda(l,k) S_{1,\lambda}(m,l)\binom{n}{m}\left(\frac{k}{2}\right)_{n-m}\right)\frac{t^n}{n!}.
\end{aligned} \tag{36}$$

On the other hand,

$$\begin{aligned}
\frac{1}{k!}t^k &= \left(\frac{t}{\log_\lambda(1+t)}\right)^k \frac{1}{k!}(\log_\lambda(1+t))^k \\
&= \sum_{l=0}^{\infty} b_{l,\lambda}^{(k)}\frac{t^l}{l!} \sum_{m=k}^{\infty} S_{1,\lambda}(m,k)\frac{t^m}{m!} \\
&= \sum_{n=k}^{\infty}\left(\sum_{m=k}^{n} S_{1,\lambda}(m,k) b_{n-m,\lambda}^{(k)}\binom{n}{m}\right)\frac{t^n}{n!}.
\end{aligned} \tag{37}$$

Therefore, by (36) and (37), we obtain the following theorem.

Theorem 4. *For $n, k \geq 0$, we have*

$$\sum_{m=k}^{n}\sum_{l=k}^{m} T_\lambda(l,k) S_{1,\lambda}(m,l)\binom{n}{m}\left(\frac{k}{2}\right)_{n-m} = \sum_{m=k}^{n} S_{1,\lambda}(m,k) b_{n-m,\lambda}^{(k)}\binom{n}{m}.$$

3. Conclusions

In this paper, we introduced the type 2 degenerate Bernoulli polynomials of the second kind and their higher-order analogues, and studied some identities and expressions for these polynomials. Specifically, we obtained a relation between the type 2 degenerate Bernoulli polynomials of the second and the degenerate Bernoulli polynomials of the second, an identity involving higher-order analogues of those polynomials and the degenerate Stirling numbers of second kind, and an expression of higher-order analogues of those polynomials in terms of the higher-order type 2 degenerate Bernoulli polynomials and the degenerate Stirling numbers of the first kind.

In addition, we obtained an identity involving the higher-order degenerate Bernoulli polynomials of the second kind, the type 2 Bernoulli polynomials and Stirling numbers of the second kind, and an identity involving the degenerate central factorial numbers of the second kind, the degenerate Stirling numbers of the first kind and the higher-order degenerate Bernoulli polynomials of the second kind.

Next, we would like to mention three possible applications of our results. The first one is their applications to identities of symmetry. For instance, in [7] by using the p-adic fermionic integrals it was possible for us to find many symmetric identities in three variables related to degenerate Euler polynomials and alternating generalized falling factorial sums.

The second one is their applications to differential equations. Indeed, in [9] we derived an infinite family of nonlinear differential equations having the generating function of the degenerate Changhee numbers of the second kind as a solution. As a result, from those differential equations we obtained an interesting identity involving the degenerate Changhee and higher-order degenerate Changhee numbers of the second kind.

The third one is their applications to probability. For example, in [19,20] we showed that both the degenerate λ-Stirling polynomials of the second and the r-truncated degenerate λ-Stirling polynomials of the second kind appear in certain expressions of the probability distributions of appropriate random variables.

These possible applications of our results require a considerable amount of work and they should appear as separate papers. We have witnessed in recent years that studying various degenerate versions of some special polynomials and numbers are very fruitful and promising [21]. It is our plan to continue to do this line of research, as one of our near future projects.

Author Contributions: T.K. and D.S.K. conceived of the framework and structured the whole paper; D.S.K. and T.K. wrote the paper; L.-C.J. and H.Y.K. checked the results of the paper; D.S.K. and T.K. completed the revision of the article. All authors have read and agreed to the published version of the manuscript.

References

1. Carlitz, L. Degenerate Stirling, Bernoulli and Eulerian numbers. *Utilitas Math.* **1979**, *15*, 51–88.
2. Carlitz, L. A degenerate Staudt-Clausen theorem. *Arch. Math.* **1956**, *7*, 28–33. [CrossRef]
3. Dolgy, D.V.; Kim, T. Some explicit formulas of degenerate Stirling numbers associated with the degenerate special numbers and polynomials. *Proc. Jangjeon Math. Soc.* **2018**, *21*, 309–317.
4. Haroon, H.; Khan, W.A. Degenerate Bernoulli numbers and polynomials associated with degenerate Hermite polynomials. *Commun. Korean Math. Soc.* **2018**, *33*, 651–669.
5. Jang, G.-W.; Kim, T. A note on type 2 degenerate Euler and Bernoulli polynomials. *Adv. Stud. Contemp. Math.* **2019**, *29*, 147–159.
6. Kim, T. A note on degenerate Stirling polynomials of the second kind. *Proc. Jangjeon Math. Soc.* **2017**, *20*, 319–331.
7. Kim, T.; Kim, D.S. Identities of symmetry for degenerate Euler polynomials and alternating generalized falling factorial sums. *Iran. J. Sci. Technol. Trans. Sci.* **2017**, *41*, 939–949. [CrossRef]
8. Kim, T.; Kim, D.S. Degenerate central factorial numbers of the second kind. *Rev. R. Acad. Cienc. Exactas Fis. Nat. Ser. A Mat. RACSAM* **2019**, 1–9. [CrossRef]
9. Kim, T.; Kim, D.S. Differential equations associated with degenerate Changhee numbers of the second kind. *Rev. R. Acad. Cienc. Exactas Fís. Nat. Ser. A Mat. RACSAM* **2019**, *113*, 1785–1793. [CrossRef]
10. Kim, T.; Yao, Y.; Kim, D.S.; Jang, G.-W. Degenerate r-Stirling numbers and r-Bell polynomials. *Russ. J. Math. Phys.* **2018**, *25*, 44–58. [CrossRef]
11. Kim, T.; Jang, G.-W. A note on degenerate gamma function and degenerate Stirling numbers of the second kind. *Adv. Stud. Contemp. Math.* **2018**, *28*, 207–214.
12. Kim, T.; Kim, D.S. Degenerate Laplace transfrom and degenerate gamma funmction. *Russ. J. Math. Phys.* **2017**, *24*, 241–248. [CrossRef]
13. Kim, T.; Kim, D.S. A note on type 2 Changhee and Daehee polynomials. *Rev. R. Acad. Cienc. Exactas Fis. Nat. Ser. A Mat. RACSAM* **2019**, *113*, 2783–2791. [CrossRef]

14. He, Y.; Araci, S. Sums of products of Apostol-Bernoulli and Apostol-Euler polynomials. *Adv. Differ. Equ.* **2014**, *2014*, 13. [CrossRef]
15. Roman, S. The umbral calculus. In *Pure and Applied Mathematics*; Academic Press Inc.: New York, NY, USA, 1984; p. 193, ISBN 0-12-594380-6.
16. Simsek, Y. Identities and relations related to combinatorial numbers and polynomials. *Proc. Jangjeon Math. Soc.* **2017**, *20*, 127–135.
17. Simsek, Y. Identities on the Changhee numbers and Apostol-type Daehee polynomials. *Adv. Stud. Contemp. Math.* **2011**, *27*, 199–212.
18. Zhang, W.; Lin, X. Identities invoving trigonometric functions and Bernoulli numbers. *Appl. Math. Comput.* **2018**, *334*, 288–294.
19. Kim, T.; Kim, D.S.; Kim, H.Y.; Kwon, J. Degenerate Stirling polynomials of the second kind and some applications. *Symmetry* **2019**, *11*, 1046. [CrossRef]
20. Kim, T.; Kim, D.S. Some identities of extended degenerate r-central Bell polynomials arising from umbral calculus. *Rev. R. Acad. Cienc. Exactas Fís. Nat. Ser. A Mat. RACSAM* **2020**, *114*, 19. [CrossRef]
21. Kim, T.; Kim, D.S.; Kim, H.Y.; Jang, L.-C. Degenerate poly-Bernoulli numbers and polynomials. *Informatica* **2020**, *31*, 2–8.

Symmetric Properties of Carlitz's Type q-Changhee Polynomials

YunJae Kim [1], Byung Moon Kim [2] and Jin-Woo Park [3,*]

1 Department of Mathematics, Dong-A University, Busan 49315, Korea; kimholzi@gmail.com
2 Department of Mechanical System Engineering, Dongguk University, Gyeongju 38066, Korea; kbm713@dongguk.ac.kr
3 Department of Mathematics Education, Daegu University, Gyeongsan 38066, Korea
* Correspondence: a0417001@knu.ac.kr.

Abstract: Changhee polynomials were introduced by Kim, and the generalizations of these polynomials have been characterized. In our paper, we investigate various interesting symmetric identities for Carlitz's type q-Changhee polynomials under the symmetry group of order n arising from the fermionic p-adic q-integral on $\mathbb{Z}_p$.

Keywords: fermionic p-adic q-integral on $\mathbb{Z}_p$; q-Euler polynomials; q-Changhee polynomials; symmetry group

MSC: 33E20; 05A30; 11B65; 11S05

---

1. Introduction

For an odd prime number p, $\mathbb{Z}_p$, $\mathbb{Q}_p$, and $\mathbb{C}_p$ denote the ring of p-adic integers, the field of p-adic rational numbers, and the completions of algebraic closure of $\mathbb{Q}_p$, respectively, throughout this paper.

The p-adic norm is normalized as $|p|_p = \frac{1}{p}$, and let q be an indeterminate in $\mathbb{C}_p$ with $|q-1|_p < p^{-\frac{1}{p-1}}$. The q-analogue of number x is defined as

$$[x]_q = \frac{1-q^x}{1-q}. \tag{1}$$

Note that $\lim_{q\to 1}[x]_q = x$ for each $x \in \mathbb{Z}_p$.

Let $C(\mathbb{Z}_p) = \{f | f : \mathbb{Z}_p \longrightarrow \mathbb{R} \text{ is continuous}\}$. Then, a fermionic p-adic q-integral of f $(\in C(\mathbb{Z}_p))$ is defined by Kim as [1–6] :

$$\begin{aligned} I_{-q}(f) &= \int_{\mathbb{Z}_p} f(x)d\mu_{-q}(x) = \lim_{N\to\infty} \sum_{x=0}^{p^N-1} f(x)\mu_{-q}\left(x+p^N\mathbb{Z}_p\right) \\ &= \lim_{N\to\infty} \frac{1}{[p^N]_{-q}} \sum_{x=0}^{p^N-1} f(x)(-q)^x \\ &= \lim_{N\to\infty} \frac{[2]_q}{2} \sum_{x=0}^{p^N-1} f(x)(-q)^x. \end{aligned} \tag{2}$$

On the other hand, it is well known that the Euler polynomial $E_n(x)$ is given by the Appell sequence with $g(t) = \frac{1}{2}(e^t + 1)$, giving the the generating function

$$\frac{2}{e^t+1}e^{xt} = \sum_{n=0}^{\infty} E_n(x)\frac{t^n}{n!},$$

(see [7–17]). In particular, if $x = 0$, $E_n = E_n(0)$ $(n \in \mathbb{N})$ is called the Euler number.

As a q-analogue of Euler polynomials, the Carlitz's type q-Euler polynomial $\mathcal{E}_{n,q}(x)$ is defined by

$$\sum_{n=0}^{\infty} \mathcal{E}_{n,q}(x)\frac{t^n}{n!} = \int_{\mathbb{Z}_p} e^{[x+y]_q t} d\mu_{-q}(y), \tag{3}$$

(see [2,13–17]). In particular, if $x = 0$, $\mathcal{E}_{n,q} = \mathcal{E}_{n,q}(0)$ is called the q-Euler number.

By (3), the Carlitz's type q-Euler polynomial $\mathcal{E}_{n,q}(x)$ is obtained as

$$\mathcal{E}_{n,q}(x) = \int_{\mathbb{Z}_p} [x+y]_q^n d\mu_{-q}(y), \ (n \geq 0). \tag{4}$$

From the fermionic p-adic q-integral on $\mathbb{Z}_p$, the degenerate q-Euler polynomial $\mathcal{E}_{n,\lambda,q}(x)$ is defined as [16]:

$$\sum_{n=0}^{\infty} \mathcal{E}_{n,\lambda,q}(x)\frac{t^n}{n!} = \int_{\mathbb{Z}_p} (1+\lambda t)^{\frac{[x+y]_q}{\lambda}} d\mu_{-q}(y). \tag{5}$$

By the binomial expansion of $(1+\lambda t)^{\frac{[x+y]_q}{\lambda}}$, we get

$$\int_{\mathbb{Z}_p} (1+\lambda t)^{\frac{[x+y]_q}{\lambda}} d\mu_{-q}(y) = \sum_{n=0}^{\infty} \lambda^n \int_{\mathbb{Z}_p} \left(\frac{[x+y]_q}{\lambda}\right)_n d\mu_{-q}(y)\frac{t^n}{n!}, \tag{6}$$

where $(\alpha)_n = \alpha(\alpha-1)\cdots(\alpha-n+1)$ for $n \in \mathbb{N}$, and by (5) and (6), we have

$$\mathcal{E}_{n,\lambda,q}(x) = \lambda^n \int_{\mathbb{Z}_p} \left(\frac{[x+y]_q}{\lambda}\right)_n d\mu_{-q}(y), \ (n \in \mathbb{N}). \tag{7}$$

Since

$$(\alpha)_n = \alpha(\alpha-1)\cdots(\alpha-n+1) = \sum_{l=0}^{n} S_1(n,l)\,\alpha^l, \tag{8}$$

$$\begin{aligned}\mathcal{E}_{n,\lambda,q}(x) =& \lambda^n \sum_{l=0}^{n} S_1(n,l) \int_{\mathbb{Z}_p} \left(\frac{[x+y]_q}{\lambda}\right)^l d\mu_{-q}(y) \\ =& \sum_{l=0}^{n} \lambda^{n-l} S_1(n,l)\mathcal{E}_{l,q}(x),\end{aligned}$$

where $S_1(n,m)$ is the Stirling number of the first kind (see [2,7,8,12,17,18]).

Now, we apply these polynomials to Changhee polynomials, introduced by Kim et al. [19]. The Changhee polynomial of the first kind $Ch_n(x)$ is defined by the generating function to be

$$\begin{aligned}\sum_{n=0}^{\infty} Ch_n(x)\frac{t^n}{n!} =& \int_{\mathbb{Z}_p} (1+t)^{x+y} d\mu_{-1}(y) \\ =& \frac{2}{2+t}(1+t)^x.\end{aligned} \tag{9}$$

(see [20,21]).

In view point of (3) and (9), Carlitz's type q-Changhee polynomial $Ch_{n,q}(x)$ is defined by

$$\sum_{n=0}^{\infty} Ch_{n,q}(x)\frac{t^n}{n!} = \int_{\mathbb{Z}_p} (1+t)^{[x+y]_q} d\mu_{-q}(y), \tag{10}$$

(see [18,22]).

By the binomial expansion of $(1+t)^{[x+y]_q}$,

$$\begin{aligned}\sum_{n=0}^{\infty} Ch_{n,q}(x)\frac{t^n}{n!} &= \int_{\mathbb{Z}_p} (1+t)^{[x+y]_q} d\mu_{-q}(y) \\ &= \sum_{n=0}^{\infty} \int_{\mathbb{Z}_p} \left([x+y]_q\right)_n d\mu_{-q}(y)\frac{t^n}{n!},\end{aligned} \tag{11}$$

and so the equation (10) and (11) yield the following:

$$Ch_{n,q}(x) = \int_{\mathbb{Z}_p} \left([x+y]_q\right)_n d\mu_{-q}(y), \tag{12}$$

(see [20,21]).

In the past decade, many different generalizations of Changhee polynomials have been studied (see [19,20,22–32]), and the relationship between important combinatorial polynomials and those polynomials was found.

Symmetric identities of special polynomials are important and interesting in number theory, pure and applied mathematics. Symmetric identities of many different polynomials were investigated in [5,10,14,16,32–39]. In particular, C. Cesarano [40] presented some techniques regarding the generating functions used, and these identities can be applicable to the theory of porous materials [41].

In current paper, we construct symmetric identities for the Carlitz's type q-Changhee polynomials under the symmetry group of order n arising from the fermionic p-adic q-integral on $\mathbb{Z}_p$, and the proof methods which was used in the Kim's previous researches are also used as good tools in this paper (see [5,10,14,16,32–39]).

2. Symmetric Identities for the Carlitz's Type q-Changhee Polynomials

Let $t \in \mathbb{C}_p$ with $|t|_p < p^{-\frac{1}{p-1}}$, and let S_n be the symmetry group of degree n. For positive integers $w_1, w_2, \ldots, w_n$ with $w_i \equiv 1 \pmod 2$ for each $i = 1, 2, \ldots n$, we consider the following integral equation for the fermionic p-adic q-integral on $\mathbb{Z}_p$;

$$\begin{aligned}&\int_{\mathbb{Z}_p} (1+t)^{\left[(\prod_{i=1}^{n-1} w_i)y+(\prod_{i=1}^{n} w_i)x+w_n \sum_{i=1}^{n-1}\left(\prod_{\substack{j=1\\ j\neq i}}^{n-1} w_j\right)k_i\right]_q} d\mu_{-q^{w_1w_2\cdots w_{n-1}}}(y) \\ &= \frac{[2]_{q^{w_1\cdots w_{n-1}}}}{2} \lim_{N\to\infty} \sum_{m=0}^{w_n-1}\sum_{y=0}^{p^N-1} (1+t)^{\left[(\prod_{i=1}^{n-1} w_i)(m+w_ny)+(\prod_{i=1}^{n} w_i)x+w_n \sum_{i=1}^{n}\left(\sum_{\substack{j=1\\ j\neq i}}^{n-1} w_j\right)k_i\right]_q} \\ &\quad \times (-1)^{m+w_ny} q^{w_1w_2\cdots w_{n-1}(m+w_ny)}.\end{aligned} \tag{13}$$

From (13), we get

$$\frac{2}{[2]_{q^{w_1 w_2 \cdots w_{n-1}}}} \prod_{m=1}^{n-1} \sum_{k_m=0}^{w_m-1} (-1)^{\sum_{i=1}^{n-1} k_i} q^{w_n \sum_{j=1}^{n-1} \left(\prod_{\substack{i=1 \\ i \neq j}}^{n-1} w_i \right) k_j}$$
$$\times \int_{\mathbb{Z}_p} (1+t)^{\left[(\prod_{i=1}^{n-1} w_i) y + (\prod_{i=1}^{n} w_i) x + w_n \sum_{i=1}^{n-1} \left(\prod_{\substack{j=1 \\ j \neq i}}^{n-1} w_j \right) k_i \right]_q} d\mu_{-q^{w_1 w_2 \cdots w_{n-1}}}(y)$$

$$= \lim_{N \to \infty} \prod_{m=1}^{n-1} \sum_{k_m=0}^{w_m-1} \sum_{l=0}^{w_n-1} \sum_{y=0}^{p^N-1} (1+t)^{\left[(\prod_{i=1}^{n-1} w_i)(m + w_n y) + (\prod_{i=1}^{n} w_i) x + w_n \sum_{i=1}^{n} \left(\sum_{\substack{j=1 \\ j \neq i}}^{n-1} w_j \right) k_i \right]_q}$$
$$\times (-1)^{\sum_{i=1}^{n-1} k_i + l + y} q^{\left(\sum_{j=1}^{n-1} \right) l + \left(\prod_{j=1}^{n} w_j \right) y + w_n \sum_{j=1}^{n-1} \left(\prod_{\substack{i=1 \\ i \neq j}} w_i \right) k_i}. \tag{14}$$

If we put

$$F(w_1, w_2, \ldots, w_n) = \frac{2}{[2]_{q^{w_1 w_2 \cdots w_{n-1}}}} \prod_{m=1}^{n-1} \sum_{k_m=0}^{w_m-1} (-1)^{\sum_{i=1}^{n-1} k_i} q^{w_n \sum_{j=1}^{n-1} \left(\prod_{\substack{i=1 \\ i \neq j}}^{n-1} w_i \right) k_j}$$
$$\times \int_{\mathbb{Z}_p} (1+t)^{\left[(\prod_{i=1}^{n-1} w_i) y + (\prod_{i=1}^{n} w_i) x + w_n \sum_{i=1}^{n-1} \left(\prod_{\substack{j=1 \\ j \neq i}}^{n-1} w_j \right) k_i \right]_q} d\mu_{-q^{w_1 w_2 \cdots w_{n-1}}}(y), \tag{15}$$

then, by (14), we know that $F(w_1, w_2, \ldots, w_n)$ is invariant for any permutation $\sigma \in S_n$.

Hence, by (14) and (15), we obtain the following theorem.

Theorem 1. *Let $w_1, w_2, \ldots, w_n$ be positive odd integers. For any $\sigma \in S_n$, $F(w_{\sigma(1)}, w_{\sigma(2)}, \ldots, w_{\sigma(n)})$ have the same value.*

By (1), we know that

$$\left[\prod_{i=1}^{n-1} w_i \right]_q \left[y + w_n x + w_n \sum_{i=1}^{n-1} \frac{k_i}{w_i} \right]_{q^{w_1 w_2 \cdots w_{n-1}}} = \left[\left(\prod_{i=1}^{n-1} w_i \right) y + \left(\prod_{i=1}^{n} w_i \right) x + w_n \sum_{i=1}^{n-1} \left(\prod_{\substack{j=1 \\ j \neq i}}^{n-1} w_j \right) k_i \right]_q. \tag{16}$$

From (5) and (16), we derive the following identities.

$$\begin{aligned}
&\int_{\mathbb{Z}_p}(1+t)^{\left[(\prod_{i=1}^{n-1}w_i)y+(\prod_{i=1}^{n}w_i)x+w_n\sum_{i=1}^{n-1}\left(\prod_{\substack{j=1\\j\neq i}}^{n-1}w_j\right)k_i\right]_q}d\mu_{-q^{w_1w_2\cdots w_{n-1}}}(y)\\
&=(1+t)^{[\prod_{i=1}^{n-1}w_i]_q}\int_{\mathbb{Z}_p}(1+t)^{\left[y+w_nx+w_n\sum_{i=1}^{n-1}\frac{k_i}{w_i}\right]_{q^{w_1w_2\cdots w_{n-1}}}}d\mu_{-q^{w_1w_2\cdots w_{n-1}}}(y)\\
&=\left(\sum_{l=0}^{\infty}\binom{[\prod_{i=1}^{n-1}w_i]_q}{l}t^l\right)\left(\sum_{m=0}^{\infty}Ch_{m,q^{w_1w_2\cdots w_{n-1}}}\left(w_nx+w_n\sum_{i=1}^{n-1}\frac{k_i}{w_i}\right)\frac{t^m}{m!}\right)\\
&=\sum_{m=0}^{\infty}\left(\sum_{r=0}^{m}\left(\left[\prod_{i=1}^{n-1}w_i\right]_q\right)_{m-r}\binom{m}{r}Ch_{r,q^{w_1w_2\cdots w_{n-1}}}\left(w_nx+w_n\sum_{i=1}^{n-1}\frac{k_i}{w_i}\right)\right)\frac{t^m}{m!},
\end{aligned}\tag{17}$$

for each positive integer n. Thus, by Theorem 1 and (17), we obtain the following corollary.

Corollary 1. *Let $w_1, w_2, \ldots, w_n$ be positive integers with $w_i \equiv 1 \pmod 2$ for each $i = 1, 2, \ldots, n$, and let m be a nonnegative integer. Then, for any permutation $\tau \in S_n$,*

$$\begin{aligned}
&\frac{2}{[2]_{q^{w_{\tau(1)}w_{\tau(2)}\cdots w_{\tau(n)}}}}\sum_{r=0}^{m}\prod_{l=1}^{n-1}\sum_{k_l=0}^{w_{\tau(l)}-1}(-1)^{\sum_{i=1}^{n-1}k_i}q^{w_{\tau(n)}\sum_{i=1}^{n-1}\left(\sum_{\substack{j=1\\j\neq i}}^{n-1}w_{\tau(j)}\right)k_i}\\
&\times\left(\left[\prod_{i=1}^{n-1}w_{\tau(i)}\right]_q\right)_{m-r}\binom{m}{r}Ch_{r,q^{w_{\tau(1)}w_{\tau(2)}\cdots w_{\tau(n-1)}}}\left(w_{\tau(n)}x+w_{\tau(n)}\sum_{i=1}^{n-1}\frac{k_i}{w_{\tau(i)}}\right)
\end{aligned}$$

have the same expressions.

Note that, by the definition of $[x]_q$,

$$\begin{aligned}
&\left[y+w_nx+w_n\sum_{i=1}^{n-1}\frac{k_i}{w_i}\right]_{q^{w_1w_2\cdots w_{n-1}}}\\
&=\frac{[w_n]_q}{\left[\prod_{i=1}^{n-1}w_i\right]_q}\left[\sum_{i=1}^{n-1}\left(\prod_{\substack{j=1\\j\neq i}}^{n-1}w_j\right)k_i\right]_{q^{w_n}}+q^{w_n\sum_{i=1}^{n-1}\left(\prod_{\substack{j=1\\j\neq i}}^{n-1}w_j\right)k_i}[y+w_nx]_{q^{w_1w_2\cdots w_{n-1}}}.
\end{aligned}\tag{18}$$

By (12), we get

$$\begin{aligned}
&Ch_{m,q^{w_1w_2\cdots w_{n-1}}}\left(w_nx+w_n\sum_{i=1}^{n-1}\frac{k_i}{w_i}\right)\\
&=\int_{\mathbb{Z}_p}\left(\left[y+w_nx+w_n\sum_{i=1}^{n-1}\frac{k_i}{w_i}\right]_{q^{w_1w_2\cdots w_{n-1}}}\right)_m d\mu_{-q^{w_1w_2\cdots w_{n-1}}},
\end{aligned}\tag{19}$$

and by (8) and (18),

$$\left(\left[y+w_n x+w_n\sum_{i=1}^{n-1}\frac{k_i}{w_i}\right]_{q^{w_1w_2\cdots w_{n-1}}}\right)_m$$

$$=\left(\frac{[w_n]_q}{\left[\prod_{i=1}^{n-1}w_i\right]_q}\left[\sum_{i=1}^{n-1}\left(\prod_{\substack{j=1\\ j\neq i}}^{n-1}w_j\right)k_i\right]_{q^{w_n}}+q^{w_n\sum_{i=1}^{n-1}\left(\prod_{j=1, j\neq i}^{n-1}w_j\right)k_i}[y+w_nx]_{q^{w_1w_2\cdots w_{n-1}}}\right)_m$$

$$=\sum_{l=0}^{m}S_1(m,l)\left(\frac{[w_n]_q}{\left[\prod_{i=1}^{n-1}w_i\right]_q}\left[\sum_{i=1}^{n-1}\left(\prod_{\substack{j=1\\ j\neq i}}^{n-1}w_j\right)k_i\right]_{q^{w_n}}+q^{w_n\sum_{i=1}^{n-1}\left(\prod_{j=1, j\neq i}^{n-1}w_j\right)k_i}[y+w_nx]_{q^{w_1w_2\cdots w_{n-1}}}\right)^l \tag{20}$$

$$=\sum_{l=0}^{m}S_1(m,l)\sum_{i=1}^{l}\binom{l}{i}\left(\frac{[w_n]_q}{\left[\prod_{i=1}^{n-1}w_i\right]_q}\right)^{l-i}\left[\sum_{i=1}^{n-1}\left(\prod_{\substack{j=1\\ j\neq i}}^{n-1}w_j\right)k_i\right]_{q^{w_n}}^{l-i}$$

$$\times q^{iw_n\sum_{i=1}^{n-1}\left(\prod_{j=1, j\neq i}^{n-1}w_j\right)k_i}[y+w_nx]^i_{q_1^{w}w_2\cdots w_{n-1}}.$$

From (4), (19) and (20), we have

$$Ch_{m,q^{w_1w_2\cdots w_{n-1}}}\left(w_nx+w_n\sum_{i=1}^{n-1}\frac{k_i}{w_i}\right)$$

$$=\sum_{l=0}^{m}S_1(m,l)\sum_{i=1}^{l}\binom{l}{i}\left(\frac{[w_n]_q}{\left[\prod_{i=1}^{n-1}w_i\right]_q}\right)^{l-i}\left[\sum_{i=1}^{n-1}\left(\prod_{\substack{j=1\\ j\neq i}}^{n-1}w_j\right)k_i\right]_{q^{w_n}}^{l-i}$$

$$\times q^{iw_n\sum_{i=1}^{n-1}\left(\prod_{j=1, j\neq i}^{n-1}w_j\right)k_i}\int_{\mathbb{Z}_p}[y+w_nx]^i_{q^{w_1w_2\cdots w_{n-1}}}\,d\mu_{-q^{w_1w_2\cdots w_{n-1}}}(y) \tag{21}$$

$$=\sum_{l=0}^{m}S_1(m,l)\sum_{i=1}^{l}\binom{l}{i}\left(\frac{[w_n]_q}{\left[\prod_{i=1}^{n-1}w_i\right]_q}\right)^{l-i}\left[\sum_{i=1}^{n-1}\left(\prod_{\substack{j=1\\ j\neq i}}^{n-1}w_j\right)k_i\right]_{q^{w_n}}^{l-i}$$

$$\times q^{iw_n\sum_{i=1}^{n-1}\left(\prod_{j=1, j\neq i}^{n-1}w_j\right)k_i}Ch_{i,q^{w_1w_2\cdots w_{n-1}}}(w_nx).$$

From (21), we have

$$\begin{aligned}
&\frac{2}{[2]_{q^{w_1w_2\cdots w_n}}}\sum_{r=0}^{m}\prod_{l=1}^{n-1}\sum_{k_l=0}^{w_l-1}(-1)^{\sum_{i=1}^{n-1}k_i}q^{w_n\sum_{i=1}^{n-1}\left(\sum_{\substack{j=1\\j\neq i}}^{n-1}w_j\right)k_i}\\
&\times\left(\left[\prod_{i=1}^{n-1}w_i\right]_q\right)_{m-r}\binom{m}{r}Ch_{r,q^{w_1w_2\cdots w_{n-1}}}\left(w_nx+w_n\sum_{i=1}^{n-1}\frac{k_i}{w_i}\right)\\
=&\frac{2}{[2]_{q^{w_1w_2\cdots w_n}}}\sum_{r=0}^{m}\prod_{l=1}^{n-1}\sum_{k_l=0}^{w_l-1}(-1)^{\sum_{i=1}^{n-1}k_i}q^{w_n\sum_{i=1}^{n-1}\left(\sum_{\substack{j=1\\j\neq i}}^{n-1}w_j\right)k_i}\left(\left[\prod_{i=1}^{n-1}w_i\right]_q\right)_{m-r}\binom{m}{r}\\
&\times\sum_{p=0}^{r}S_1(r,p)\sum_{i=1}^{l}\binom{l}{i}\left(\frac{[w_n]_q}{\left[\prod_{i=1}^{n-1}w_i\right]_q}\right)^{p-i}\left[\sum_{i=1}^{n-1}\left(\prod_{\substack{j=1\\j\neq i}}^{n-1}w_j\right)k_i\right]_{q^{w_n}}^{p-i}\\
&\times q^{iw_n\sum_{i=1}^{n-1}\left(\prod_{\substack{j=1\\j\neq i}}^{n-1}w_j\right)k_i}Ch_{i,q^{w_1w_2\cdots w_{n-1}}}(w_nx)\\
=&\sum_{r=0}^{m}\sum_{l=0}^{r}\sum_{i=1}^{l}S_1(r,l)\binom{m}{r}\binom{l}{i}\left(\frac{[w_n]_q}{\left[\prod_{i=1}^{n-1}w_i\right]_q}\right)^{p-i}\left(\left[\prod_{i=1}^{n-1}w_i\right]_q\right)_{m-r}Ch_{i,q^{w_1w_2\cdots w_{n-1}}}(w_nx)\\
&\times\frac{2}{[2]_{q^{w_1w_2\cdots w_n}}}\prod_{l=1}^{n-1}\sum_{k_l=0}^{w_l-1}(-1)^{\sum_{i=1}^{n-1}k_i}q^{(1+i)w_n\sum_{i=1}^{n-1}\left(\sum_{\substack{j=1\\j\neq i}}^{n-1}w_j\right)k_i}\left[\sum_{i=1}^{n-1}\left(\prod_{\substack{j=1\\j\neq i}}^{n-1}w_j\right)k_i\right]_{q^{w_n}}^{p-i}\\
=&\sum_{r=0}^{m}\sum_{l=0}^{r}\sum_{i=1}^{l}S_1(r,l)\binom{m}{r}\binom{l}{i}\\
&\times\left(\frac{[w_n]_q}{\left[\prod_{i=1}^{n-1}w_i\right]_q}\right)^{p-i}\left(\left[\prod_{i=1}^{n-1}w_i\right]_q\right)_{m-r}Ch_{i,q^{w_1w_2\cdots w_{n-1}}}(w_nx)F_{n,q^{w_n}}(w_1,\ldots,w_{n-1}|i+1),
\end{aligned}\tag{22}$$

where

$$F_{n,q}(w_1,\ldots,w_{n-1}|i)=\frac{2}{[2]_{q^{w_1w_2\cdots w_n}}}\prod_{l=1}^{n-1}\sum_{k_l=0}^{w_l-1}(-1)^{\sum_{i=1}^{n-1}k_i}q^{i\sum_{i=1}^{n-1}\left(\sum_{\substack{j=1\\j\neq i}}^{n-1}w_j\right)k_i}\left[\sum_{t=1}^{n-1}\left(\prod_{\substack{j=1\\j\neq t}}^{n-1}w_j\right)k_t\right]_q^{p-i-1}$$

Theorem 2. *For each nonnegative odd integers $w_1, w_2, \ldots, w_n$ and for any permutation σ in the symmetry group of degree n, the expressions*

$$\begin{aligned}
&\sum_{r=0}^{m}\sum_{l=0}^{r}\sum_{i=1}^{l}S_1(r,l)\binom{m}{r}\binom{l}{i}\left(\frac{[w_{\sigma(n)}]_q}{\left[\prod_{i=1}^{n-1}w_{\sigma(i)}\right]_q}\right)^{p-i}\left(\left[\prod_{i=1}^{n-1}w_{\sigma(i)}\right]_q\right)_{m-r}\\
&\times Ch_{i,q^{w_{\sigma(1)}w_{\sigma(2)}\cdots w_{\sigma(n-1)}}}\left(w_{\sigma(n)}x\right)F_{n,q^{w_{\sigma(n)}}}\left(w_{\sigma(1)},\ldots,w_{\sigma(n-1)}|i+1\right)
\end{aligned}$$

3. Conclusion

The Changhee numbers are closely related with the Euler numbers, the Stirling numbers of the first kind and second kind and the harmonic numbers, and so on. Throughout this paper, we investigate that the function $F(w_{\sigma(1)}, w_{\sigma(2)}, \ldots, w_{\sigma(n)})$ for the Carlitz's type q-Changhee polynomials is invariant under the symmetry group $\sigma \in S_n$. From the invariance of $F(w_{\sigma(1)}, w_{\sigma(2)}, \ldots, w_{\sigma(n)})$, $\sigma \in S_n$, we construct symmetric identities of the Carlitz's type q-Changhee polynomials from the fermionic p-adic q-integral on $\mathbb{Z}_p$. As Bernoulli and Euler polynomials, our properties on the Carlitz's type q-Changhee polynomials play an crucial role in finding identities for numbers in algebraic number theory.

Author Contributions: All authors contributed equally to this work; All authors read and approved the final manuscript.

Acknowledgments: The authors would like to thank the referees for their valuable and detailed comments which have significantly improved the presentation of this paper.

References

1. Kim, T. q-Euler numbers and polynomials associated with p-adic q-integrals. *J. Nonlinear Math. Phys.* **2007**, *14*, 15–27. [CrossRef]
2. Kim, T. Some identities on the q-Euler polynomials of higher order and q-Stirling numbers by the fermionic p-adic integral on $\mathbb{Z}_p$. *Russ. J. Math. Phys.* **2009**, *16*, 484–491. [CrossRef]
3. Kim, T. On q-analogue of the p-adic log gamma functions and related integral. *J. Number Theory* **1999**, *76*, 320–329. [CrossRef]
4. Kim, T. q-Volkenborn integration. *Russ. J. Math. Phys.* **2002**, *9*, 288–299.
5. Kim, T. Symmetry of power sum polynomials and multivariate fermionic p-adic invariant integral on $\mathbb{Z}_p$. *Russ. J. Math. Phys.* **2009**, *16*, 93–96. [CrossRef]
6. Kim, D.S.; Kim, T. Some p-adic integrals on $\mathbb{Z}_p$ associated with trigonometric Functions. *Russ. J. Math. Phys.* **2018**, *25*, 300–308. [CrossRef]
7. Kim, T. A study on the q-Euler numbers and the fermionic q-integrals of the product of several type q-Bernstein polynomials on $\mathbb{Z}_p$. *Adv. Stud. Contemp. Math.* **2013**, *23*, 5–11.
8. Bayad, A.; Kim, T. Identities involving values of Bernstein, q-Bernoulli, and q-Euler polynomials. *Russ. J. Math. Phys.* **2011**, *18*, 133–143. [CrossRef]
9. Gaboury, S.; Tremblay, R.; Fugere, B.-J. Some explicit formulas for certain new classes of Bernoulli, Euler and Genocchi polynomials. *Proc. Jangjeon Math. Soc.* **2014**, *17*, 115–123.
10. Kim, D.S.; Kim, T. Some symmetric identites for the higher-order q-Euler polynomials related to symmetry group S_3 arising from p-adic q-fermionic integral on $\mathbb{Z}_p$. *Filomat* **2016**, *30*, 1717–1721. [CrossRef]
11. Sharma, A. q-Bernoulli and Euler numbers of higher order. *Duke Math. J.* **1958**, *25*, 343–353. [CrossRef]
12. Srivastava, H. Some generalizations and basic (or q-)extensions of the Bernoulli, Euler and Genocchi polynomials. *Appl. Math. Inf. Sci.* **2011**, *5*, 390–444.
13. Zhang, Z.; Yang, H. Some closed formulas for generalized Bernoulli-Euler numbers and polynomials. *Proc. Jangjeon Math. Soc.* **2008**, *11*, 191–198.
14. Kim, T.; Kim, D.S. Identities of symmetry for degenerate Euler polynomials and alternating generalized falling factorial sums. *Iran. J. Sci. Technol. Trans. A Sci.* **2017**, *41*, 939–949. [CrossRef]
15. Carlitz, L. q-Bernoulli and Eulerian numbers. *Trans. Amer. Math. Soc.* **1954**, *76*, 332–350.
16. Kim, D.S.; Kim, T. Symmetric identities of higher-order degenerate q-Euler polynomials. *J. Nonlinear Sci. Appl.* **2016**, *9*, 443–451. [CrossRef]
17. Comtet, L. *Advanced Combinatorics*; Reidel: Dordrecht, The Netherlands, 1974.
18. Dolgy, D.V.; Jang, G.W.; Kwon, H.I.; Kim, T. A note on Carlitzs type q-Changhee numbers and polynomials. *Adv. Stud. Contemp. Math.* **2017**, *27*, 451–459.
19. Kim, D.S.; Kim, T.; Seo, J.J. A note on Changhee polynomials and numbers. *Adv. Stud. Theor. Phys.* **2013**, *7*, 993–1003. [CrossRef]
20. Kim, T.; Kim, D.S. A note on nonlinear Changhee differential equations. *Russ. J. Math. Phys.* **2016**, *23*, 88–92. [CrossRef]

21. Kim, T.; Kim, D.S. Identities for degenerate Bernoulli polynomials and Korobov polynomials of the first kind. *Sci. China Math.* **2018**. [CrossRef]
22. Kim, B.M.; Jang, L.C.; Kim, W.; Kwon, H.I. On Carlitz's Type Modified Degenerate Changhee Polynomials and Numbers. *Discrete Dyn. Nat. Soc.* **2018**, *2018*, 9520269. [CrossRef]
23. Kim, T.; Kim, D.S. Differential equations associated with degenerate Changhee numbers of the second kind. *Rev. R. Acad. Cienc. Exactas Fis. Nat. Ser. A Math. RACSAM* **2018**. [CrossRef]
24. Moon, E.J.; Park, J.W. A note on the generalized q-Changhee numbers of higher order. *J. Comput. Anal. Appl.* **2016**, *20*, 470–479.
25. Kwon, J.; Park, J.W. On modified degenerate Changhee polynomials and numbers. *J. Nonlinear Sci. Appl.* **2015**, *18*, 295–305. [CrossRef]
26. Kim, T.; Kwon, H.I.; Seo, J.J. Degenerate q-Changhee polynomials. *J. Nonlinear Sci. Appl.* **2016**, *9*, 2389–2393. [CrossRef]
27. Kwon, H.I.; Kim, T.; Seo, J.J. A note on degenerate Changhee numbers and polynomials. *Proc. Jangjeon Math. Soc.* **2015**, *18*, 295–305.
28. Arici, S.; Ağyüz, E.; Acikgoz, M. On a q-analogue of some numbers and polynomials. *J. Ineqal. Appl.* **2015**, *2015*, 19. [CrossRef]
29. Pak, H.K.; Jeong, J.; Kang, D.J.; Rim, S.H. Changhee-Genocchi numbers and their applications. *ARS Combin.* **2018**, *136*, 153–159.
30. Kim, B.M.; Jeong, J.; Rim, S.H. Some explicit identities on Changhee-Genocchi polynomials and numbers. *Adv. Differ. Equ.* **2016**, *2016*, 202. [CrossRef]
31. Rim, S.H.; Park, J.W.; Pyo, S.S.; Kwon, J. The n-th twisted Changhee polynomials and numbers. *Bull. Korean Math. Soc.* **2015**, *52*, 741–749. [CrossRef]
32. Simsek, Y. Identities on the Changhee numbers and Apostol-type Daehee polynomials. *Adv. Stud. Contemp. Math.* **2017**, *27*, 199–212.
33. Kim, T.; Kim, D.S. Degenerate Bernstein polynomials. *Rev. R. Acad. Cienc. Exactas Fis. Nat. Ser. A Math. RACSAM* **2018**. [CrossRef]
34. Liu, C.; Bao, W. Application of Probabilistic Method on Daehee Sequences. *Eur. J. Pure Appl. Math.* **2018**, *11*, 69–78. [CrossRef]
35. Kim, D.S.; Lee, N.; Na, J.; Park, K.H. Abundant symmetry for higher-order Bernoulli polynomials (I). *Adv. Stud. Contemp. Math.* **2013**, *23*, 461–482.
36. Kim, D.S.; Lee, N.; Na, J.; Pak, K.H. Identities of symmetry for higher-order Euler polynomials in three variables (I). *Adv. Stud. Contemp. Math.* **2012**, *22*, 51–74. [CrossRef]
37. Kim, D.S.; Lee, N.; Na, J.; Pak, K.H. Identities of symmetry for higher-order Bernoulli polynomials in three variables (II). *Proc. Jangjeon Math. Soc.* **2013**, *16*, 359–378.
38. Kim, D.S. Symmetry identities for generalized twisted Euler polynomials twisted by unramified roots of unity. *Proc. Jangjeon Math. Soc.* **2012**, *15*, 303–316.
39. Kim, T.; Kim, D.S. An identity of symmetry for the degenerate Frobenius-Euler polynomials. *Math. Slovaca* **2008**, *68*, 239–243. [CrossRef]
40. Cesarano, C. Operational methods and new identities for Hermit polynomials. *Math. Model. Nat. Phenom.* **2017**, *12*, 44–50. [CrossRef]
41. Marin, M. Weak solutions in elasticity of dipolar porous materials. *Math. Probl. Eng.* **2008**, *2008*, 158908. [CrossRef]

Representation by Chebyshev Polynomials for Sums of Finite Products of Chebyshev Polynomials

Taekyun Kim [1], Dae San Kim [2], Lee-Chae Jang [3,*] and Dmitry V. Dolgy [4]

1 Department of Mathematics, Kwangwoon University, Seoul 139-701, Korea; tkkim@kw.ac.kr
2 Department of Mathematics, Sogang University, Seoul 121-742, Korea; dskim@sogang.ac.kr
3 Graduate School of Education, Konkuk University, Seoul 139-701, Korea
4 Hanrimwon, Kwangwoon University, Seoul 139-701, Korea; d_dol@mail.ru
* Correspondence: lcjang@konkuk.ac.kr

Abstract: In this paper, we consider sums of finite products of Chebyshev polynomials of the first, third, and fourth kinds, which are different from the previously-studied ones. We represent each of them as linear combinations of Chebyshev polynomials of all kinds whose coefficients involve some terminating hypergeometric functions ${}_2F_1$. The results may be viewed as a generalization of the linearization problem, which is concerned with determining the coefficients in the expansion of the product of two polynomials in terms of any given sequence of polynomials. These representations are obtained by explicit computations.

Keywords: Chebyshev polynomials of the first, second, third, and fourth kinds; sums of finite products; representation

---

1. Introduction and Preliminaries

We first fix some notations that will be used throughout this paper. For any nonnegative integer n, the falling factorial sequence $(x)_n$ and the rising factorial sequence $<x>_n$ are respectively given by:

$$(x)_n = x(x-1)\cdots(x-n+1), \quad (n \geq 1), \quad (x)_0 = 1, \tag{1}$$

$$<x>_n = x(x+1)\cdots(x+n-1), \quad (n \geq 1), \quad <x>_0 = 1. \tag{2}$$

Then, we easily see that the two factorial sequences are related by:

$$(-1)^n (x)_n = <-x>_n . \tag{3}$$

The Gauss hypergeometric function ${}_2F_1(a,b;c;x)$ is defined by:

$${}_2F_1(a,b;c;x) = \sum_{n=0}^{\infty} \frac{<a>_n <b>_n}{<c>_n} \frac{x^n}{n!}, \quad (|x| < 1). \tag{4}$$

In this paper, we only need very basic facts about Chebyshev polynomials of the first, second, third, and fourth kinds, which we recall briefly in the following. The Chebyshev polynomials belong to the family of orthogonal polynomials. We let the interested reader refer to [1–4] for more details on these.

In terms of generating functions, the Chebyshev polynomials of the first, second, third, and fourth kinds are respectively given by:

$$F_1(t,x) = \frac{1-xt}{1-2xt+t^2} = \sum_{n=0}^{\infty} T_n(x)t^n, \tag{5}$$

$$F_2(t,x) = \frac{1}{1-2xt+t^2} = \sum_{n=0}^{\infty} U_n(x)t^n, \tag{6}$$

$$F_3(t,x) = \frac{1-t}{1-2xt+t^2} = \sum_{n=0}^{\infty} V_n(x)t^n, \tag{7}$$

$$F_4(t,x) = \frac{1+t}{1-2xt+t^2} = \sum_{n=0}^{\infty} W_n(x)t^n. \tag{8}$$

They are also explicitly given by the following expressions:

$$\begin{aligned} T_n(x) &= {}_2F_1\left(-n, n; \frac{1}{2}; \frac{1-x}{2}\right) \\ &= \frac{n}{2}\sum_{l=0}^{[\frac{n}{2}]}(-1)^l \frac{1}{n-l}\binom{n-l}{l}(2x)^{n-2l}, \quad (n \geq 1), \end{aligned} \tag{9}$$

$$\begin{aligned} U_n(x) &= (n+1){}_2F_1\left(-n, n+2; \frac{3}{2}; \frac{1-x}{2}\right) \\ &= \sum_{l=0}^{[\frac{n}{2}]}(-1)^l \binom{n-l}{l}(2x)^{n-2l}, \quad (n \geq 0), \end{aligned} \tag{10}$$

$$\begin{aligned} V_n(x) &= {}_2F_1\left(-n, n+1; \frac{1}{2}; \frac{1-x}{2}\right) \\ &= \sum_{l=0}^{n}\binom{n+l}{2l}2^l(x-1)^l, \quad (n \geq 0), \end{aligned} \tag{11}$$

$$\begin{aligned} W_n(x) &= (2n+1){}_2F_1\left(-n, n+1; \frac{3}{2}; \frac{1-x}{2}\right) \\ &= (2n+1)\sum_{l=0}^{n}\frac{2^l}{2l+1}\binom{n+l}{2l}(x-1)^l, \quad (n \geq 0), \end{aligned} \tag{12}$$

The Chebyshev polynomials of all four kinds are also expressed by the Rodrigues formulas, which are given by:

$$T_n(x) = \frac{(-1)^n 2^n n!}{(2n)!}(1-x^2)^{\frac{1}{2}}\frac{d^n}{dx^n}(1-x^2)^{n-\frac{1}{2}}, \tag{13}$$

$$U_n(x) = \frac{(-1)^n 2^n (n+1)!}{(2n+1)!}(1-x^2)^{-\frac{1}{2}}\frac{d^n}{dx^n}(1-x^2)^{n+\frac{1}{2}}, \tag{14}$$

$$\begin{aligned} &(1-x)^{-\frac{1}{2}}(1+x)^{\frac{1}{2}}V_n(x) \\ &= \frac{(-1)^n 2^n n!}{(2n)!}\frac{d^n}{dx^n}(1-x)^{n-\frac{1}{2}}(1+x)^{n+\frac{1}{2}}, \end{aligned} \tag{15}$$

$$\begin{aligned} &(1-x)^{\frac{1}{2}}(1+x)^{-\frac{1}{2}}W_n(x) \\ &= \frac{(-1)^n 2^n n!}{(2n)!}\frac{d^n}{dx^n}(1-x)^{n+\frac{1}{2}}(1-x)^{n-\frac{1}{2}}. \end{aligned} \tag{16}$$

They satisfy orthogonalities with respect to various weight functions as given in the following:

$$\int_{-1}^{1}(1-x^2)^{-\frac{1}{2}}T_n(x)T_m(x)dx = \frac{\pi}{\varepsilon_n}\delta_{n,m}, \tag{17}$$

where:

$$\varepsilon_n = \begin{cases} 1, & \text{if} \quad n = 0, \\ 2, & \text{if} \quad n \geq 1, \end{cases} \tag{18}$$

$$\delta_{n,m} = \begin{cases} 0, & \text{if} \quad n \neq m, \\ 1, & \text{if} \quad n = m. \end{cases} \tag{19}$$

$$\int_{-1}^{1} (1-x^2)^{\frac{1}{2}} U_n(x) U_m(x) dx = \frac{\pi}{2} \delta_{n,m}, \tag{20}$$

$$\int_{-1}^{1} \left(\frac{1+x}{1-x}\right)^{\frac{1}{2}} V_n(x) V_m(x) dx = \pi \delta_{n,m}, \tag{21}$$

$$\int_{-1}^{1} \left(\frac{1-x}{1+x}\right)^{\frac{1}{2}} W_n(x) W_m(x) dx = \pi \delta_{n,m}. \tag{22}$$

For convenience, we let:

$$\alpha_{m,r}(x) = \sum_{i_1+\cdots+i_{r+1}=m} T_{i_1}(x) \cdots T_{i_{r+1}}(x), \quad (m, r \geq 0), \tag{23}$$

$$\beta_{m,r}(x) = \sum_{i_1+\cdots+i_{r+1}=m} V_{i_1}(x) \cdots V_{i_{r+1}}(x), \quad (m, r \geq 0), \tag{24}$$

$$\gamma_{m,r}(x) = \sum_{i_1+\cdots+i_{r+1}=m} W_{i_1}(x) \cdots W_{i_{r+1}}(x), \quad (m, r \geq 0), \tag{25}$$

Here, all the sums in (23)–(25) are over all nonnegative integers $i_1, \cdots, i_{r+1}$, with $i_1 + i_2 + \cdots + i_{r+1} = m$. Furthermore, note here that $\alpha_{m,r}(x)$, $\beta_{m,r}(x)$, $\gamma_{m,r}(x)$ all have degree m.

Further, let us put:

$$\begin{aligned} &\sum_{l=0}^{m} \sum_{i_1+\cdots+i_{r+1}=m-l} \binom{r+l}{r} x^l T_{i_1}(x) \cdots T_{i_{r+1}}(x) \\ &- \sum_{l=0}^{m-2} \sum_{i_1+\cdots+i_{r+1}=m-l-2} \binom{r+l}{r} x^l T_{i_1}(x) \cdots T_{i_{r+1}}(x), \quad (m \geq 2, r \geq 1), \end{aligned} \tag{26}$$

$$\sum_{l=0}^{m} \sum_{i_1+\cdots+i_{r+1}=l} \binom{r-1+m-l}{r-1} V_{i_1}(x) \cdots V_{i_{r+1}}(x), \quad (m \geq 0, r \geq 1), \tag{27}$$

$$\sum_{l=0}^{m} \sum_{i_1+\cdots+i_{r+1}=l} (-1)^{m-l} \binom{r-1+m-l}{r-1} W_{i_1}(x) \cdots W_{i_{r+1}}(x), \quad (m \geq 0, r \geq 1). \tag{28}$$

We considered the expression (26) in [5] and (27) and (28) in [6] and were able to express each of them in terms of the Chebyshev polynomials of all four kinds. It is amusing to note that in such expressions, some terminating hypergeometric functions ${}_2F_1$ and ${}_3F_2$ appear respectively for (26)–(28). We came up with studying the sums in (26)–(28) by observing that they are respectively equal to $\frac{1}{2^{r-1}r!} T_{m+r}^{(r)}(x)$, $\frac{1}{2^r r!} V_{m+r}^{(r)}(x)$, and $\frac{1}{2^r r!} W_{m+r}^{(r)}(x)$. Actually, these easily follow by differentiating the generating functions in (5), (7), and (8).

In this paper, we consider the expressions $\alpha_{m,r}(x)$, $\beta_{m,r}(x)$, and $\gamma_{m,r}(x)$ in (23)–(25), which are sums of finite products of Chebyshev polynomials of the first, third, and fourth kinds, respectively. Then, we express each of them as linear combinations of $T_n(x)$, $U_n(x)$, $V_n(x)$, and $W_n(x)$. Here, we remark that $\alpha_{m,r}(x)$, $\beta_{m,r}(x)$, and $\gamma_{m,r}(x)$ are expressed in terms of $U_{m-j+r}^{(r)}(x)$, $(j = 0, 1, \cdots, m)$ (see Lemmas 2 and 3) by making use of the generating function in (6). This is unlike the previous works for (26)–(28) (see [5,6]), where we showed they are respectively equal to $\frac{1}{2^{r-1}r!}T^{(r)}_{m+r}(x)$, $\frac{1}{2^r r!}V^{(r)}_{m+r}(x)$, and $\frac{1}{2^r r!}W^{(r)}_{m+r}(x)$ by exploiting the generating functions in (5), (7) and (8). Then, our results for $\alpha_{m,r}(x)$, $\beta_{m,r}(x)$, and $\gamma_{m,r}(x)$ will be found by making use of Lemmas 1 and 2, the general formulas in Propositions 1 and 2, and integration by parts. As we can notice here, generating functions play important roles in the present and the previous works in [5,6]. We would like to remark here that the technique of generating functions has been widely used not only in mathematics, but also in physics and biology. For this matter, we recommend the reader to refer to [7–9]. The next three theorems are our main results.

Theorem 1. *For any nonnegative integers m, r, the following identities hold true.*

$$\sum_{i_1+\cdots+i_{r+1}=m} T_{i_1}(x)\cdots T_{i_{r+1}}(x) = \frac{1}{r!}\sum_{s=0}^{[\frac{m}{2}]}\sum_{l=0}^{s}\frac{\varepsilon_{m-2s}(-1)^l(m+r-l)!}{l!(m-s-l)!(s-l)!}\,{}_2F_1\left(2l-m,-r-1;l-m-r;\frac{1}{2}\right)T_{m-2s}(x) \tag{29}$$

$$= \frac{1}{r!}\sum_{s=0}^{[\frac{m}{2}]}\sum_{l=0}^{s}\frac{(-1)^l(m-2s+1)(m+r-l)!}{l!(m-s+1-l)!(s-l)!}\,{}_2F_1\left(2l-m,-r-1;l-m-r;\frac{1}{2}\right)U_{m-2s}(x) \tag{30}$$

$$= \frac{1}{r!}\sum_{s=0}^{m}\sum_{l=0}^{[\frac{s}{2}]}\frac{(-1)^l(m+r-l)!}{l!\left(m-[\frac{s}{2}]-l\right)!\left([\frac{s}{2}]-l\right)!}\,{}_2F_1\left(2l-m,-r-1;l-m-r;\frac{1}{2}\right)V_{m-s}(x) \tag{31}$$

$$= \frac{1}{r!}\sum_{s=0}^{m}\sum_{l=0}^{[\frac{s}{2}]}\frac{(-1)^{s+l}(m+r-l)!}{l!\left(m-[\frac{s}{2}]-l\right)!\left([\frac{s}{2}]-l\right)!}\,{}_2F_1\left(2l-m,-r-1;l-m-r;\frac{1}{2}\right)W_{m-s}(x). \tag{32}$$

Theorem 2. *For any nonnegative integers m, r, we have the following identities.*

$$\sum_{i_1+\cdots+i_{r+1}=m} V_{i_1}(x)\cdots V_{i_{r+1}}(x) = \frac{1}{r!}\sum_{k=0}^{m}\sum_{l=0}^{[\frac{m-k}{2}]}\frac{(-1)^{m-k}\varepsilon_k(k+2s+r)!}{(s+k)!s!}\binom{r+1}{m-k-2s}{}_2F_1(-s,-s-k;-k-2s-r;1)\,T_k(x) \tag{33}$$

$$= \frac{1}{r!}\sum_{k=0}^{m}\sum_{s=0}^{[\frac{m-k}{2}]}\frac{(-1)^{m-k}(k+1)(k+2s+r)!}{(s+k+1)!s!}\binom{r+1}{m-k-2s}{}_2F_1(-s,-s-k-1;-k-2s-r;1)\,U_k(x) \tag{34}$$

$$= \frac{1}{r!}\sum_{k=0}^{m}\sum_{s=0}^{m-k}\frac{(-1)^{m-k-s}(k+r+s)!}{(k+[\frac{s+1}{2}])![\frac{s}{2}]!}\binom{r+1}{m-k-s}{}_2F_1\left(-\left[\frac{s}{2}\right],-\left[\frac{s+1}{2}\right]-k;-k-s-r;1\right)V_k(x) \tag{35}$$

$$= \frac{1}{r!}\sum_{k=0}^{m}\sum_{s=0}^{m-k}\frac{(-1)^{m-k}(k+r+s)!}{(k+[\frac{s+1}{2}])![\frac{s}{2}]!}\binom{r+1}{m-k-s}{}_2F_1\left(-\left[\frac{s}{2}\right],-\left[\frac{s+1}{2}\right]-k;-k-s-r;1\right)W_k(x) \tag{36}$$

Theorem 3. *For any nonnegative integers m, r, the following identities are valid.*

$$\sum_{i_1+\cdots+i_{r+1}=m} W_{i_1}(x)\cdots W_{i_{r+1}}(x) = \frac{1}{r!}\sum_{k=0}^{m}\sum_{l=0}^{[\frac{m-k}{2}]}\frac{\varepsilon_k(k+2s+r)!}{(s+k)!s!}\binom{r+1}{m-k-2s}{}_2F_1(-s,-s-k;-k-2s-r;1)\,T_k(x) \tag{37}$$

$$= \frac{1}{r!}\sum_{k=0}^{m}\sum_{s=0}^{[\frac{m-k}{2}]}\frac{(k+1)(k+2s+r)!}{(s+k+1)!s!}\binom{r+1}{m-k-2s}{}_2F_1(-s,-s-k-1;-k-2s-r;1)\,U_k(x) \tag{38}$$

$$= \frac{1}{r!} \sum_{k=0}^{m} \sum_{s=0}^{m-k} \frac{(-1)^s (k+r+s)!}{\left(k+\left[\frac{s+1}{2}\right]\right)! \left[\frac{s}{2}\right]!} \binom{r+1}{m-k-s} {}_2F_1\left(-\left[\frac{s}{2}\right], -\left[\frac{s+1}{2}\right]-k; -k-s-r; 1\right) V_k(x) \quad (39)$$

$$= \frac{1}{r!} \sum_{k=0}^{m} \sum_{s=0}^{m-k} \frac{(k+r+s)!}{\left(k+\left[\frac{s+1}{2}\right]\right)! \left[\frac{s}{2}\right]!} \binom{r+1}{m-k-s} {}_2F_1\left(-\left[\frac{s}{2}\right], -\left[\frac{s+1}{2}\right]-k; -k-s-r; 1\right) W_k(x) \quad (40)$$

Before moving on to the next section, we would like to say a few words on the previous works that are associated with the results in the present paper. In terms of Bernoulli polynomials, quite a few sums of finite products of some special polynomials are expressed. They include Chebyshev polynomials of all four kinds, and Bernoulli, Euler, Genocchi, Legendre, Laguerre, Fibonacci, and Lucas polynomials (see [10–16]). All of these expressions in terms of Bernoulli polynomials have been derived from the Fourier series expansions of the functions closely related to each such polynomials. Further, as for Chebyshev polynomials of all four kinds and Legendre, Laguerre, Fibonacci, and Lucas polynomials, certain sums of finite products of such polynomials are also expressed in terms of all four kinds of Chebyshev polynomials in [5,6,17,18]. Finally, the reader may want to look at [19–21] for some applications of Chebyshev polynomials.

2. Proof of Theorem 1

In this section, we will prove Theorem 1. In order to do this, we first state Propositions 1 and 2 that are needed in proving Theorems 1–3. Here, we note that the facts (a), (b), (c), and (d) in Proposition 1 are stated respectively in the Equations (24) of [22], (36) of [22], (23) of [23], and (38) of [23]. All of them follow easily from the orthogonality relations in (17) and (20)–(22), Rodrigues' formulas in (13)–(16), and integration by parts.

Proposition 1. *For any polynomial $q(x) \in \mathbb{R}[x]$ of degree n, we have the following formulas.*

(a) $q(x) = \sum_{k=0}^{n} C_{k,1} T_k(x)$, *where:*

$$C_{k,1} = \frac{(-1)^k 2^k k! \varepsilon_k}{(2k)! \pi} \int_{-1}^{1} q(x) \frac{d^k}{dx^k} (1-x^2)^{k-\frac{1}{2}} dx,$$

(b) $q(x) = \sum_{k=0}^{n} C_{k,2} U_k(x)$, *where:*

$$C_{k,2} = \frac{(-1)^k 2^{k+1} (k+1)!}{(2k+1)! \pi} \int_{-1}^{1} q(x) \frac{d^k}{dx^k} (1-x^2)^{k+\frac{1}{2}} dx,$$

(c) $q(x) = \sum_{k=0}^{n} C_{k,3} V_k(x)$, *where:*

$$C_{k,3} = \frac{(-1)^k k! 2^k}{(2k)! \pi} \int_{-1}^{1} q(x) \frac{d^k}{dx^k} (1-x)^{k-\frac{1}{2}} (1+x)^{k+\frac{1}{2}} dx,$$

(d) $q(x) = \sum_{k=0}^{n} C_{k,4} W_k(x)$, *where,*

$$C_{k,4} = \frac{(-1)^k k! 2^k}{(2k)! \pi} \int_{-1}^{1} q(x) \frac{d^k}{dx^k} (1-x)^{k+\frac{1}{2}} (1+x)^{k+\frac{1}{2}} dx,$$

The next proposition is stated and proven in [17].

Proposition 2. *For any nonnegative integers m, k, we have the following formulas:*

(a)
$$\int_{-1}^{1}(1-x^2)^{k-\frac{1}{2}}x^m dx = \begin{cases} 0, & \text{if } m \equiv 1 \pmod 2, \\ \frac{m!(2k)!\pi}{2^{m+2k}\left(\frac{m}{2}+k\right)!\left(\frac{m}{2}\right)!k!}, & \text{if } m \equiv 0 \pmod 2. \end{cases}$$

(b)
$$\int_{-1}^{1}(1-x^2)^{k+\frac{1}{2}}x^m dx = \begin{cases} 0, & \text{if } m \equiv 1 \pmod 2, \\ \frac{m!(2k+2)!\pi}{2^{m+2k+2}\left(\frac{m}{2}+k+1\right)!\left(\frac{m}{2}\right)!(k+1)!}, & \text{if } m \equiv 0 \pmod 2. \end{cases}$$

(c)
$$\int_{-1}^{1}(1-x)^{k-\frac{1}{2}}(1+x)^{k+\frac{1}{2}}x^m dx = \begin{cases} \frac{(m+1)!(2k)!\pi}{2^{m+2k+1}\left(\frac{m+1}{2}+k\right)!\left(\frac{m+1}{2}\right)!k!}, & \text{if } m \equiv 1 \pmod 2, \\ \frac{m!(2k)!\pi}{2^{m+2k}\left(\frac{m}{2}+k\right)!\left(\frac{m}{2}\right)!k!}, & \text{if } m \equiv 0 \pmod 2. \end{cases}$$

(d)
$$\int_{-1}^{1}(1-x)^{k+\frac{1}{2}}(1+x)^{k-\frac{1}{2}}x^m dx = \begin{cases} -\frac{(m+1)!(2k)!\pi}{2^{m+2k+1}\left(\frac{m+1}{2}+k\right)!\left(\frac{m+1}{2}\right)!k!}, & \text{if } m \equiv 1 \pmod 2, \\ \frac{m!(2k)!\pi}{2^{m+2k}\left(\frac{m}{2}+k\right)!\left(\frac{m}{2}\right)!k!}, & \text{if } m \equiv 0 \pmod 2. \end{cases}$$

The following lemma was shown in [24] and can be derived by differentiating [23].

Lemma 1. *For any nonnegative integers n, r, the following identity holds:*

$$\sum_{i_1+\cdots+i_{r+1}=n} U_{i_1}(x)\cdots U_{i_{r+1}}(x) = \frac{1}{2^r r!}U_{n+k}^{(r)}(x), \tag{41}$$

where the sum is over all nonnegative integers $i_1, \cdots, i_{r+1}$, with $i_1+\cdots+i_{r+1}=n$.

Further, Equation (41) is equivalent to:

$$\left(\frac{1}{1-2xt+t^2}\right)^{r+1} = \frac{1}{2^r r!}\sum_{n=0}^{\infty} U_{n+r}^{(r)}(x)t^n. \tag{42}$$

In reference [24], the following lemma is stated for $m \geq r+1$. However, it holds for any nonnegative integer m, under the usual convention $\binom{r+1}{j}=0$, for $j > r+1$. Therefore, we are going to give a proof for the next lemma.

Lemma 2. *Let m, r be any nonnegative integers. Then, the following identity holds.*

$$\begin{aligned} &\sum_{i_1+\cdots+i_{r+1}=m} T_{i_1}(x)\cdots T_{i_{r+1}}(x) \\ &= \frac{1}{2^r r!}\sum_{j=0}^{m}(-1)^j\binom{r+1}{j}x^j U_{m-j+r}^{(r)}(x), \end{aligned} \tag{43}$$

where $\binom{r+1}{j}=0$, for $j > r+1$.

Proof. By making use of (42), we have:

$$
\begin{aligned}
&\sum_{m=0}^{\infty}\left(\sum_{i_1+\cdots+i_{r+1}=m} T_{i_1}(x)\cdots T_{i_{r+1}}(x)\right)t^m \\
&=\left(\frac{1}{1-2xt+t^2}\right)^{r+1}(1-xt)^{r+1} \\
&=\frac{1}{2^r r!}\sum_{n=0}^{\infty}U_{n+r}^{(r)}(x)t^n\sum_{j=0}^{r+1}\binom{r+1}{j}(-x)^j t^j \\
&=\frac{1}{2^r r!}\sum_{m=0}^{\infty}\left(\sum_{j=0}^{\min\{m,r+1\}}(-1)^j\binom{r+1}{j}x^jU_{m-j+r}^{(r)}(x)\right)t^m \\
&=\frac{1}{2^r r!}\sum_{m=0}^{\infty}\left(\sum_{j=0}^{m}(-1)^j\binom{r+1}{j}x^jU_{m-j+r}^{(r)}(x)\right)t^m.
\end{aligned}
\tag{44}
$$

Now, by comparing both sides of (44), we have the desired result. □

From (10), we see that the rth derivative of $U_n(x)$ is given by:

$$
U_n^{(r)}(x)=\sum_{l=0}^{\left[\frac{n-r}{2}\right]}(-1)^l\binom{n-l}{l}(n-2l)_r 2^{n-2l}x^{n-2l-r}. \tag{45}
$$

Especially, we have:

$$
x^jU_{m-j+r}^{(r)}(x)=\sum_{l=0}^{\left[\frac{m-j}{2}\right]}(-1)^l\binom{m-j+r-l}{l}(m-j+r-2l)_r 2^{m-j+r-2l}x^{m-2l}. \tag{46}
$$

In this section, we will show (29) and (31) of Theorem 1 and leave similar proofs for (30) and (32) as exercises to the reader. As in (23), let us put:

$$
\alpha_{m,r}(x)=\sum_{i_1+\cdots+i_{r+1}=m}T_{i_1}(x)\cdots T_{i_{r+1}}(x),
$$

and set:

$$
\alpha_{m,r}(x)=\sum_{k=0}^{m}C_{k,1}T_k(x). \tag{47}
$$

Then, we can now proceed as follows by using (a) of Proposition 1, (43) and (46), and integration by parts k times.

$$
\begin{aligned}
C_{k,1} &= \frac{(-1)^k 2^k k! \varepsilon_k}{(2k)!\pi} \int_{-1}^{1} \alpha_{m,r}(x) \frac{d^k}{dx^k}(1-x^2)^{k-\frac{1}{2}} dx \\
&= \frac{(-1)^k 2^k k! \varepsilon_k}{(2k)!\pi 2^r r!} \sum_{j=0}^{m} (-1)^j \binom{r+1}{j} \int_{-1}^{1} x^j U_{m-j+r}^{(r)}(x) \frac{d^k}{dx^k}(1-x^2)^{k-\frac{1}{2}} dx \\
&= \frac{(-1)^k 2^k k! \varepsilon_k}{(2k)!\pi 2^r r!} \sum_{j=0}^{m} (-1)^j \binom{r+1}{j} \sum_{l=0}^{\left[\frac{m-j}{2}\right]} (-1)^l \binom{m-j+r-l}{l} \\
&\quad \times (m-j+r-2l)_r 2^{m-j+r-2l} \int_{-1}^{1} x^{m-2l} \frac{d^k}{dx^k}(1-x^2)^{k-\frac{1}{2}} dx. \\
&= \frac{2^k k! \varepsilon_k}{(2k)!\pi 2^r r!} \sum_{j=0}^{m} (-1)^j \binom{r+1}{j} \sum_{l=0}^{\left[\frac{m-j}{2}\right]} (-1)^l \binom{m-j+r-l}{l} \\
&\quad \times (m-j+r-2l)_r 2^{m-j+r-2l} (m-2l)_k \int_{-1}^{1} x^{m-k-2l} (1-x^2)^{k-\frac{1}{2}} dx \\
&= \frac{2^k k! \varepsilon_k}{(2k)!\pi 2^r r!} \sum_{l=0}^{\left[\frac{m-k}{2}\right]} \sum_{j=0}^{m-2l} (-1)^j \binom{r+1}{j} (-1)^l \binom{m-j+r-l}{l} \\
&\quad \times (m-j+r-2l)_r 2^{m-j+r-2l} (m-2l)_k \int_{-1}^{1} x^{m-k-2l} (1-x^2)^{k-\frac{1}{2}} dx.
\end{aligned} \tag{48}
$$

Now, from (a) of Proposition 2 and after some simplifications, we see that:

$$
\begin{aligned}
\alpha_{m,r}(x) &= \frac{1}{r!} \sum_{0 \le k \le m,\, k \equiv m (\mathrm{mod} 2)} \sum_{l=0}^{\left[\frac{m-k}{2}\right]} \sum_{j=0}^{m-2l} \varepsilon_k (-1)^j \binom{r+1}{j} 2^{-j} \\
&\quad \times \frac{(-1)^l (m-j+r-l)!(m-2l)!}{l!(m-j-2l)! \left(\frac{m+k}{2}-l\right)! \left(\frac{m-k}{2}-l\right)!} T_k(x) \\
&= \frac{1}{r!} \sum_{s=0}^{\left[\frac{m}{2}\right]} \sum_{l=0}^{s} \frac{\varepsilon_{m-2s} (-1)^l (m-2l)!}{l!(m-s-l)!(s-l)!} \\
&\quad \times \sum_{j=0}^{m-2l} \frac{2^{-j}(-1)^j (m+r-l-j)!(r+1)_j}{j!(m-2l-j)!} T_{m-2s}(x) \\
&= \frac{1}{r!} \sum_{s=0}^{\left[\frac{m}{2}\right]} \sum_{l=0}^{s} \frac{\varepsilon_{m-2s} (-1)^l (m+r-l)!}{l!(m-s-l)!(s-l)!} \\
&\quad \times \sum_{j=0}^{m-2l} \frac{2^{-j}(-1)^j (m-2l)_j (r+1)_j}{j!(m+r-l)_j} T_{m-2s}(x) \\
&= \frac{1}{r!} \sum_{s=0}^{\left[\frac{m}{2}\right]} \sum_{l=0}^{s} \frac{\varepsilon_{m-2s} (-1)^l (m+r-l)!}{l!(m-s-l)!(s-l)!} \\
&\quad \times \sum_{j=0}^{m-2l} \frac{2^{-j} < 2l-m >_j < -r-1 >_j}{j! < l-m-r >_j} T_{m-2s}(x) \\
&= \frac{1}{r!} \sum_{s=0}^{\left[\frac{m}{2}\right]} \sum_{l=0}^{s} \frac{\varepsilon_{m-2s} (-1)^l (m+r-l)!}{l!(m-s-l)!(s-l)!} \\
&\quad \times {}_2F_1\left(2l-m, -r-1; l-m-r; \frac{1}{2}\right) T_{m-2s}(x),
\end{aligned} \tag{49}
$$

where we note that we made the change of variables $m-k=2s$.

This completes the proof for (29). Next, we let:

$$\alpha_{m,r}(x) = \sum_{k=0}^{m} C_{k,3} V_k(x). \tag{50}$$

Then, we can obtain the following by making use of (c) of Proposition 1, (43) and (46), and integration by parts k times.

$$\begin{aligned} C_{k,3} = \frac{k!2^k}{(2k)!\pi 2^r r!} \sum_{l=0}^{\left[\frac{m-k}{2}\right]} \sum_{j=0}^{m-2l} (-1)^j \binom{r+1}{j} (-1)^l \binom{m-j+r-l}{l} (m-j+r-2l)_r 2^{m-j+r-2l} \\ \times (m-2l)_k \int_{-1}^{1} x^{m-2l-k} (1-x)^{k-\frac{1}{2}} (1+x)^{k+\frac{1}{2}} dx. \end{aligned} \tag{51}$$

where we note from (c) of Proposition 2 that:

$$\begin{aligned} &\int_{-1}^{1} x^{m-2l-k} (1-x)^{k-\frac{1}{2}} (1+x)^{k+\frac{1}{2}} dx \\ &= \begin{cases} \frac{(m-2l-k+1)!(2k)!\pi}{2^{m+k-2l+1}\left(\frac{m+k+1}{2}-l\right)!\left(\frac{m-k+1}{2}-l\right)!k!}, & \text{if } k \not\equiv m \pmod 2, \\ \frac{(m-2l-k)!(2k)!\pi}{2^{m+k-2l}\left(\frac{m+k}{2}-l\right)!\left(\frac{m-k}{2}-l\right)!k!}, & \text{if } k \equiv m \pmod 2. \end{cases} \end{aligned} \tag{52}$$

From (50)–(52), and after some simplifications, we get:

$$\alpha_{m,r}(x) = \sum\nolimits_1 + \sum\nolimits_2, \tag{53}$$

where:

$$\begin{aligned} \sum\nolimits_1 &= \frac{1}{r!} \sum_{0 \le k \le m,\, k \not\equiv m \pmod 2} \sum_{l=0}^{\left[\frac{m-k}{2}\right]} \sum_{j=0}^{m-2l} (-1)^j \binom{r+1}{j} 2^{-j-1} \\ &\quad \times \frac{(-1)^l (m-j+r-l)!(m-2l)!(m-2l-k+1)}{l!(m-j-2l)!\left(\frac{m+k+1}{2}-l\right)!\left(\frac{m-k+1}{2}-l\right)!} V_k(x), \\ \sum\nolimits_2 &= \frac{1}{r!} \sum_{0 \le k \le m,\, k \equiv m \pmod 2} \sum_{l=0}^{\left[\frac{m-k}{2}\right]} \sum_{j=0}^{m-2l} (-1)^j \binom{r+1}{j} 2^{-j} \\ &\quad \times \frac{(-1)^l (m-j+r-l)!(m-2l)!}{l!(m-j-2l)!\left(\frac{m+k}{2}-l\right)!\left(\frac{m-k}{2}-l\right)!} V_k(x). \end{aligned} \tag{54}$$

Proceeding analogously to the case of (29), we observe from (54) that:

$$\begin{aligned} \sum\nolimits_1 &= \frac{1}{r!} \sum_{s=0}^{\left[\frac{m-1}{2}\right]} \sum_{l=0}^{s} \frac{(-1)^l (m+r-l)!}{l!(m-s-l)!(s-l)!} \\ &\quad \times \sum_{j=0}^{m-2l} \frac{2^{-j}(-1)^j (r+1)_j (m-2l)_j}{j!(m+r-l)_j} V_{m-2s-1}(x) \\ &= \frac{1}{r!} \sum_{s=0}^{\left[\frac{m-1}{2}\right]} \sum_{l=0}^{s} \frac{(-1)^l (m+r-l)!}{l!(m-s-l)!(s-l)!} \\ &\quad \times {}_2F_1\left(2l-m, -r-1; l-m-r; \frac{1}{2}\right) V_{m-2s-1}(x), \end{aligned} \tag{55}$$

$$\begin{aligned}\sum_2 &= \frac{1}{r!}\sum_{s=0}^{\left[\frac{m}{2}\right]}\sum_{l=0}^{s}\frac{(-1)^l(m+r-l)!}{l!(m-s-l)!(s-l)!}\\ &\quad\times\sum_{j=0}^{m-2l}\frac{2^{-j}(-1)^j(r+1)_j(m-2l)_j}{j!(m+r-l)_j}V_{m-2s}(x)\\ &= \frac{1}{r!}\sum_{s=0}^{\left[\frac{m}{2}\right]}\sum_{l=0}^{s}\frac{(-1)^l(m+r-l)!}{l!(m-s-l)!(s-l)!}\\ &\quad\times{}_2F_1\left(2l-m,-r-1;l-m-r;\frac{1}{2}\right)V_{m-2s}(x).\end{aligned} \tag{56}$$

We now obtain the result in (31) from (53), (55) and (56).

3. Proofs of Theorems 2 and 3

In this section, we will show (34) and (36) for Theorem 2, leaving (33) and (35) as exercises to the reader, and note that Theorem 3 follows from (33)–(36) by simple observation. The next lemma can be shown analogously to Lemma 1.

Lemma 3. *For any nonnegative integers m, r, the following identities are valid.*

$$\begin{aligned}&\sum_{i_1+\cdots+i_{r+1}=m} V_{i_1}(x)\cdots V_{i_{r+1}}(x)\\ &= \frac{1}{2^r r!}\sum_{j=0}^{m}(-1)^j\binom{r+1}{j}U_{m-j+r}^{(r)}(x),\end{aligned} \tag{57}$$

$$\begin{aligned}&\sum_{i_1+\cdots+i_{r+1}=m} W_{i_1}(x)\cdots W_{i_{r+1}}(x)\\ &= \frac{1}{2^r r!}\sum_{j=0}^{m}\binom{r+1}{j}U_{m-j+r}^{(r)}(x),\end{aligned} \tag{58}$$

where $\binom{r+1}{j}=0$, for $j>r+1$.

As in (24), let us set:

$$\beta_{m,r}(x) = \sum_{i_1+\cdots+i_{r+1}=m} V_{i_1}(x)\cdots V_{i_{r+1}}(x),$$

and put:

$$\beta_{m,r}(x) = \sum_{k=0}^{m} C_{k,2}U_k(x). \tag{59}$$

First, we note:

$$U_{m-j+r}^{(r+k)}(x) = \sum_{l=0}^{\left[\frac{m-j-k}{2}\right]}(-1)^l\binom{m-j+r-l}{l}(m-j+r-2l)_{r+k}2^{m-j+r-2l}x^{m-j-k-2l}. \tag{60}$$

Then, we have the following by exploiting (b) of Proposition 1, (57) and (60), and integration by parts k times.

$$\begin{aligned}C_{k,2} &= \frac{(-1)^k 2^{k+1}(k+1)!}{(2k+1)!\pi}\int_{-1}^{1}\beta_{m,r}(x)\frac{d^k}{dx^k}(1-x^2)^{k+\frac{1}{2}}dx \\ &= \frac{(-1)^k 2^{k+1}(k+1)!}{(2k+1)!\pi 2^r r!}\sum_{j=0}^{m}(-1)^j\binom{r+1}{j}\int_{-1}^{1}U_{m-j+r}^{(r)}(x)\frac{d^k}{dx^k}(1-x^2)^{k+\frac{1}{2}}dx \\ &= \frac{2^{k+1}(k+1)!}{(2k+1)!\pi 2^r r!}\sum_{j=0}^{m-k}(-1)^j\binom{r+1}{j}\int_{-1}^{1}U_{m-j+r}^{(r+k)}(x)(1-x^2)^{k+\frac{1}{2}}dx \\ &= \frac{2^{k+1-r}(k+1)!}{(2k+1)!\pi r!}\sum_{j=0}^{m-k}(-1)^j\binom{r+1}{j}\sum_{l=0}^{\left[\frac{m-j-k}{2}\right]}(-1)^l\binom{m-j+r-l}{l} \\ &\quad\times (m-j+r-2l)_{r+k}2^{m-j+r-2l}\int_{-1}^{1}x^{m-j-k-2l}(1-x^2)^{k+\frac{1}{2}}dx\end{aligned} \tag{61}$$

where we note from (b) of Proposition 2 that:

$$\begin{aligned}&\int_{-1}^{1}x^{m-j-k-2l}(1-x^2)^{k+\frac{1}{2}}dx \\ &= \begin{cases} 0, & \text{if } j \not\equiv m-k \pmod 2, \\ \frac{(m-j-k-2l)!(2k+2)!\pi}{2^{m-j+k-2l+2}\left(\frac{m-j+k}{2}+1-l\right)!\left(\frac{m-j-k}{2}-l\right)!(k+1)!}, & \text{if } j \equiv m-k \pmod 2. \end{cases}\end{aligned} \tag{62}$$

From (59), (61) and (62), and after some simplifications, we obtain:

$$\begin{aligned}\beta_{m,r}(x) &= \frac{1}{r!}\sum_{k=0}^{m}\sum_{0\le j\le m-k, j\equiv m-k (\mathrm{mod} 2)}\sum_{l=0}^{\left[\frac{m-k-j}{2}\right]}(-1)^j\binom{r+1}{j}(k+1) \\ &\quad\times\frac{(-1)^l(m-j+r-l)!}{l!\left(\frac{m-j+k}{2}+1-l\right)!\left(\frac{m-j-k}{2}-l\right)!}U_k(x) \\ &= \frac{1}{r!}\sum_{k=0}^{m}\sum_{s=0}^{\left[\frac{m-k}{2}\right]}\frac{(-1)^{m-k}(k+1)(k+2s+r)!}{(s+k+1)!s!}\binom{r+1}{m-k-2s} \\ &\quad\times\sum_{l=0}^{s}\frac{(-1)^l(s+k+1)_l(s)_l}{l!(k+2s+r)_l}U_k(x) \\ &= \frac{1}{r!}\sum_{k=0}^{m}\sum_{s=0}^{\left[\frac{m-k}{2}\right]}\frac{(-1)^{m-k}(k+1)(k+2s+r)!}{(s+k+1)!s!}\binom{r+1}{m-k-2s} \\ &\quad\times\sum_{l=0}^{s}\frac{<-s>_l<-s-k-1>_l}{l!<-k-2s-r>_l}U_k(x)\end{aligned} \tag{63}$$

$$\begin{aligned}&= \frac{1}{r!}\sum_{k=0}^{m}\sum_{s=0}^{\left[\frac{m-k}{2}\right]}\frac{(-1)^{m-k}(k+1)(k+2s+r)!}{(s+k+1)!s!}\binom{r+1}{m-k-2s} \\ &\quad\times{}_2F_1\left(-s,-s-k-1;-k-2s-r;1\right)U_k(x).\end{aligned}$$

This completes the proof for (34). Next, we let:

$$\beta_{m,r}(x) = \sum_{k=0}^{m}C_{k,4}W_k(x). \tag{64}$$

Then, from (d) of Proposition 1, (57) and (60), and integration by parts k times, we have:

$$C_{k,4} = \frac{k!2^{k-r}}{(2k)!\pi r!}\sum_{j=0}^{m-k}(-1)^j\binom{r+1}{j}\sum_{l=0}^{\left[\frac{m-j-k}{2}\right]}(-1)^l\binom{m-j+r-l}{l} \\ \times (m-j+r-2l)_{r+k}2^{m-j+r-2l}\int_{-1}^{1}x^{m-j-k-2l}(1-x)^{k+\frac{1}{2}}(1-x)^{k-\frac{1}{2}}dx \tag{65}$$

From (d) of Proposition 2, we observe that:

$$\int_{-1}^{1}x^{m-j-k-2l}(1-x)^{k+\frac{1}{2}}(1-x)^{k-\frac{1}{2}}dx \\ = \begin{cases} -\frac{(m-j-k-2l+1)!(2k)!\pi}{2^{m-j+k-2l+1}\left(\frac{m-j+k+1}{2}-l\right)!\left(\frac{m-j-k+1}{2}-l\right)!k!}, & \text{if } j \not\equiv m-k \pmod 2, \\ \frac{(m-j-k-2l)!(2k)!\pi}{2^{m-j+k-2l}\left(\frac{m-j+k}{2}-l\right)!\left(\frac{m-j-k}{2}-l\right)!k!}, & \text{if } j \equiv m-k \pmod 2. \end{cases} \tag{66}$$

By (64)–(66), and after some simplifications, we get:

$$\beta_{m,r}(x) = -\frac{1}{2r!}\sum_{k=0}^{m}\sum_{0\le j\le m-k, j\not\equiv m-k(\mathrm{mod}2)}(-1)^j\binom{r+1}{j}\sum_{l=0}^{\left[\frac{m-j-k}{2}\right]} \\ \times \frac{(-1)^l(m-j+r-l)!(m-j-k-2l+1)}{l!\left(\frac{m-j+k+1}{2}-l\right)!\left(\frac{m-j-k+1}{2}-l\right)!}W_k(x) \\ +\frac{1}{r!}\sum_{k=0}^{m}\sum_{0\le j\le m-k, j\equiv m-k(\mathrm{mod}2)}(-1)^j\binom{r+1}{j}\sum_{l=0}^{\left[\frac{m-j-k}{2}\right]} \\ \times \frac{(-1)^l(m-j+r-l)!}{l!\left(\frac{m-j+k}{2}-l\right)!\left(\frac{m-j-k}{2}-l\right)!}W_k(x) \\ = \frac{1}{r!}\sum_{k=0}^{m}\sum_{s=0}^{\left[\frac{m-k-1}{2}\right]}(-1)^{m-k}\binom{r+1}{m-k-2s-1}\frac{(k+2s+r+1)!}{(s+k+1)!s!} \\ \times \sum_{l=0}^{s}\frac{(-1)^l(s+k+1)_l(s)_l}{l!(k+2s+r+1)_l}W_k(x) \\ +\frac{1}{r!}\sum_{k=0}^{m}\sum_{s=0}^{\left[\frac{m-k}{2}\right]}(-1)^{m-k}\binom{r+1}{m-k-2s}\frac{(k+2s+r)!}{(s+k)!s!} \\ \times \sum_{l=0}^{s}\frac{(-1)^l(s+k)_l(s)_l}{l!(k+2s+r)_l}W_k(x). \tag{67}$$

Further modification of (67) gives us:

$$\begin{aligned}\beta_{m,r}(x) &= \frac{1}{r!}\sum_{k=0}^{m}\sum_{s=0}^{\left[\frac{m-k-1}{2}\right]}(-1)^{m-k}\frac{(k+2s+r+1)!}{(s+k+1)!s!}\binom{r+1}{m-k-2s-1}\\ &\times {}_2F_1(-s,-s-k-1;-k-2s-r-1;1)W_k(x)\\ &+\frac{1}{r!}\sum_{k=0}^{m}\sum_{l=0}^{\left[\frac{m-k}{2}\right]}(-1)^{m-k}\frac{(k+2s+r)!}{(s+k)!s!}\binom{r+1}{m-k-2s}\\ &\times {}_2F_1(-s,-s-k;-k-2s-r;1)W_k(x)\\ &=\frac{1}{r!}\sum_{k=0}^{m}\sum_{s=0}^{m-k}\frac{(-1)^{m-k}(k+r+s)!}{\left(k+\left[\frac{s+1}{2}\right]\right)!\left[\frac{s}{2}\right]!}\binom{r+1}{m-k-s}\\ &\times {}_2F_1\left(-\left[\frac{s}{2}\right],-\left[\frac{s+1}{2}\right]-k;-k-s-r;1\right)W_k(x).\end{aligned} \tag{68}$$

This finishes up the proof for (36).

Remark 1. *We note from* (57) *and* (58) *that the only difference between $\beta_{m,r}(x)$ and $\gamma_{m,r}(x)$ (see* (24) *and* (25)*) is the alternating sign $(-1)^j$ in their sums, which corresponds to $(-1)^{m-k}$ in* (33)–(36)*. This remark gives the results in* (37)–(40) *of Theorem 3.*

4. Conclusions

Our paper can be viewed as a generalization of the linearization problem, which is concerned with determining the coefficients in the expansion $a_n(x)b_m(x) = \sum_{k=0}^{n+m} c_k(nm)p_k(x)$ of the product $a_n(x)b_m(x)$ of two polynomials $a_n(x)$ and $b_m(x)$ in terms of an arbitrary polynomial sequence $\{p_k(x)\}_{k\geq 0}$. Our pursuit of this line of research can also be justified from another fact; namely, the famous Faber–Pandharipande–Zagier and Miki identities follow by expressing the sum $\sum_{k=1}^{m-1}\frac{1}{k(m-k)}B_k(x)B_{m-k}(x)$ as a linear combination of Bernoulli polynomials. For some details on this, we let the reader refer to the Introduction of [15]. Here, we considered sums of finite products of the Chebyshev polynomials of the first, third, and fourth kinds and represented each of those sums of finite products as linear combinations of $T_n(x)$, $U_n(x)$, $V_n(x)$, and $W_n(x)$, which involve some terminating hypergeometric function ${}_2F_1$. Here, we remark that $\alpha_{m,r}(x)$, $\beta_{m,r}(x)$, and $\gamma_{m,r}(x)$ are expressed in terms of $U_{m-j+r}^{(r)}(x)$, $(j=0,1,\cdots,m)$ (see Lemmas 2 and 3) by making use of the generating function in (6). This is unlike the previous works for (26)–(28) (see [5,6]), where we showed they are respectively equal to $\frac{1}{2^{r-1}r!}T_{m+r}^{(r)}(x)$, $\frac{1}{2^r r!}V_{m+r}^{(r)}(x)$, and $\frac{1}{2^r r!}W_{m+r}^{(r)}(x)$ by exploiting the generating functions in (5), (7) and (8). Then, our results for $\alpha_{m,r}(x)$, $\beta_{m,r}(x)$, and $\gamma_{m,r}(x)$ were found by making use of Lemmas 1 and 2, the general formulas in Propositions 1 and 2, and integration by parts. It is certainly possible to represent such sums of finite products by other orthogonal polynomials, which is one of our ongoing projects. More generally, along the same line as the present paper, we are planning to consider some sums of finite products of many special polynomials and want to find their applications.

Author Contributions: T.K. and D.S.K. conceived of the framework and structured the whole paper; T.K. wrote the paper; L.-C.J. and D.V.D. checked the results of the paper; D.S.K. and T.K. completed the revision of the article.

References

1. Andrews, G.E.; Askey, R.; Roy, R. *Special Functions*; Encyclopedia of Mathematics and Its Applications 71; Cambridge University Press: Cambridge, UK, 1999.
2. Beals, R.; Wong, R. *Special Functions and Orthogonal Polynomials*; Cambridge Studies in Advanced Mathematics 153; Cambridge University Press: Cambridge, UK, 2016.

3. Kim, T.; Kim, D.S.; Jang, G.-W.; Jang, L.C. Fourier series of functions involving higher-order ordered Bell polynomials. *Open Math.* **2017**, *15*, 1606–1617. [CrossRef]
4. Mason, J.C.; Handscomb, D.C. *Chebyshev Polynomials*; Chapman&Hall/CRC: Boca Raton, FC, USA, 2003.
5. Kim, T.; Kim, D.S.; Dolgy, D.V.; Kwon, J. Representing sums of finite products of chebyshev polynomials of the first kind and lucas polynomials by chebyshev polynomials. *Math. Comput. Sci.* **2018**. [CrossRef]
6. Kim, T.; Kim, D.S.; Dolgy, D.V.; Ryoo, C.S. Representing sums of finite products of Chebyshev polynomials of the third and fourth kinds by Chebyshev polynomials. *Symmetry* **2018**, *10*, 258. [CrossRef]
7. Shang, Y. Unveiling robustness and heterogeneity through percolation triggered by random-link breakdown. *Phys. Rev. E* **2014**, *90*, 032820. [CrossRef] [PubMed]
8. Shang, Y. Effect of link oriented self-healing on resilience of networks. *J. Stat. Mech. Theory Exp.* **2016**, *2016*, 083403. [CrossRef]
9. Shang, Y. Modeling epidemic spread with awareness and heterogeneous transmission rates in networks. *J. Biol. Phys.* **2013**, *39*, 489–500. [PubMed]
10. Agarwal, R.P.; Kim, D.S.; Kim, T.; Kwon, J. Sums of finite products of Bernoulli functions. *Adv. Differ. Equ.* **2017**, *2017*, 237. [CrossRef]
11. Kim, T.; Kim, D.S.; Dolgy, D.V.; Kwon, J. Sums of finite products of Chebyshev polynomials of the third and fourth kinds. *Adv. Differ. Equ.* **2018**, *2018*, 283. [CrossRef]
12. Kim, T.; Kim, D.S.; Dolgy, D.V.; Park, J.-W. Sums of finite products of Chebyshev polynomials of the second kind and of Fibonacci polynomials. *J. Inequal. Appl.* **2018**, *2018*, 148. [CrossRef] [PubMed]
13. Kim, T.; Kim, D.S.; Dolgy, D.V.; Park, J.-W. Sums of finite products of Legendre and Laguerre polynomials. *Adv. Differ. Equ.* **2018**, *2018*, 277. [CrossRef]
14. Kim, T.; Kim, D.S.; Jang, G.-W.; Kwon, J. Sums of finite products of Euler functions. In *Advances in Real and Complex Analysis with Applications*; Trends in Mathematics; Birkhäuser: Basel, Switzerland, 2017; pp. 243–260.
15. Kim, T.; Kim, D.S.; Jang, L.C.; Jang, G.-W. Sums of finite products of Genocchi functions. *Adv. Differ. Equ.* **2017**, *2017*, 268. [CrossRef]
16. Kim, T.; Kim, D.S.; Jang, L.C.; Jang, G.-W. Fourier series for functions related to Chebyshev polynomials of the first kind and Lucas polynomials. *Mathematics* **2018**, *6*, 276. [CrossRef]
17. Kim, T.; Dolgy, D.V.; Kim, D.S. Representing sums of finite products of Chebyshev polynomials of the second kind and Fibonacci polynomials in terms of Chebyshev polynomials. *Adv. Stud. Contemp. Math.* **2018**, *28*, 321–335.
18. Kim, T.; Kim, D.S.; Kwon, J.; Jang, G.-W. Sums of finite products of Legendre and Laguerre polynomials by Chebyshev polynomials. *Adv. Stud. Contemp. Math.* **2018**, *28*, 551–565.
19. Doha, E.H.; Abd-Elhameed, W.M.; Alsuyuti, M.M. On using third and fourth kinds Chebyshev polynomials for solving the integrated forms of high odd-order linear boundary value problems. *J. Egypt. Math. Soc.* **2015**, *23*, 397–405. [CrossRef]
20. Eslahchi, M.R.; Dehghan, M.; Amani, S. The third and fourth kinds of Chebyshev polynomials and best uniform approximation. *Math. Comput. Model.* **2012**, *55*, 1746–1762. [CrossRef]
21. Mason, J.C. Chebyshev polynomials of the second, third and fourth kinds in approximation, indefinite integration, and integral transforms. *J. Comput. Appl. Math.* **1993**, *49*, 169–178. [CrossRef]
22. Kim, D.S.; Kim, T.; Lee, S.-H. Some identities for Bernoulli polynomials involving Chebyshev polynomials. *J. Comput. Anal. Appl.* **2014**, *16*, 172–180.
23. Kim, D.S.; Dolgy, D.V.; Kim, T.; Rim, S.-H. Identities involving Bernoulli and Euler polynomials arising from Chebyshev polynomials. *Proc. Jangjeon Math. Soc.* **2012**, *15*, 361–370.
24. Zhang, W. Some identities involving the Fibonacci numbers and Lucas numbers. *Fibonacci Q.* **2004**, *42*, 149–154.

The Power Sums Involving Fibonacci Polynomials and their Applications

Li Chen [1] **and Xiao Wang** [1,*]

[1] School of Mathematics, Northwest University, Xi'an 710127, Shaanxi, China; cl1228@stumail.nwu.edu.cn
[*] Correspondence: wangxiao_0606@stumail.nwu.edu.cn;

Abstract: The Girard and Waring formula and mathematical induction are used to study a problem involving the sums of powers of Fibonacci polynomials in this paper, and we give it interesting divisible properties. As an application of our result, we also prove a generalized conclusion proposed by R. S. Melham.

Keywords: Fibonacci polynomials; Lucas polynomials; sums of powers; divisible properties; R. S. Melham's conjectures

MSC: 11B39

---

1. Introduction

For any integer $n \geq 0$, the famous Fibonacci polynomials $\{F_n(x)\}$ and Lucas polynomials $\{L_n(x)\}$ are defined as $F_0(x) = 0$, $F_1(x) = 1$, $L_0(x) = 2$, $L_1(x) = x$ and $F_{n+2}(x) = xF_{n+1}(x) + F_n(x)$, $L_{n+2}(x) = xL_{n+1}(x) + L_n(x)$ for all $n \geq 0$. Now, if we let $\alpha = \frac{x+\sqrt{x^2+4}}{2}$ and $\beta = \frac{x-\sqrt{x^2+4}}{2}$, then it is easy to prove that

$$F_n(x) = \frac{1}{\alpha - \beta}\left(\alpha^n - \beta^n\right) \text{ and } L_n(x) = \alpha^n + \beta^n \text{ for all } n \geq 0.$$

If $x = 1$, we have that $\{F_n(x)\}$ turns into Fibonacci sequences $\{F_n\}$, and $\{L_n(x)\}$ turns into Lucas sequences $\{L_n\}$. If $x = 2$, then $F_n(2) = P_n$, the nth Pell numbers, they are defined by $P_0 = 0$, $P_1 = 1$ and $P_{n+2} = 2P_{n+1} + P_n$ for all $n \geq 0$. In fact, $\{F_n(x)\}$ is a second-order linear recursive polynomial, when x takes a different value x_0, then $F_n(x_0)$ can become a different sequence.

Since the Fibonacci numbers and Lucas numbers occupy significant positions in combinatorial mathematics and elementary number theory, they are thus studied by plenty of researchers, and have gained a large number of vital conclusions; some of them can be found in References [1–15]. For example, Yi Yuan and Zhang Wenpeng [1] studied the properties of the Fibonacci polynomials, and proved some interesting identities involving Fibonacci numbers and Lucas numbers. Ma Rong and Zhang Wenpeng [2] also studied the properties of the Chebyshev polynomials, and obtained some meaningful formulas about the Chebyshev polynomials and Fibonacci numbers. Kiyota Ozeki [3] got some identity involving sums of powers of Fibonacci numbers. That is, he proved that

$$\sum_{k=1}^{n} F_{2k}^{2m+1} = \frac{1}{5^m} \sum_{j=0}^{m} \frac{(-1)^j}{L_{2m+1-2j}} \binom{2m+1}{j} \left(F_{(2m+1-2j)(2n+1)} - F_{2m+1-2j}\right).$$

Helmut Prodinger [4] extended the result of Kiyota Ozeki [3].

In addition, regarding many orthogonal polynomials and famous sequences, Kim et al. have done a lot of important research work, obtaining a series of interesting identities. Interested readers can refer to References [16–22]; we will not list them one by one.

In this paper, our main purpose is to care about the divisibility properties of the Fibonacci polynomials. This idea originated from R. S. Melham. In fact, in [5], R. S. Melham proposed two interesting conjectures as follows:

Conjecture 1. *If $m \geq 1$ is a positive integer, then the summation*

$$L_1L_3L_5\cdots L_{2m+1}\sum_{k=1}^{n}F_{2k}^{2m+1}$$

can be written as $(F_{2n+1}-1)^2\,P_{2m-1}(F_{2n+1})$, where $P_{2m-1}(x)$ is an integer coefficients polynomial with degree $2m-1$.

Conjecture 2. *If $m \geq 0$ is an integer, then the summation*

$$L_1L_3L_5\cdots L_{2m+1}\sum_{k=1}^{n}L_{2k}^{2m+1}$$

can be written as $(L_{2n+1}-1)\,Q_{2m}(L_{2n+1})$, where $Q_{2m}(x)$ is an integer coefficients polynomial with degree $2m$.

Wang Tingting and Zhang Wenpeng [6] solved Conjecture 2 completely. They also proved a weaker conclusion for Conjecture 1. That is,

$$L_1L_3L_5\cdots L_{2m+1}\sum_{k=1}^{n}F_{2k}^{2m+1}$$

can be expressed as $(F_{2n+1}-1)\,P_{2m}(F_{2n+1})$, where $P_{2m}(x)$ is a polynomial of degree $2m$ with integer coefficients.

Sun et al. [7] solved Conjecture 1 completely. In fact, Ozeki [3] and Prodinger [4] indicated that the odd power sum of the first several consecutive Fibonacci numbers of even order is equivalent to the polynomial estimated at a Fibonacci number of odd order. Sun et al. in [7] proved that this polynomial and its derivative both disappear at 1, and it can be an integer polynomial when a product of the first consecutive Lucas numbers of odd order multiplies it. This presents an affirmative answer to Conjecture 1 of Melham.

In this paper, we are going to use a new and different method to study this problem, and give a generalized conclusion. That is, we will use the Girard and Waring formula and mathematical induction to prove the conclusions in the following:

Theorem 1. *If n and h are positive integers, then we have the congruence*

$$L_1(x)L_3(x)\cdots L_{2n+1}(x)\sum_{m=1}^{h}F_{2m}^{2n+1}(x)\equiv 0 \bmod (F_{2h+1}(x)-1)^2.$$

Taking $x=1$ and $x=2$ in Theorem 1, we can instantly infer the two corollaries:

Corollary 1. *Let F_n and L_n be Fibonacci numbers and Lucas numbers, respectively. Then, for any positive integers n and h, we have the congruence*

$$L_1L_3L_5\cdots L_{2n+1}\sum_{m=1}^{h}F_{2m}^{2n+1}\equiv 0 \bmod (F_{2h+1}-1)^2.$$

Corollary 2. *Let P_n be nth Pell numbers. Then, for any positive integers n and h, we have the congruence*

$$L_1(2)L_3(2)L_5(2)\cdots L_{2n+1}(2)\sum_{m=1}^{h} P_{2m}^{2n+1} \equiv 0 \bmod (P_{2h+1}-1)^2,$$

where $L_n(2) = \left(1+\sqrt{2}\right)^n + \left(1-\sqrt{2}\right)^n$ *is called nth Pell–Lucas numbers.*

It is clear that our Corollary 1 gave a new proof for Conjecture 1.

2. Several Lemmas

In this part, we will give four simple lemmas, which are essential to prove our main results.

Lemma 1. *Let h be any positive integer; then, we have*

$$\left(x^2+4, F_{2h+1}(x)-1\right) = 1,$$

where x^2+4 and $F_{2h+1}(x)-1$ are said to be relatively prime.

Proof. From the definition of $F_n(x)$ and binomial theorem, we have

$$\begin{aligned} F_{2h+1}(x) &= \frac{1}{2^{2h+1}\sqrt{x^2+4}}\sum_{k=0}^{2h+1}\binom{2h+1}{k}x^k\left(x^2+4\right)^{\frac{2h+1-k}{2}} \\ &\quad -\frac{1}{2^{2h+1}\sqrt{x^2+4}}\sum_{k=0}^{2h+1}\binom{2h+1}{k}x^k(-1)^{2h+1-k}\left(x^2+4\right)^{\frac{2h+1-k}{2}} \\ &= \frac{1}{4^h}\sum_{k=0}^{h}\binom{2h+1}{2k}x^{2k}\left(x^2+4\right)^{h-k}. \end{aligned} \tag{1}$$

Thus, from Equation (1), we have the polynomial congruence

$$\begin{aligned} 4^h F_{2h+1}(x) &= \sum_{k=0}^{h}\binom{2h+1}{2k}x^{2k}\left(x^2+4\right)^{h-k} \equiv (2h+1)x^{2h} \\ &\equiv (2h+1)\left(x^2+4-4\right)^h \equiv (2h+1)(-4)^h \bmod (x^2+4) \end{aligned}$$

or

$$F_{2h+1}(x)-1 \equiv (2h+1)(-1)^h - 1 \bmod (x^2+4). \tag{2}$$

Since x^2+4 is an irreducible polynomial of x, and $(2h+1)(-1)^h-1$ is not divisible by (x^2+4) for all integer $h \geq 1$, so, from (2), we can deduce that

$$\left(x^2+4, F_{2h+1}(x)-1\right) = 1.$$

Lemma 1 is proved. □

Lemma 2. *Let h and n be non-negative integers with $h \geq 1$; then, we have*

$$(x^2+4)F_{(2h+1)(2n+1)}(x) - L_{2n}(x) - L_{2n+2}(x) \equiv 0 \bmod (F_{2h+1}(x)-1).$$

Proof. We use mathematical induction to calculate the polynomial congruence for n. Noting $L_0(x) = 2$, $L_1(x) = x$, $L_2(x) = x^2 + 2$. Thus, if $n = 0$, then

$$\begin{aligned} & (x^2+4)F_{(2h+1)(2n+1)}(x) - L_{2n}(x) - L_{2n+2}(x) \\ = \; & (x^2+4)F_{2h+1}(x) - 2 - x^2 - 2 \\ = \; & (x^2+4)\left(F_{2h+1}(x) - 1\right) \equiv 0 \bmod \left(F_{2h+1}(x) - 1\right). \end{aligned}$$

If $n = 1$, then $L_2(x) + L_4(x) = x^2 + 2 + x^4 + 4x^2 + 2 = x^4 + 5x^2 + 4$. Note that the identity $F^3_{2h+1}(x) = \frac{1}{x^2+4}\left(F_{3(2h+1)}(x) + 3F_{2h+1}(x)\right)$, so we obtain the congruence

$$\begin{aligned} & (x^2+4)F_{(2h+1)(2n+1)}(x) - L_{2n}(x) - L_{2n+2}(x) \\ = \; & (x^2+4)F_{3(2h+1)}(x) - x^4 - 5x^2 - 4 \\ = \; & (x^2+4)\left[(x^2+4)F^3_{2h+1}(x) - 3F_{2h+1}(x)\right] - x^4 - 5x^2 - 4 \\ = \; & (x^2+4)^2\left[F^3_{2h+1}(x) - F_{2h+1}(x)\right] + (x^2+4)(x^2+1)F_{2h+1}(x) - x^4 - 5x^2 - 4 \\ \equiv \; & (x^2+4)^2\left(F^2_{2h+1}(x) + F_{2h+1}(x)\right)\left(F_{2h+1}(x) - 1\right) \equiv 0 \bmod \left(F_{2h+1}(x) - 1\right), \end{aligned}$$

which means that Lemma 2 is correct for $n = 0$ and 1.

Assume Lemma 2 is right for all integers $n = 0, 1, 2, \cdots, k$. Namely,

$$(x^2+4)F_{(2h+1)(2n+1)}(x) - L_{2n}(x) - L_{2n+2}(x) \equiv 0 \bmod \left(F_{2h+1}(x) - 1\right), \tag{3}$$

where $0 \le n \le k$.

Thus, $n = k + 1 \ge 2$, and we notice that

$$L_{2(2h+1)}(x)F_{(2h+1)(2k+1)}(x) = F_{(2h+1)(2k+3)}(x) + F_{(2h+1)(2k-1)}(x),$$

$$L_{2k+2}(x) + L_{2k+4}(x) = (x^2+2)L_{2k}(x) + (x^2+2)L_{2k+2}(x) - \left(L_{2k-2}(x) + L_{2k}(x)\right)$$

and

$$L_{2(2h+1)}(x) = (x^2+4)F^2_{2h+1}(x) - 2 \equiv x^2 + 2 \bmod \left(F_{2h+1}(x) - 1\right).$$

From inductive assumption (3), we have

$$\begin{aligned} & (x^2+4)F_{(2h+1)(2n+1)}(x) - L_{2n}(x) - L_{2n+2}(x) \\ = \; & (x^2+4)F_{(2h+1)(2k+3)}(x) - L_{2k+2}(x) - L_{2k+4}(x) \\ = \; & (x^2+4)L_{2(2h+1)}(x)F_{(2h+1)(2k+1)}(x) - (x^2+4)F_{(2h+1)(2k-1)} - L_{2k+2}(x) - L_{2k+4}(x) \\ \equiv \; & (x^2+4)(x^2+2)F_{(2h+1)(2k+1)}(x) - (x^2+2)L_{2k}(x) - (x^2+2)L_{2k+2}(x) \\ & -(x^2+4)F_{(2h+1)(2k-1)}(x) + L_{2k-2}(x) + L_{2k}(x) \\ \equiv \; & (x^2+2)\left[(x^2+4)F_{(2h+1)(2k+1)}(x) - L_{2k}(x) - L_{2k+2}(x)\right] \\ & - \left[(x^2+4))F_{(2h+1)(2k-1)}(x) - L_{2k-2}(x) - L_{2k}(x)\right] \\ \equiv \; & 0 \bmod \left(F_{2h+1}(x) - 1\right). \end{aligned}$$

Now, we have achieved the results of Lemma 2. □

Lemma 3. *Let h and n be non-negative integers with $h \geq 1$; then, we have the polynomial congruence*

$$\begin{aligned} & L_1(x)L_3(x)\cdots L_{2n+1}(x)\sum_{m=1}^{h}\left[F_{2m(2n+1)}(x)-(2n+1)F_{2m}(x)\right] \\ \equiv\ & 0 \bmod \left(F_{2h+1}(x)-1\right)^2. \end{aligned}$$

Proof. For positive integer n, first note that $\alpha\beta=-1$, $L_n(x)=\alpha^n+\beta^n$,

$$\begin{aligned} & \sum_{m=1}^{h} F_{2m(2n+1)}(x) = \frac{1}{\sqrt{x^2+4}}\sum_{m=1}^{h}\left[\alpha^{2m(2n+1)}-\beta^{2m(2n+1)}\right] \\ =\ & \frac{1}{\sqrt{x^2+4}}\left[\frac{\alpha^{2(2n+1)}\left(\alpha^{2h(2n+1)}-1\right)}{\alpha^{2(2n+1)}-1}-\frac{\beta^{2(2n+1)}\left(\beta^{2h(2n+1)}-1\right)}{\beta^{2(2n+1)}-1}\right] \\ =\ & \frac{1}{L_{2n+1}(x)}\left[F_{(2h+1)(2n+1)}(x)-F_{2n+1}(x)\right] \end{aligned} \tag{4}$$

and

$$\sum_{m=1}^{h} F_{2m}(x) = \frac{1}{\sqrt{x^2+4}}\sum_{m=1}^{h}\left[\alpha^{2m}-\beta^{2m}\right] = \frac{1}{L_1(x)}\left[F_{(2h+1)}(x)-1\right]. \tag{5}$$

Thus, from Labels (4) and (5), we know that, to prove Lemma 3, now we need to obtain the polynomial congruence

$$\begin{aligned} & L_1(x)\left(F_{(2h+1)(2n+1)}(x)-F_{2n+1}(x)\right)-(2n+1)L_{2n+1}(x)\left(F_{2h+1}(x)-1\right) \\ \equiv\ & 0 \bmod \left(F_{2h+1}(x)-1\right)^2. \end{aligned} \tag{6}$$

Now, we prove (6) by mathematical induction. If $n=0$, then it is obvious that (6) is correct. If $n=1$, we notice that $L_1(x)=x$, $F_{3(2h+1)}(x)=(x^2+4)F_{2h+1}^3(x)-3F_{2h+1}(x)$ and $F_{2h+1}^3(x)\equiv (F_{2h+1}(x)-1+1)^3\equiv 3F_{2h+1}(x)-2 \bmod (F_{2h+1}(x)-1)^2$ we have

$$\begin{aligned} & L_1(x)F_{(2h+1)(2n+1)}(x)-L_1(x)F_{2n+1}(x)-(2n+1)L_{2n+1}(x)\left(F_{2h+1}(x)-1\right) \\ =\ & xF_{3(2h+1)}(x)-xF_3(x)-3L_3(x)\left(F_{2h+1}(x)-1\right) \\ =\ & x(x^2+4)F_{2h+1}^3(x)-3xF_{2h+1}(x)-x(x^2+1)-3(x^3+3x)\left(F_{2h+1}(x)-1\right) \\ \equiv\ & (x^3+4x)\left(3F_{2h+1}(x)-2\right)-3xF_{2h+1}(x)-(x^3+x)-3(x^3+3x)\left(F_{2h+1}(x)-1\right) \\ \equiv\ & 3(x^3+3x)\left(F_{2h+1}(x)-1\right)-3(x^3+3x)\left(F_{2h+1}(x)-1\right) \\ \equiv\ & 0 \bmod \left(F_{2h+1}(x)-1\right)^2. \end{aligned}$$

Thus, $n=1$ is fit for (6). Assume that (6) is correct for all integers $n=0,1,2,\cdots,k$. Namely,

$$\begin{aligned} & L_1(x)\left(F_{(2h+1)(2n+1)}(x)-F_{2n+1}(x)\right)-(2n+1)L_{2n+1}(x)\left(F_{2h+1}(x)-1\right) \\ \equiv\ & 0 \bmod \left(F_{2h+1}(x)-1\right)^2 \end{aligned} \tag{7}$$

for all $n=0,1,\cdots,k$.

Where $n=k+1\geq 2$, we notice

$$L_{2(2h+1)}(x)F_{(2h+1)(2k+1)}(x)=F_{(2h+1)(2k+3)}(x)+F_{(2h+1)(2k-1)}(x)$$

and

$$\begin{aligned} & L_{2(2h+1)}(x) = (x^2+4)F^2_{2h+1}(x) - 2 = (x^2+4)\left(F_{2h+1}(x) - 1 + 1\right)^2 - 2 \\ = & \ (x^2+4)\left[(F_{2h+1}(x)-1)^2 + 2(F_{2h+1}(x)-1)\right] + x^2 + 2 \\ \equiv & \ 2(x^2+4)(F_{2h+1}(x)-1) + x^2 + 2 \bmod (F_{2h+1}(x)-1)^2. \end{aligned}$$

From inductive assumption (7) and Lemma 2, we have

$$\begin{aligned} & xF_{(2h+1)(2n+1)}(x) - xF_{2n+1}(x) - (2n+1)L_{2n+1}(x)\left(F_{2h+1}(x)-1\right) \\ = & \ xF_{(2h+1)(2k+3)}(x) - xF_{2k+3}(x) - (2k+3)L_{2k+3}(x)\left(F_{2h+1}(x)-1\right) \\ = & \ xL_{2(2h+1)}(x)F_{(2h+1)(2k+1)}(x) - xF_{(2h+1)(2k-1)}(x) - xF_{2k+3}(x) \\ & -(2k+3)L_{2k+3}(x)\left(F_{2h+1}(x)-1\right) \\ \equiv & \ 2x(x^2+4)(F_{2h+1}(x)-1)F_{(2h+1)(2k+1)}(x) + x(x^2+2)F_{(2h+1)(2k+1)}(x) \\ & -xF_{(2h+1)(2k-1)}(x) - x(x^2+2)F_{2k+1}(x) + xF_{2k-1}(x) \\ & -(x^2+2)(2k+1)L_{2k+1}(x)\left(F_{2h+1}(x)-1\right) + (2k-1)L_{2k-1}(x)\left(F_{2h+1}(x)-1\right) \\ & -2x\left(L_{2k}(x) + L_{2k+2}(x)\right)\left(F_{2h+1}(x)-1\right) \\ \equiv & \ 2x(F_{2h+1}(x)-1)\left[(x^2+4)F_{(2h+1)(2k+1)}(x) - L_{2k}(x) - L_{2k+2}(x)\right] \\ & +(x^2+2)\left[xF_{(2h+1)(2k+1)}(x) - xF_{2k+1}(x) - (2k+1)L_{2k+1}(x)\left(F_{2h+1}(x)-1\right)\right] \\ & -\left[xF_{(2h+1)(2k-1)}(x) - xF_{2k-1}(x) - (2k-1)L_{2k-1}(x)\left(F_{2h+1}(x)-1\right)\right] \\ \equiv & \ 2x(F_{2h+1}(x)-1)\left[(x^2+4)F_{(2h+1)(2k+1)}(x) - L_{2k}(x) - L_{2k+2}(x)\right] \\ \equiv & \ 0 \bmod (F_{2h+1}(x)-1)^2. \end{aligned}$$

Now, we attain Lemma 3 by mathematical induction. □

Lemma 4. *For all non-negative integers u and real numbers X, Y, we have the identity*

$$X^u + Y^u = \sum_{k=0}^{\left[\frac{u}{2}\right]} (-1)^k \frac{u}{u-k}\binom{u-k}{k}(X+Y)^{u-2k}(XY)^k,$$

in which $[x]$ denotes the greatest integer $\leq x$.

Proof. This formula due to Waring [15]. It can also be found in Girard [14]. □

3. Proof of the Theorem

We will achieve the theorem by these lemmas. Taking $X = \alpha^{2m}$, $Y = -\beta^{2m}$ and $U = 2n+1$ in Lemma 4, we notice that $XY = -1$, from the expression of $F_n(x)$

$$\begin{aligned} F_{2m(2n+1)}(x) &= \sum_{k=0}^{n} (-1)^k \frac{2n+1}{2n+1-k}\binom{2n+1-k}{k}(x^2+4)^{n-k}F_{2m}^{2n+1-2k}(x)(-1)^k \\ &= \sum_{k=0}^{n} \frac{2n+1}{2n+1-k}\binom{2n+1-k}{k}(x^2+4)^{n-k}F_{2m}^{2n+1-2k}(x). \end{aligned} \quad (8)$$

For any integer $h \geq 1$, from (8), we get

$$\begin{aligned}&\sum_{m=1}^{h}\left[F_{2m(2n+1)}(x)-(2n+1)F_{2m}(x)\right]\\&=\sum_{k=0}^{n-1}\frac{2n+1}{2n+1-k}\binom{2n+1-k}{k}(x^2+4)^{n-k}\sum_{m=1}^{h}F_{2m}^{2n+1-2k}(x).\end{aligned}\tag{9}$$

If $n = 1$, then, from (9), we can get

$$L_1(x)L_3(x)\sum_{m=1}^{h}(F_{6m}(x)-3F_{2m}(x))=L_1(x)L_3(x)(x^2+4)\sum_{m=1}^{h}F_{2m}^3(x).\tag{10}$$

From Lemma 1, we know that $(x^2+4, F_{2h+1}(x)-1) = 1$, so, applying Lemma 3 and (10), we deduce that

$$L_1(x)L_3(x)\sum_{m=1}^{h}F_{2m}^3(x)\equiv 0 \bmod (F_{2h+1}(x)-1)^2.\tag{11}$$

This means that Theorem 1 is suitable for $n = 1$.

Assume that Theorem 1 is correct for all integers $n = 1, 2, \cdots, s$. Then,

$$L_1(x)L_3(x)\cdots L_{2n+1}(x)\sum_{m=1}^{h}F_{2m}^{2n+1}(x)\equiv 0 \bmod (F_{2h+1}(x)-1)^2\tag{12}$$

for all integers $1 \leq n \leq s$.

When $n = s + 1$, from (9), we obtain

$$\begin{aligned}&\sum_{m=1}^{h}\left(F_{2m(2s+3)}(x)-(2s+3)F_{2m}(x)\right)\\&=\sum_{k=0}^{s}\frac{2s+3}{2s+3-k}\binom{2s+3-k}{k}(x^2+4)^{s+1-k}\sum_{m=1}^{h}F_{2m}^{2s+3-2k}(x)\\&=\sum_{k=1}^{s}\frac{2s+3}{2s+3-k}\binom{2s+3-k}{k}(x^2+4)^{s+1-k}\sum_{m=1}^{h}F_{2m}^{2s+3-2k}(x)\\&\quad+(x^2+4)^{s+1}\sum_{m=1}^{h}F_{2m}^{2s+3}(x).\end{aligned}\tag{13}$$

From Lemma 3, we have

$$\begin{aligned}&L_1(x)L_3(x)\cdots L_{2s+3}(x)\sum_{m=1}^{h}\left[F_{2m(2s+3)}(x)-(2s+3)F_{2m}(x)\right]\\&\equiv 0 \bmod (F_{2h+1}(x)-1)^2.\end{aligned}\tag{14}$$

Applying inductive hypothesis (12), we obtain

$$\begin{aligned}&L_1(x)L_3(x)\cdots L_{2s+1}(x)\sum_{k=1}^{s}\frac{2s+3}{2s+3-k}\binom{2s+3-k}{k}\\&\times(x^2+4)^{s+1-k}\sum_{m=1}^{h}F_{2m}^{2s+3-2k}(x)\equiv 0 \bmod (F_{2h+1}(x)-1)^2.\end{aligned}\tag{15}$$

Combining (13), (14), (15) and Lemma 3, we have the conclusion

$$\begin{aligned} & L_1(x)L_3(x)\cdots L_{2s+3}(x)\cdot(x^2+4)^{s+1}\sum_{m=1}^{h}F_{2m}^{2s+3}(x) \\ \equiv\ & 0 \bmod \left(F_{2h+1}(x)-1\right)^2. \end{aligned} \tag{16}$$

Note that $\left(x^2+4, F_{2h+1}(x)-1\right)=1$, so (16) indicates the conclusion

$$L_1(x)L_3(x)\cdots L_{2s+3}(x)\cdot\sum_{m=1}^{h}F_{2m}^{2s+3}(x)\equiv 0 \bmod \left(F_{2h+1}(x)-1\right)^2.$$

Now, we apply mathematical induction to achieve Theorem 1.

Author Contributions: Conceptualization, L.C.; methodology, L.C and X.W.; validation, L.C. and X.W.; formal analysis, L.C.; investigation, X.W.; resources, L.C.; writing—original draft preparation, L.C.; writing—review and editing, X.W.; visualization,L.C.; supervision, L.C.; project administration, X.W.; all authors have read and approved the final manuscript.

Acknowledgments: The authors would like to thank the referees for their very helpful and detailed comments, which have significantly improved the presentation of this paper.

References

1. Yi, Y.; Zhang, W.P. Some identities involving the Fibonacci polynomials. *Fibonacci Q.* **2002**, *40*, 314–318.
2. Ma, R.; Zhang, W.P. Several identities involving the Fibonacci numbers and Lucas numbers. *Fibonacci Q.* **2007**, *45*, 164–170.
3. Ozeki, K. On Melham's sum. *Fibonacci Q.* **2008**, *46*, 107–110.
4. Prodinger, H. On a sum of Melham and its variants. *Fibonacci Q.* **2008**, *46*, 207–215.
5. Melham, R.S. Some conjectures concerning sums of odd powers of Fibonacci and Lucas numbers. *Fibonacci Q.* **2009** , *4*, 312–315.
6. Wang, T.T.; Zhang, W.P. Some identities involving Fibonacci, Lucas polynomials and their applications. *Bull. Math. Soc. Sci. Math. Roumanie* **2012**, *55*, 95–103.
7. Sun, B.Y.; Xie, M.H.Y.; Yang, A.L.B. Melham's Conjecture on Odd Power Sums of Fibonacci Numbers. *Quaest. Math.* **2016**, *39*, 945–957. [CrossRef]
8. Duncan, R.I. Application of uniform distribution to the Fibonacci numbers. *Fibonacci Q.* **1967** , *5*, 137–140.
9. Kuipers, L. Remark on a paper by R. L. Duncan concerning the uniform distribution mod 1 of the sequence of the logarithms of the Fibonacci numbers. *Fibonacci Q.* **1969** , *7*, 465–466.
10. Chen, L.; Zhang, W.P. Chebyshev polynomials and their some interesting applications. *Adv. Differ. Equ.* **2017**, *2017*, 303.
11. Li, X.X. Some identities involving Chebyshev polynomials, Mathematical Problems in Engineering. *Math. Probl. Eng.* **2015**, *2015*. [CrossRef]
12. Ma, Y.K; Lv, X.X. Several identities involving the reciprocal sums of Chebyshev polynomials. *Math. Probl. Eng.* **2017**, *2017*. [CrossRef]
13. Clemente, C. Identities and generating functions on Chebyshev polynomials *Georgian Math. J.* **2012**, *39*, 427–440.
14. Gould, H.W. The Girard-Waring power sum formulas for symmetric functions and Fibonacci sequences. *Fibonacci Q.* **1999**, *37*, 135–140.
15. Waring, E. *Miscellanea Analytica: De Aequationibus Algebraicis, Et Curvarum Proprietatibus*; Nabu Press: Charleston, SC, USA, 2010.
16. Kim, D.S.; Dolgy, D.V.; Kim, D.; Kim, T. Representing by Orthogonal Polynomials for Sums of Finite Products of Fubini Polynomials. *Mathematics* **2019**, *7*, 319. [CrossRef]

17. Kim, T.; Hwang, K.-W.; Kim, D.S.; Dolgy, D.V. Connection Problem for Sums of Finite Products of Legendre and Laguerre Polynomials. *Symmetry* **2019**, *11*, 317. [CrossRef]
18. Kizilates, C.; Cekim, B.; Tuglu, N.; Kim, T. New Families of Three-Variable Polynomials Coupled with Well-Known Polynomials and Numbers. *Symmetry* **2019**, *11*, 264. [CrossRef]
19. Kim, T.; Kim, D.S.; Dolgy, D.V.; Park, J.-W. Sums of finite products of Chebyshev polynomials of the second kind and of Fibonacci polynomials. *J. Inequal. Appl.* **2018**, *2018*, 148. [CrossRef]
20. Kim, T.; Dolgy, D.V.; Kim, D.S.; Seo, J.J. Convolved Fibonacci numbers and their applications. *Ars Combin.* **2017**, *135*, 119–131.
21. Kim, T.; Kim, D.S.; Dolgy, D.V.; Kwon, J. Representing Sums of Finite Products of Chebyshev Polynomials of the First Kind and Lucas Polynomials by Chebyshev Polynomials. *Mathematics* **2019**, *7*, 26. [CrossRef]
22. Kim, T.; Kim, D.S.; Kwon, J.; Dolgy, D.V. Expressing Sums of Finite Products of Chebyshev Polynomials of the Second Kind and of Fibonacci Polynomials by Several Orthogonal Polynomials. *Mathematics* **2018**, *6*, 210. [CrossRef]

Permissions

All chapters in this book were first published in MDPI; hereby published with permission under the Creative Commons Attribution License or equivalent. Every chapter published in this book has been scrutinized by our experts. Their significance has been extensively debated. The topics covered herein carry significant findings which will fuel the growth of the discipline. They may even be implemented as practical applications or may be referred to as a beginning point for another development.

The contributors of this book come from diverse backgrounds, making this book a truly international effort. This book will bring forth new frontiers with its revolutionizing research information and detailed analysis of the nascent developments around the world.

We would like to thank all the contributing authors for lending their expertise to make the book truly unique. They have played a crucial role in the development of this book. Without their invaluable contributions this book wouldn't have been possible. They have made vital efforts to compile up to date information on the varied aspects of this subject to make this book a valuable addition to the collection of many professionals and students.

This book was conceptualized with the vision of imparting up-to-date information and advanced data in this field. To ensure the same, a matchless editorial board was set up. Every individual on the board went through rigorous rounds of assessment to prove their worth. After which they invested a large part of their time researching and compiling the most relevant data for our readers.

The editorial board has been involved in producing this book since its inception. They have spent rigorous hours researching and exploring the diverse topics which have resulted in the successful publishing of this book. They have passed on their knowledge of decades through this book. To expedite this challenging task, the publisher supported the team at every step. A small team of assistant editors was also appointed to further simplify the editing procedure and attain best results for the readers.

Apart from the editorial board, the designing team has also invested a significant amount of their time in understanding the subject and creating the most relevant covers. They scrutinized every image to scout for the most suitable representation of the subject and create an appropriate cover for the book.

The publishing team has been an ardent support to the editorial, designing and production team. Their endless efforts to recruit the best for this project, has resulted in the accomplishment of this book. They are a veteran in the field of academics and their pool of knowledge is as vast as their experience in printing. Their expertise and guidance has proved useful at every step. Their uncompromising quality standards have made this book an exceptional effort. Their encouragement from time to time has been an inspiration for everyone.

The publisher and the editorial board hope that this book will prove to be a valuable piece of knowledge for researchers, students, practitioners and scholars across the globe.

List of Contributors

Dae San Kim
Department of Mathematics, Sogang University, Seoul 121-742, Korea
Department of Mathematics, Sogang University, Seoul 04107, Korea

Taekyun Kim
School of Science, Xi'an Technological University, Xi'an 710021, China
Department of Mathematics, Kwangwoon University, Seoul 139-701, Korea
Department of Mathematics, College of Science, Tianjin Polytechnic University, Tianjin 300160, China
Department of Mathematics, Kwangwoon University, Seoul 01897, Korea
Department of Mathematics, Tianjin Polytechnic University, Tianjin 300387, China

Hyunseok Lee
Department of Mathematics, Kwangwoon University, Seoul 139-701, Korea

Farah Jawad and Andriy Zagorodnyuk
Department of Mathematical and Functional Analysis, Vasyl Stefanyk Precarpathian National University, 57 Shevchenka Str., 76018 Ivano-Frankivsk, Ukraine

Cheon Seoung Ryoo
Department of Mathematics, Hanman University, Daejeon 10216, Korea
Department of Mathematics, Hannam University, Daejeon 306-791, Korea

Jung Yoog Kang
Department of Mathematics Education, Silla University, Busan 469470, Korea

Serkan Araci
Department of Economics, Faculty of Economics, Administrative and Social Sciences, Hasan Kalyoncu University, TR-27410 Gaziantep, Turkey

Waseem Ahmad Khan
Department of Mathematics, Faculty of Science, Integral University, Lucknow-226026, India

Kottakkaran Sooppy Nisar
Department of Mathematics, College of Arts and Science-Wadi Aldawaser, Prince Sattam bin Abdulaziz University, 11991 Riyadh Region, Kingdom of Saudi Arabia

Dmitry V. Dolgy
Kwangwoon Institute for Advanced Studies, Kwangwoon University, Seoul 139-701, Korea
Hanrimwon, Kwangwoon University, Seoul 139-701, Korea
Institute of Natural Sciences, Far Eastern Federal University, 690950 Vladivostok, Russia

Jongkyum Kwon
Department of Mathematics Education and ERI, Gyeongsang National University, Jinju, Gyeongsangnamdo 52828, Korea

Lee-Chae Jang
Graduate School of Education, Konkuk University, Seoul 143-701, Korea
Graduate School of Education, Konkuk University, Seoul 139-701, Korea
Graduate School of Education, Konkuk University, Seoul 05029, Korea

Dmitry Victorovich Dolgy
Kwangwoon Institute for Advanced Studies, Kwangwoon University, Seoul 139-701, Korea
Institute of National Sciences, Far Eastern Federal University, 690950 Vladivostok, Russia

Kyung-Won Hwang
Department of Mathematics, Dong-A University, Busan 49315, Korea

Younjin Kim
Institute of Mathematical Sciences, Ewha Womans University, Seoul 03760, Korea

Naeem N. Sheikh
School of Sciences and Engineering, Al Akhawayn University, 53000 Ifrane, Morocco

Gwan-Woo Jang
Department of Mathematics, Kwangwoon University, Seoul 139-701, Korea

Dug Hun Hong
Department of Mathematics, Myongji University, Yongin 449-728, Kyunggido, Korea

Lee Jinwoo
Department of Mathematics, Kwangwoon University, Seoul 01897, Korea

Wenpeng Zhang
School of Mathematics and Statistics, Kashgar University, Xinjiang 844006, China
School of Mathematics, Northwest University, Xi'an 710127, Shaanxi, China

Zhuoyu Chen
School of Mathematics, Northwest University, Xi'an 710127, China

Lan Qi
School of Mathematics and Statistics, Yulin University, Yulin 719000, China

YunJae Kim
Department of Mathematics, Dong-A University, Busan 49315, Korea

Byung Moon Kim
Department of Mechanical System Engineering, Dongguk University, Gyungju-si, Gyeongsangbukdo 38066, Korea

Ran Duan and Shimeng Shen
School of Mathematics, Northwest University, Xi'an 710127, China

Tingting Wang and Liang Qiao
College of Science, Northwest A&F University, Yangling 712100, China

Yonghong Yao
Department of Mathematics, Tianjin Polytechnic University, Tianjin 300387, China
Institute of Fundamental and Frontier Sciences, University of Electronic Science and Technology of China, Chengdu 610054, China

Han Young Kim
Department of Mathematics, Kwangwoon University, Seoul 139-701, Korea

Jin-Woo Park
Department of Mathematics Education, Daegu University, Gyeongsan 38066, Korea

Li Chen and Xiao Wang
School of Mathematics, Northwest University, Xi'an 710127, Shaanxi, China

Index

Printed in the USA
CPSIA information can be obtained
at www.ICGtesting.com
JSHW061708180324
59435JS00005B/117

9 781647 254629